Statistical Methods for QTL Mapping

CHAPMAN & HALL/CRC
Mathematical and Computational Biology Series

Aims and scope:

This series aims to capture new developments and summarize what is known over the entire spectrum of mathematical and computational biology and medicine. It seeks to encourage the integration of mathematical, statistical, and computational methods into biology by publishing a broad range of textbooks, reference works, and handbooks. The titles included in the series are meant to appeal to students, researchers, and professionals in the mathematical, statistical and computational sciences, fundamental biology and bioengineering, as well as interdisciplinary researchers involved in the field. The inclusion of concrete examples and applications, and programming techniques and examples, is highly encouraged.

Series Editors

N. F. Britton
Department of Mathematical Sciences
University of Bath

Xihong Lin
Department of Biostatistics
Harvard University

Hershel M. Safer
School of Computer Science
Tel Aviv University

Maria Victoria Schneider
European Bioinformatics Institute

Mona Singh
Department of Computer Science
Princeton University

Anna Tramontano
Department of Biochemical Sciences
University of Rome La Sapienza

Proposals for the series should be submitted to one of the series editors above or directly to:
CRC Press, Taylor & Francis Group
3 Park Square, Milton Park
Abingdon, Oxfordshire OX14 4RN
UK

Published Titles

Published Titles (continued)

Meta-analysis and Combining Information in Genetics and Genomics
Rudy Guerra and Darlene R. Goldstein

Methods in Medical Informatics: Fundamentals of Healthcare Programming in Perl, Python, and Ruby
Jules J. Berman

Modeling and Simulation of Capsules and Biological Cells
C. Pozrikidis

Niche Modeling: Predictions from Statistical Distributions
David Stockwell

Normal Mode Analysis: Theory and Applications to Biological and Chemical Systems
Qiang Cui and Ivet Bahar

Optimal Control Applied to Biological Models
Suzanne Lenhart and John T. Workman

Pattern Discovery in Bioinformatics: Theory & Algorithms
Laxmi Parida

Python for Bioinformatics
Sebastian Bassi

Quantitative Biology: From Molecular to Cellular Systems
Sebastian Bassi

Spatial Ecology
Stephen Cantrell, Chris Cosner, and Shigui Ruan

Spatiotemporal Patterns in Ecology and Epidemiology: Theory, Models, and Simulation
Horst Malchow, Sergei V. Petrovskii, and Ezio Venturino

Statistical Methods for QTL Mapping
Zehua Chen

Statistics and Data Analysis for Microarrays Using R and Bioconductor, Second Edition
Sorin Drăghici

Stochastic Modelling for Systems Biology, Second Edition
Darren J. Wilkinson

Structural Bioinformatics: An Algorithmic Approach
Forbes J. Burkowski

The Ten Most Wanted Solutions in Protein Bioinformatics
Anna Tramontano

Chapman & Hall/CRC Mathematical and Computational Biology Series

Statistical Methods for QTL Mapping

Zehua Chen

National University of Singapore

CRC Press
Taylor & Francis Group
Boca Raton   London   New York

CRC Press is an imprint of the
Taylor & Francis Group, an **informa** business

A CHAPMAN & HALL BOOK

CRC Press
Taylor & Francis Group
6000 Broken Sound Parkway NW, Suite 300
Boca Raton, FL 33487-2742

© 2014 by Taylor & Francis Group, LLC
CRC Press is an imprint of Taylor & Francis Group, an Informa business

No claim to original U.S. Government works

Printed on acid-free paper
Version Date: 20130830

International Standard Book Number-13: 978-1-4398-6830-0 (Hardback)

Library of Congress Cataloging-in-Publication Data

Chen, Zehua, 1956-
   Statistical methods for QTL mapping / Zehua Chen.
      pages cm. -- (Chapman & Hall/CRC mathematical and computational biology
      series)
   "A CRC title."
   Includes bibliographical references and index.
   ISBN 978-1-4398-6830-0 (hardcover : alk. paper)
   1. Gene mapping. 2. Genetic markers. I. Title. II. Title: Statistical methods for
quantitative trait loci mapping.

QH445.2.C44 2014
572.8'633--dc23   2013034937

Visit the Taylor & Francis Web site at
http://www.taylorandfrancis.com

and the CRC Press Web site at
http://www.crcpress.com

Contents

List of Figures

List of Tables

Preface

Phenotypic traits that vary continuously are referred to as quantitative traits. Stretches of DNA containing or linked to the genes that underlie a quantitative trait are called quantitative trait loci (QTL). QTL mapping is a scientific undertaking to identify QTL using information of biological markers. As biotechnology progresses, more and more biological markers have become available in many organisms, and the genotyping of the markers has become less and less expensive. QTL mapping has now been used on a daily basis in genetic research as well as in medical studies. Appropriate and efficient statistical methodologies are crucial for QTL mapping to be fruitful. As the data complexity of QTL mapping evolves due to the progress of bio-technology, many advanced statistical methods have been developed in recent years. There is a need for a book that provides a summary of these methods.

There are a few books available on QTL mapping such as [162], [21] and [202]. In [162], the concentration is on the genome scans based on a one-QTL model and the statistical setting is quite simple. The emphasis of [21] is more on the practical issues and on the details of the R/qtl package. A wide range of genetic experiments and practical problems related to QTL mapping are covered in [202] but important more advanced statistical methods are only briefly discussed. A volume for a comprehensive coverage of the recently developed statistical methods is still a void. The current book aims at filling this void.

This book provides the most up-to-date coverage of statistical methods for QTL mapping from simple one-QTL model approaches to advanced genome-wide multiple-QTL model approaches. A balanced treatment on the theoretical development and practical implementation of these methods is adopted in this book in order to serve the needs of both the theoreticians and practitioners. Necessary biological and statistical backgrounds are provided so that the book is as self-contained as possible, and also serves as a bridge between the geneticists and statisticians involved in the joint venture of QTL mapping.

The book contains eight chapters.

Chapter 1 covers the basic topics of genetics which are necessary for the study of QTL mapping. These topics include the basic elements of genetics, mitosis and meiosis, crossover and recombination, gene expression, mapping functions, genetic maps and experimental crosses.

Chapter 2 covers some selected topics in statistics. It contains traditional topics such as estimation, hypothesis testing, confidence interval, linear and generalized linear models, Bayesian analysis, and so on. It also contains some more advanced topics such as mixture model, EM algorithm, MCMC, small-n-large-p high dimensional data analysis etc. The statistical principles rather than the details of these topics are elucidated so that the reader can have a deeper understanding of the essence of statistical analysis.

Chapter 3 discusses the principles of quantitative genetics and general statistical issues on QTL mapping. The following topics are covered: genetic models of QTL effects, genetic variances of quantitative traits, inference on unknown QTL genotypes from marker genotypes in experimental crosses, and general statistical models for quantitative traits.

Chapter 4 covers commonly used one-dimensional QTL mapping approaches, including single-marker analysis, single interval mapping and composite interval mapping. It also covers some related topics such as the determination of threshold value, the determination of sample size and selective genotyping.

Chapter 5 covers the topics of multiple interval mapping. Mixture linear and generalized linear models for Gaussian and non-Gaussian traits and the related EM algorithms are developed in detail in this chapter. Multiple interval mapping methods for mapping categorical traits and traits with a spike in their distribution are also developed in this chapter.

Chapter 6 concerns with QTL mapping with dense markers. The QTL mapping problem is dealt with by a feature selection approach. Newly developed penalized likelihood methods and sequential methods together with variable selection criterion for small-n-large-p high-dimensional feature space are used to tackle the problem. Strategies are developed to treat both the case of non-epistatic QTL effects and the case of epistatic QTL effects.

Chapter 7 provides comprehensive coverage on Bayesian approaches for QTL mapping. Various Bayesian models are presented, including the ones treating the number of QTL as fixed, the ones treating the number of QTL as a random variable, the ones taking into account epistatic QTL effects, the ones for ordinal polytomous traits, etc. The MCMC algorithms for these Bayesian models are described in detail. Both inbred line crosses and outbred crosses are considered.

Chapter 8 summarizes the methods for multi-trait QTL mapping and eQTL mapping. The methods covered include approaches based on single-QTL models such as one-trait-at-a-time methods, meta-trait methods and multivariate composite interval mapping, and more sophisticated approaches such as multivariate sparse partial least square regression, multivariate sequential procedures, and multivariate Bayesian approaches.

Chapters 4, 5, 6 and 7 are quite independent from each other. They can be read in any order or selectively read according to the needs of the reader. However, to better understand Chapter 8, the knowledge of Chapters 4, 5 and 6 is necessary.

The book is at an advanced level. A lot of emphasis has been put on the statistical issues and modern statistical methodology for QTL mapping.

Its primary audience is the geneticists and statisticians specializing in QTL mapping. It can be used as a supplementary material for graduate courses on QTL mapping. It can also be used by PhD students working on QTL mapping projects.

I would like to thank Professor Howell Tong, one of the Chapman & Hall/ CRC series editors, who initiated the project of writing this book. Without his recommendation and encouragement, this book would not be in existence. I would also like to express my gratitude to the anonymous reviewers who provided valuable suggestions on the original proposal of the book. Finally, my appreciation goes to my wife, Dr. Liang Shen, my son, Xingjian, and my daughters Ruoxi and Jingxuan, for their understanding and encouragement while the book was being written.

Zehua Chen
National University of Singapore
Singapore
May 2013

Chapter 1

Biological Background

The basic topics of genetics which are necessary for the study of QTL mapping are discussed in this chapter. First, we give a brief sketch of the history of genetics. Then, we describe the basic elements of genetics, elucidate two types of cell division: mitosis and meiosis, and explain the phenomena of crossover and recombination. Finally, we discuss topics most closely related to QTL mapping: genetic maps, mapping functions and experimental crosses. Some of the basic materials in this chapter are adapted from the text book by Robert J. Brooker [23]. For a deeper immersion in genetics, the reader is referred to this book. For a quick and general overview on genetics, *The Cartoon Guide to Genetics* [74] is recommended.

1.1 A Brief Sketch of the History of Genetics and Mendel's Pea Plant Experiments

Why do cats only produce cats, orange trees only bear oranges? Why do tall parents tend to have tall offspring? How do the characteristics of a species pass from one generation to the next? These questions are answered by the chromosome theory of inheritance established only in the 20th century. But questions like these attracted human attention even in ancient times. There was a guess or belief in the ancient Greek era that all parts of the body of a living thing collectively produced seeds which were transmitted to the offspring. In the long history of human beings, two schools of inheritance were developed from this primitive belief: spermists and ovists. The two schools have one viewpoint in common; that is, the offspring was grown from a miniature being transmitted from only one of the parents. They differ in that the miniature being is contained in the sperm of the male parent according to the spermists, but in the egg of the female parent according to the ovists. The idea that both parents contribute equally to the offspring came about around the middle of the 18th century. The idea is now common sense, but it was not firmly verified and confirmed until Gregor J. Mendel established *Mendel's law of segregation* through pea plant experiments which was reported in 1865. Mendel's work demarcated a watershed in the history of genetics.

Mendel was a monk in the Augustinian monastery of St. Thomas in Moravia which was then a part of Austria [86]. He began his pea plant experiments in 1856 and continued for seven years. Seven characteristics of pea plant were selected in his experiments: flower color — purple and white, flower position — axial and terminal, seed color — yellow and green, seed shape — round and wrinkled, pod shape — inflated and constricted, pod color — green and yellow and stem length — tall and dwarf. These characteristics are called traits.

Mendel first obtained true breeding lines by self-fertilization. A true breeding line is one that always produces the same offspring in terms of a given trait. For example, a true breeding line with purple flower always produces purple flower offspring from generation to generation. Mendel experimented with the seven traits one at a time. For each trait, the two true breeding lines were crossed to generate the F_1 (first filial) generation. The F_1 generation was then self-fertilized to generate the F_2 (second filial) generation.

Mendel counted the number of different forms of the trait in each generation. In the case of flower color, for example, he found that all the plants in the F_1 generation had purple flowers and that, in the F_2 generation, the ratio of the plants with purple flower and those with white flower was 3.15:1. He discovered the same phenomenon in all the other traits: in F_1 generation, all plants showed only one form of the trait, in F_2 generation, the form present in F_1 and the form absent in F_1 had a ratio around 3:1. This ratio is easy to explain now even for a fresh undergraduate. But what this ratio means was still mysterious to Mendel at the time.

Mendel then continued self-fertilizing the F_2 generation and found that the ratio was about 5:3 in the F_3 generation. Mendel had been trained in combinatorics. He used combinatorics to analyze what he had found. In the case of flower color, he let A represent the genetic element (the word gene was not coined yet) causing purple color, and a, white color. He formulated his theory as follows. The true breeding lines carried AA and aa (he in fact used a single A for AA, similarly for aa) respectively. The parents contributed equally to the offspring. Thus, in the F_1 generation, every plant was a hybrid carrying Aa, but A is dominant and a is recessive and hence every plant had purple flowers. In the F_2 generation, there should be three combinations of A and a, i.e., AA, Aa and aa. While self-fertilized, the AA and aa plants should breed true, i.e., producing purple flower and white flower respectively. The Aa plants, like their parents, should have bred the purple and white flowers in a ratio 3:1. Denote by r the proportion of Aa plants in F_2. Then $\frac{r}{4} + \frac{1}{4} = \frac{3}{8}$ and hence $r = \frac{1}{2}$. Mendel then refined the ratio 3:1 to 1:2:1.

Eventually, Mendel arrived at his law of segregation: each individual contains genetic elements from both parents, the genetic element from the father and that from the mother segregate from each other during transmission from the individual to the next offspring and each of the parents' genetic elements is transmitted at random with equal probability.

Mendel's law of segregation remained obscure until around 1900 when it was rediscovered independently by Hugh de Vries, Carl Correns and Erich von Tschermak from Holland, Germany and Austria respectively. Since then many exciting events have happened. The chromosome theory of inheritance was proposed in 1903 independently by Walter Sutton and Theodore Boveri and finally established in the early years of the 20th century. Around the 1940s, it was shown that the chromosomes are DNA (deoxyribonucleic acid) sequences. In the 1950s, the double helix structure of DNA sequences was elucidated. Following that the function and structure of DNA at cytological and molecular level were extensively explored. Most remarkably the Human Genome Project started in 1990 was completed in the first decade of this century. Now scientists from many fields including biologists, geneticists, chemists, computer scientists, statisticians, mathematicians and the like are joining their hands together in genetic study and advancing human's knowledge in genetics at an unprecedented pace.

1.2 Basic Elements of Genetics

This section covers basic elements of genetics such as cells, chromosomes, DNA sequences, genes, genetic traits, gene effects and genetic modes etc..

1.2.1 Cells, Chromosomes and DNA Sequences

Living things consist of cells. Cell is the basic unit of life. Genetic materials are contained in chromosomes which are in turn contained in cells. There are two types of cells: *prokaryotic* and *eukaryotic*. In a prokaryotic cell, there is no nucleus. The genetic materials are contained in a single type of chromosome in a circular form residing in the cytoplasm of the cell. In a eukaryotic cell, there is a separate nucleus within the cell. The genetic materials are contained in mutiple types of chromosomes with a linear structure, and the chromosomes reside within the nucleus of the cell.

A species consisting of prokaryotic cells is referred to as a *prokaryote*. Prokaryotes are mainly bacteria. A species consisting of eukaryotic cells is referred to as a *eukaryote*. Eukaryotes include plants, animals and human beings. Eukaryotes are mostly diploid, i.e., each type of chromosomes exists in a *homologous pair*. One member of the pair is inherited from the male parent, and the other from the female parent. When passing the genetic materials to the next generation, the two members of the pair segregate and are transmitted at random with equal probability, as formulated in Mendel's law of segregation.

Each organism has a different set of chromosomes. The particular set of chromosomes of an organism is referred to as its *genome*. For example, the human genome consists of 46 chromosomes: 22 pairs of homologous chromosomes and one pair of non-homologous sex chromosomes. The homologous chromosomes are also called *autosomes*.

A chromosome is a double helix of two strands of DNA sequences. Each strand is formed by four types of base nucleotides: *adenine (A), cytosine (C), guanine (G) and thymine (T)*, which are arranged somehow in a random manner. The two strands are paired off in a complementary way to form a double helix: A in one strand is paired off with T in the other strand, and C in one strand is paired off with G in the other strand. For example, if GATACCT is a segment on one strand then the complementary segment on the other strand is CTATGGA. Each complementary pair is referred to as a base pair. The double helix structure of DNA is illustrated in Figure 1.1.

Because of the complementary nature of the two strands, the double helix of DNA sequences is denoted by the nucleotide sequence on any one of its two strands. The four base nucleotides are like the letters in an alphabet. They can

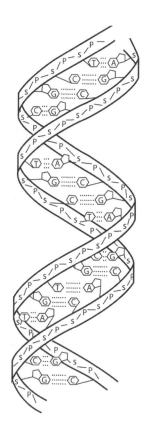

FIGURE 1.1: The double helix structure of DNA sequence.

form meaningful words. The words in the DNA sequence are all three letters long. For example, CCA is a word, CCG is another word. There are all together $4^3 = 64$ words. Each word codes the information for a particular amino acid. There are 20 distinct amino acids. Each amino acid is coded by more than one word. The words coding for the same amino acid can be regarded as acronyms. The three letter words in DNA sequences are referred to as *codons*. Like the words in a language that can form sentences of particular meanings, the codons can form segments carrying particular information. There are two types of such segments: those which code information for producing proteins are referred to as *exons* and those which do not are referred to as *introns*.

1.2.2 Genes

In genetics, the question of most concern is how the characteristics or traits of an organism are affected by genes. Genes are what Mendel called genetic elements, though Mendel at his time did not know the substance and structure of the genes. Gene is the unit of heredity. A gene consists of a sequence of exons and introns which are transmitted as a whole from generation to generation.

As chromosomes are in pairs, a gene always has two homologous forms unless it is located on sex chromosomes. The homologous forms of a gene are called *alleles*. Alleles could differ in their contents. In the population of an organism, a gene has two or more distinct alleles. But, for each individual, a gene always consists of two alleles. The combination of two alleles of a gene is called its *genotype*. The two alleles in a genotype could be the same or different. If the two alleles are the same, the genotype is said to be *homozygous*. If the two alleles are distinct, the genotype is said to be *heterozygous*. For example, the gene determining the blood type of human beings has three alleles in the population, denoted by A, B and O. There are six different genotypes: AA, AB, AO, BB, BO and OO. Each individual has one of these six genotypes. The genotypes AA, BB and OO are homozygous and AB, AO and BO are heterozygous. If one can tell which allele is from which parental chromosome in a genotype, the genotype is said to be phase-known. In other words, the phase of a genotype identifies the parental sources of its alleles.

The case of genes on sex chromosomes is slightly different. In the X-Y system (there are other systems such as X-O and Z-W), a male has one X and one Y, a female has two X's. Most genes on sex chromosomes are located on the X chomosome. They are called X-linked genes. For an X-linked gene, there is only one allele in the genotype of a male while there are still two alleles in the genotype of a female. The genotype of a female can also be homozygous or heterozygous.

1.2.3 Genetic Traits and Phenotypes

An organism displays many characteristics which are affected by genes. These characteristics are referred to as *genetic traits* or simply traits. There are *morphological traits* that relate to the appearance of an organism. For example, the seven traits of the pea plants in Mendel's experiment are all morphological traits. There are *physiological traits* which are related to the ability of an organism to perform certain functions. Examples of physiological traits include the IQ of a person, the rate of metabolization of certain food, the resistance of a crop to herbicide. Most of the morphological traits are categorical. Most of physiological traits are quantitative.

Both morphological and physiological traits are also affected by environment in addition to genes. But the extent to which they are affected by enviroment is different. For categorical traits, while the effect of environment is negligible, there is a deterministic relationship between the traits and the genotypes. In this case, the observable categories of a trait are called the *phenotypes* of the gene that affects the trait. For example, in Mendel's pea plant experiments, purple and white are two phenotypes of the gene determining the flower color, axial and terminal are two phenotypes of the gene determining the position of the flower, etc.. For quantitative traits, the environmental effect usually masks the relationship between the traits and genotypes. The relationship between quantitative traits and genes will be discussed in more detail in Chapter 3.

1.2.4   Gene Effects and Genetic Modes

The phenotypes are affected jointly by the alleles of the genes. The ways which the alleles of a gene affect the phenotypes are referred to as *genetic modes*. There are four basic types of genetic modes: dominant, recessive, codominant and overdominant. An allele in a genotype does not necessarily always manifest itself in the phenotype. If the existence of one allele in a genotype masks the existence of the other, i.e., only this allele manifest itself but the other allele does not, then the allele that masks the existence of the other is said to be *dominant* to the other allele, and the other allele is said to be *recessive* to the dominant allele. If both alleles in a genotype manifest themselves in the phenotype, the two alleles are said to be *codominant* to each other. If the existence of two different alleles in a genotype enhances the functionality of both alleles, i.e., the heterozygous genotype has a better functionality than both homozygous genotypes of the alleles, then the allele in the homozygote that is superior than the homozygote with the other allele is said to be *overdominant* to the other allele. This situation is referred to as *heterozygote advantage*.

The blood type gene of human beings provides a good example for the illustration of genetic modes. In this gene, each of the A, B and O alleles controls a distinct surface antigen on the red blood cells which can be recognized by the antibodies of the immune system. By an abuse of notation, denote the distinct antigens still by A, B and O. If an individual has genotype AO or BO, then only the antigen A or B can be recognized in the individual's blood. The existence of either A or B masks the existence of O. Therefore, A and

B are dominant to O, or O is recessive to A and B. But if an individual has genotype AB, both antigens A and B can be recognized in the individual's blood. Thus A and B are codominant to each other.

A phenotype could be affected by multiple genes. This phenomenon is referred to as *gene interaction*. The terminology "interaction" has a different meaning in genetics than its meaning in statistics. In statistics, interaction refers to the situation that the two genes do not act independently, or their effects are not additive, more precisely, the effect of one gene while the other gene is fixed at one genotype is different from the effect of the gene while the other gene is fixed at other genotypes. The statistical interaction is referred to as *epistasis* in genetics.

The relationship between genes and traits reveals that the variation in the genotypes of the gene causes the variation in the phenotypes of the gene. This provides the basis for gene mapping, especially for QTL mapping, which is the theme of this book. QTL mapping is essentially a search for genes whose variation in genotypes can explain the variation in the quantitative trait of concern.

1.3 Cell Division, Crossover and Recombination

We already know from the chromosome theory that, in diploid organisms, each type of chromosome exists in a homologous pair in an individual's cells, one is inherited from the male parent and the other from the female parent. The parents in turn have their homologous pairs inherited from their parents, i.e., the grandparents. A parent only transmits one homologous chromosome in the pair to the offspring. However, the transmitted chromosome is in general not purely the one from either the grandfather or the grandmother. It is usually a combination of segments from both grandparents. This gives rise to recombination of genes. Recombination is a major force behind biological diversity. Understanding of recombination is crucial for the study of QTL mapping.

Recombination is caused by *crossover*, an event occuring during cell division to produce *gametes*. A gamete is either a sperm cell or an egg cell. There are two processes of cell division: *mitosis* and *meiosis*. Mitosis is not directly related to heredity. It is meiosis that tells the story of genetics. However, an understanding of mitosis helps the understanding of meiosis. In this section, we describe the processes of mitosis and meiosis and elucidate the mechanisms of crossover and recombination.

1.3.1   Mitosis

Mitosis is a process that divides an original cell (the mother cell) into two daughter cells. Each of the daughter cells contains the same number and type of chromosomes as the mother cell. Mitosis is responsible for growth.

In a cell cycle involving mitosis, there are two major stages called *interphase* and *mitotic phase*. The interphase is subdivided into two growth phases and a synthesis phase. Molecular changes are accumulated in the first growth phase. The first growth phase is followed by the synthesis phase during which each chromosome is replicated into two duplicate copies. The two duplicate copies are called *sister chromatids* (they are not only homologous but also consist of exactly the same molecules). The two sister chromatids are connected together at a structure called *centromere*. Then the cell undergoes the second growth phase during which materials are accumulated for cell division. The mitotic phase distributes the sister chromatids and divides the cell into two daughter cells. The mitotic phase is further divided into four sub phases: *prophase, metaphase, anaphase and telophase*. The process of mitosis is illustrated in Figure 1.2 with an animal cell containing four pairs of sister chromatids.

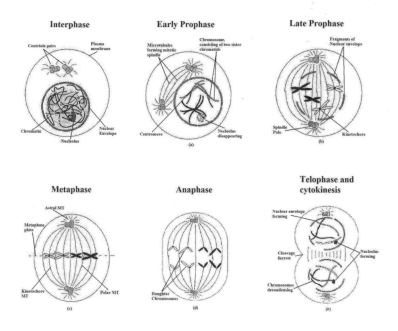

FIGURE 1.2: The process of mitosis.

During the prophase, a spindle of *microtubules* is formed from two *centrosomes* that provide the poles of the spindle, and the sister chromatids condense (Figure 1.2-*a* and *b*). In the metaphase, the sister chromatids line up along the *metaphase plate* in a single row, and each pair of sister chromatids is attached to both poles of the spindle by *kinetochore microtubule* (Figure 1.2-*c*). During the anaphase, the connection holding sister chromatids together is broken and the chromatids move towards the poles in the way that, for each pair of sister chromatids, one member moves towards one pole, the other moves towards the other pole (Figure 1.2-*d*). During the telophase, all the chromosomes have reached the poles of the spindle, two separate nuclei are formed, and each contains the same type and the same number of chromosomes as the mother cell (Figure 1.2-*e*).

1.3.2   Meiosis

Unlike mitosis, meiosis is responsible for reproduction. During meiosis, a mother cell is divided into four gamete cells (sperm cells in male, egg cells in female), each containing the same type but half the number of chromosomes of the mother cell.

Similar to mitosis, meiosis happens after the interphase of the cell cycle during which chromosomes have replicated themselves. There is a major difference between mitosis and meiosis. The meiosis involves two cell divisions while there is only one cell division in mitosis. The two cell division processes are called meiosis I and meiosis II. After the two phases of meiosis, the mother cell eventually divides into four gamete cells (a gamete cell contains only one chromosome for each type) that contain only half of the materials of the mother cell. Like mitosis, each phase of meiosis I and meiosis II is divided into the same four sub phases. The process of meiosis is illustrated in Figure 1.3.

The different features of meiosis are elaborated in the following. During the prophase of meiosis I, the homologous chromosomes recognize each other. They are first "zipped" together along their entire length in a process called *synapsis* to form *bivalents*. A bivalent is a bundle of two pairs of sister chromatids. During the synapsis, the nonsister chromatids in a bivalent exchange their materials at some points called *chiasmata* (singular, chiasma). This event is called *crossover*. At each chiasma, only one chromatid from each sister pair participates the crossover. After the synapsis, the homologous chromosomes are "unzipped" but still remain attached at chiasmata. When they break apart, at each chiasma, the segments of the chromatid-participating crossover in one sister pair are connected with the opposite segments of the chromatid-participating crossover in the other sister pair. Each pair of sister chromatids in a bivalent is then attached to a single pole of the spindle by kinetochore microtubule, (compared with mitosis in which each pair of sister chromatids is attached to both poles). During the metaphase of meiosis I, the sister chromatids line up along the metaphase plate in two rows, (compared with mitosis in which the sister chromatids line up in a single row). During

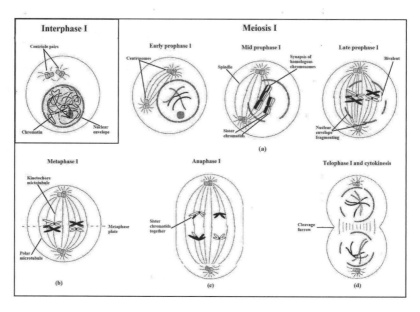

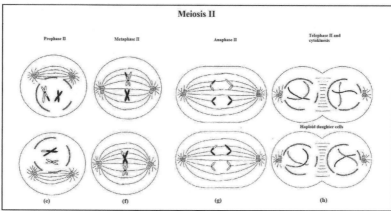

FIGURE 1.3: The process of meiosis.

anaphase and telophase, the same number and type of pairs of sister chromatids go to each pole and two nuclei are formed. Each of the newly formed nuclei contains only half of the original chromatids. These two nuclei undergo a further division during meiosis II. The process of meiosis II is the same as mitosis, except that the number of chromosomes involved is only half of that in mitosis. At the end of meiosis II, the original mother cell is eventually divided into four gamete cells, or simply four gametes.

1.3.3 Recombination

As described above, in the process of meiosis, the mother cell is divided into four gamete cells. The gamete cells are haploid; that is, chromosomes in gamete cells are no longer in homologous pairs; for each type, there is only one chromosome. Due to crossover during meiosis, some of the chromosomes in the gamete cells are combinations of paternal and maternal homologous chromosomes.

Figure 1.4 illustrates a bundle of four homologous sister chromatids with two chiasmata and its four gametes. The black ones stand for the paternal chromosomes and the white ones for the maternal chromosomes. The alleles on the paternal chromosomes are denoted by capital letters and those on the maternal chromosomes are denoted by small case letters. At the two chiasmata, the same chromatids participated in crossover, which resulted in gametes 2 and 3. Gametes 2 and 3 are formed by materials from both paternal and maternal chromosomes. Gametes 1 and 4 remain pure in the source of their genetic materials.

Recombination only concerns any two loci on a gamete. If the alleles at two loci are not from the same parental source, it is said that there is a recombination of these two loci. The two loci are called a recombinant. For example, on gamete 2, A and b, b and C are recombinants; on gamete 3, a

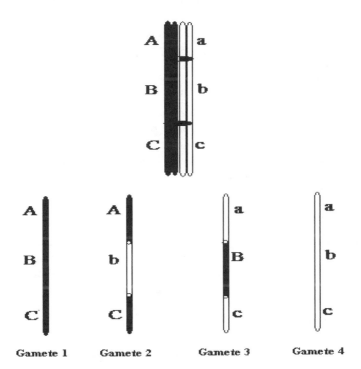

FIGURE 1.4: Illustration of recombination.

and B, B and c are recombinants. But A and C on gamete 2, a and c on gamete 3 are not recombinants. Another way to identify whether two loci are a recombinant is to count the number of crossovers between them. If the number of crossovers between them is odd, the two loci are a recombinant, otherwise they are not a recombinant.

For any two fixed loci, the proportion of recombinants among all gametes produced by meiosis in the population is called the recombination fraction between them. For a randomly selected gamete, the recombination fraction is the probability that the gamete is a recombinant. Recombinants can be observed in recombination experiments and the recombination fraction can be estimated.

1.4   Gene Expression

The traits of an organism are essentially determined by the characteristics of its cells. The characteristics of a cell are, in turn, determined by the types of proteins which it makes. The DNA sequence of the genome stores the information for the synthesis of cellular proteins. It provides a blueprint for the production of proteins in correct cells, at proper time and in suitable amount. The process of converting the information stored by DNA into functional proteins is called gene expression. In this section, we give a brief description of gene expression.

Gene expression is a two-stage process. The first stage is called transcription and the second stage translation. In the stage of transcription, a DNA sequence is copied into an RNA sequence called messenger RNA (mRNA). An mRNA sequence consists of four nucleotides: *adenine (A), cytosine (C), guanine (G) and uracil (U)*. In the stage of translation, the mRNA sequence is translated into an amino acid sequence of a polypeptide. After the translation is complete, the sequence of polypeptide folds to form a functional protein.

1.4.1   Transcription

As mentioned in §1.2.2, gene is the unit of heredity. A gene consists of different types of nucleotide sequences. These nucleotide sequences perform their specific functions during gene expression. Roughly, a gene has a promoter sequence, a terminator sequence and a middle sequence between them consisting of exons and introns. The promoter sequence signals the beginning of the transcription and the terminator sequence signals the end of the transcription. The middle sequence is copied into an mRNA sequence.

Transcription occurs in three steps: initiation, elongation and termination. In the initiation step, proteins of a particular type called transcription factors recognize the promoter site of the gene and enable an enzyme called RNA polymerase to bind to the site. After binding, the part of DNA bound with the RNA polymerase is denatured into a bubble called the open complex and the elongation step then starts. In the elongation step, the RNA polymerase slides along the middle sequence of the DNA. While the RNA polymerase moves, the DNA helix unwinds and its two strands separate within the open complex. One of the strands is used as a template for RNA synthesis, which is called the template strand. The other strand is called the coding strand which is not used in the RNA synthesis. The mRNA sequence is synthesized according to the template strand of the DNA by the rule: $T_{DNA} \rightarrow A_{RNA}$, $A_{DNA} \rightarrow U_{RNA}$, $C_{DNA} \rightarrow G_{RNA}$, $G_{DNA} \rightarrow C_{RNA}$. The mRNA sequence is a single strand sequence. It is exactly the same as the DNA coding strand except that the T nucleotide is replaced by the U nucleotide. Behind the open complex, the two strands of DNA rewind back into the double helix and the RNA sequence dissociates from the DNA. When the RNA polymerase reaches the terminator of the gene, the RNA polymerase and the whole RNA transcript dissociate from the DNA, and the process of transcription is completed.

1.4.2 Translation

After the mRNA sequence is synthesized, it is translated into a sequence of amino acids within a polypeptide. This process is called translation. In prokaryotic species, the whole mRNA sequence is usually translated. There is a one-to-one correspondence between the DNA coding strand and the amino acid sequence. However, in eukaryotic species, there is no longer such one-to-one correspondence. The mRNA sequence is not completely translated. As mentioned earlier, the mRNA sequence is a copy of the middle sequence of a gene which consists of exons and introns. In eukaryotic species, as well as some prokaryotic species, before the translation process, the sequences in the mRNA corresponding to the introns are cut out, and the exons are then spliced together. The spliced mRNA sequence is a sequence of codons, i.e., the three letter words such as UCA, CAG, AGA, GGU and so on. The codon AUG, which is the start codon, appears in the beginning segment of the spliced mRNA sequence. The three codons, UAG, UAA and UGA, which are stop codons, appear in the end segment of the spliced mRNA sequence.

In the translation process, tRNAs, a specific type of RNA coded by tRNA genes, participate in the synthesis of amino acid sequences. Each tRNA carries an anticodon. An anticodon is the complementary sequence of a codon. For example, the anticodon AAG is complementary to a UUC codon. In the synthesis, each tRNA recognizes the codon in the mRNA which is complementary to the anticodon it carries, and the anticodon binds to the codon to form an amino acid molecule. These and many additional events which are necessary in the synthesis of an amino acid sequence occur on the surface of a large macromolecular complex known as the ribosome. The role of ribosomes in translation is similar to the role of RNA polymerase in the transcription. It provides an arena for the translation.

Like the transcription, the process of translation also occur in three steps: initiation, elongation and termination. In the initiation step, an initiator tRNA recognizes and binds to the start codon AUG in the mRNA, then the mRNA, initiator tRNA and ribosomal subunits assemble to form a complex. In the elongation step, the ribosome slides along the mRNA sequence. While the ribosome moves over the codons, tRNA molecules sequentially bind to the codons of the mRNA and the ribosome and form the appropriate amino acids. Once an amino acid is formed, it is carried away and linked to the growing amino acid sequence. The order of the amino acid in the sequence is dictated by the codon sequence of the mRNA. Eventually, when a stop codon is reached, disassembly occurs and the newly made amino acid sequence is released, then the translation is terminated.

1.4.3   Gene Regulation

The rate at which gene expression occurs is referred to as gene expression level. For some genes, the expression level is essentially a constant in all conditions over time. But, for the majority of genes, the expression level varies under different conditions. The variation of gene expression levels is caused by gene regulation. Gene regulation is the action of regulatory proteins in conjunction with small effector molecules. Gene regulation can occur at any stage of gene expression. Roughly, there are transcriptional, translational and post translational regulation types. Here we focus on the transcriptional regulation.

There are two common types of regulatory proteins: activator and repressor. An activator binds to the promotor site of a gene to increase its rate of transcription. A repressor binds to the promotor site to inhibit the gene transcription.

The small effector molecules effect the gene expression through their influences on the regulatory proteins. There are three types of such effector molecules: inducer, corepressor and inhibitor. An inducer causes the rate of transcription to increase. It either binds to an activator protein and causes it bind to the DNA, or binds to a repressor to prevent it from binding to the DNA. A corepressor binds to a repressor protein and causes the latter to bind to the DNA. An inhibitor binds to an activator protein to prevent it from binding to the DNA.

The regulatory proteins and small effectors are encoded by separate genes. Usually, the regulatory genes are located near the gene which is being regulated. But there are also regulatory genes which are located far away from the regulated gene. A regulatory gene might regulate only one gene and might also regulate many genes. The gene regulation causes the gene expression of an organism to vary over time or under a different environment. The variation in the regulatory genes in different individuals also causes the gene expression to differ from individual to individual.

1.5　Genetic Maps and Mapping Functions

Genetic maps and mapping functions play important roles in genetic studies. There are three types of genetic maps: cytogenetic maps, linkage maps and physical maps. Among these three types, cytogenetic and physical maps are not directly relevant to QTL mapping. However linkage maps are crucial in the study of QTL mapping. We give a brief description of cytogenetic and physical maps first, then we concentrate on linkage maps and mapping functions which are closely related to linkage maps.

1.5.1　Cytogenetic Maps

A cytogenetic map describes the locations of particular genes relative to banding patterns along chromosomes. Each chromosome has a centromere and two arms extending from it: a long arm (denoted by the letter q) and a short arm (denoted by the letter p). For each chromosome, a particular banding pattern is identified by a certain biotechnology. Each arm is divided into a number of distinguishable bands which in turn are divided into a number of distinguishable regions. The bands and regions are numbered from the centromere towards the telomere of the arm, i.e., the end of the chromosome on the arm. A cytogenetic map describes locations of particular genes in terms of chromosome type, arm, band and region. For example, in the human genome, the location of a gene suspected of being related to prostate cancer is denoted as 8q24, i.e., the 4th region of the second band on the long arm of chromosome 8 [209].

1.5.2　Physical Maps

A physical map describes genes in linear order on the chromosomes and their physical distances. The physical distance between two loci on a chromosome is the number of base pairs (bp) between them. The genomes of different organisms have different total lengths in terms of base pairs. The construction of a physical map requires the sequencing of the whole genome. Usually, eukaryotic genomes are huge. The construction of a physical map is a stupendous undertaking. For example, the human genome has a total length about 3 billion bp. The Human Genome Project which involved many laboratories in many countries took more than ten years to finish.

1.5.3　Genetic Linkage Maps

A genetic linkage map shows genetic markers (simply markers) or genes in linear order on each chromosome and describes the length of chromosome

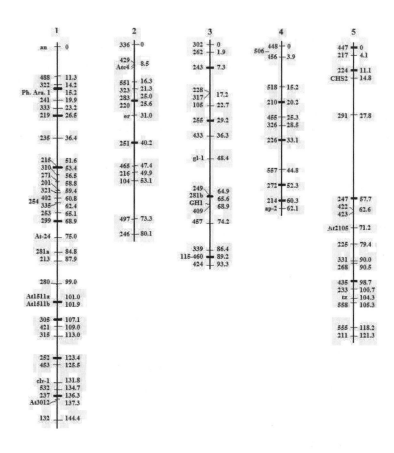

FIGURE 1.5: An RFLP linkage map of *Arabidopsis thaliana*.

segments by genetic distance. For example, Figure 1.5 displays a linkage map of *Arabidopsis thaliana*.

A genetic marker is a DNA fragment that is uniquely recognizable and varies from individual to individual but does not necessarily encode genes. The terminology used for genes such as alleles and genotypes are also used for markers, though a marker is in general not a gene. There are different types of genetic markers. Common genetic markers include *restriction fragment length polymorphisms (RFLPs)* and *single nucleotide polymorphisms (SNPs)*. RFLPs are DNA segments where the numbers and locations of the sites of a particular restriction enzyme differ from individual to individual. For example, a restriction enzyme called *EcoRI* can recognize the DNA fragment GAATTC which appears along a chromosome once in every 4,096 bp on average. The EcoRI can cut chromosomes into pieces of different lengths at the sites consisting of the fragment. But if at a site the fragment GAATTC has mutated and become unrecognizable by EcoRI, the site cannot be cut by EcoRI. Then there is variation in the number of cutting sites in the DNA segment containing this site. This gives rise to an RFLP. An SNP is a DNA sequence variation occurring when a single nucleotide differs from individual to individual. For example, suppose in a segment of DNA sequence different individuals have different fragments, say, AAGCCTA or AAGCTTA. These two fragments contain a difference in a single nucleotide. This gives rise to a SNP. A SNP has two different alleles in the population. In the example, the two alleles are C and T. For other types of markers, the reader is referred to [121].

On a linkage map, the length of a chromosome segment between two markers is measured by genetic distance. Genetic distance is in units of Morgans. It is defined as the expected number of crossovers on a single gamete between the two markers. For example, the genetic distance is a Morgan if the expected number of crossovers is 1. The unit of centiMorgan (cM) is also used. A centiMorgan is one hundredth of a Morgan. The genetic distance is loosely related to physical distance. On average, a genetic distance of 1 cM corresponds approximately to 1 million base pairs. However, the exact number varies from organism to organism, even from region to region in the genome of the same organism.

Genetic distances in general cannot be directly measured since crossovers are usually not observable. However, because crossovers cause recombination and the recombination fraction between two markers among experimental subjects can be observed in recombination experiments, the genetic distance can be inferred from recombination fraction through their intrinsic relationship. Linkage maps are constructed through recombination experiments. The exact intrinsic relationship between genetic distance and recombination fraction, which is complicated and impossible to reveal, is approximated by a mapping function. A mapping function is a monotone function $\theta = M(d)$ that relates the recombination fraction θ and the genetic distance d in units of Morgans. Various mapping functions are discussed in the remainder of this section.

1.5.4 Crossover Interference

Before we consider mapping functions, we need to understand crossover interference. As discussed in Section 1.3, crossovers occur at chiasmata that are formed along the bundle of two pairs of sister chromatids during meiosis. At each chiasma, only one chromatid from each sister pair participates in crossover. The chiasmata, in general, do not occur at random. The chance that a chiasma occures in a region is affected by whether or not there are chiasmata occuring in the vicinity of the region. This is referred to as *chiasmata interference*. Among the two chromatids in a sister pair, the chance that a particular chromatid participates in crossover at one chiasma, in general, depends on whether or not this chromatid has participated in crossover at other nearby chiasmata. This is referred to as *chromatid interference*. Crossover interference concerns a single gamete. Crossover interference refers to the situation in which the occurrence of crossover at a locus affects the chance of crossover in the vicinity of the locus. If it suppresses crossover in the vicinity, it is said there is *positive interference*. If it encourages crossover in the vicinity, it is said there is *negative interference*. If it does not affect the occurrence of crossover in the vicinity, it is said there is no crossover interference. In general, crossover interference is caused jointly by chiasmata interference and chromatid interference. If there is no chromatid interference, which is a common assumption made for deriving almost all mapping functions, crossover interference is solely caused by chiasmata interference. The two interferences are then equivalent in a certain sense.

The degree of crossover interference is quantified by a *coincidence coefficient*. Let $\mathbf{A}$, $\mathbf{B}$ and $\mathbf{C}$ be three ordered loci on a chromosome. Let p_{ij}, $i, j = 0$ or 1, denote the joint probability that there is i recombination between $\mathbf{A}$ and $\mathbf{B}$ and there is j recombination between $\mathbf{B}$ and $\mathbf{C}$. The coincidence coefficient is defined as

$$C = \frac{p_{11}}{(p_{01} + p_{11})(p_{10} + p_{11})}. \tag{1.1}$$

Note that $p_{10} + p_{11}$ is the marginal probability of recombination between $\mathbf{A}$ and $\mathbf{B}$, and $p_{01} + p_{11}$ is the marginal probability of recombination between $\mathbf{B}$ and $\mathbf{C}$. Rewrite (1.1) as follows:

$$C(p_{01} + p_{11}) = \frac{p_{11}}{p_{10} + p_{11}}. \tag{1.2}$$

On the left hand side of (1.2), it is the product of the coincidence coefficient and the marginal probability of recombination between $\mathbf{B}$ and $\mathbf{C}$. On the right hand side of (1.2), it is the conditional probability of recombination between $\mathbf{B}$ and $\mathbf{C}$ given that there is a recombination between $\mathbf{A}$ and $\mathbf{B}$. If $C = 1$, the marginal probability and the conditional probability are the same, which implies that the recombination between $\mathbf{A}$ and $\mathbf{B}$ and the recombination between $\mathbf{B}$ and $\mathbf{C}$ are independent. Hence there is no crossover interference. If $C < 1$, the conditional probability is smaller than the marginal probability; that is, the occurrence of recombination between $\mathbf{A}$ and $\mathbf{B}$ decreases the probability of recombination between $\mathbf{B}$ and $\mathbf{C}$ and therefore there is positive interference. If $C > 1$, it is the other way around and there is negative interference.

1.5.5 A General Form of Mapping Functions

The marginal probabilities of recombination are in fact what we call recombination fractions. Let θ_{AB}, θ_{BC} and θ_{AC} denote, respectively, the recombination fractions between $\mathbf{A}$ and $\mathbf{B}$, between $\mathbf{B}$ and $\mathbf{C}$, and between $\mathbf{A}$ and $\mathbf{C}$. Then

$$\theta_{AB} = p_{10} + p_{11}, \quad \theta_{BC} = p_{01} + p_{11}.$$

It is easy to derive that

$$p_{11} = \frac{1}{2}(\theta_{AB} + \theta_{BC} - \theta_{AC}).$$

Thus, we can rewrite (1.1) as

$$\theta_{AC} = \theta_{AB} + \theta_{BC} - 2C\theta_{AB}\theta_{BC}. \tag{1.3}$$

Suppose that the genetic distance between **A** and **B** is d, and between **B** and **C** is Δd. Since the coincidence coefficient C indeed depends on the genetic distances, we denote C by $c(d, \Delta d)$ to make this dependence explicit. Let $\theta = M(d)$ be any mapping function. Then equation (1.3) can be expressed as

$$M(d + \Delta d) - M(d) = M(\Delta d) - 2c(d, \Delta d)M(d)M(\Delta d).$$

Dividing both sides of the above equation by Δd and then letting Δd go to zero yields

$$M'(d) = \lim_{\Delta d \to 0} \frac{M(\Delta d)}{\Delta d} - 2M(d) \lim_{\Delta d \to 0} c(d, \Delta d) \lim_{\Delta d \to 0} \frac{M(\Delta d)}{\Delta d},$$

where $M'(d)$ denotes the derivative of M at d. Suppose $M'(0) = 1$, which implies that if the genetic distance is close to zero then it approximately equals the recombination fraction. Let $c_0(d) = \lim_{\Delta d \to 0} c(d, \Delta d)$ which can be considered as the instantaneous coincidence coefficient associated with a chromosome segment of genetic distance d. We arrive at the following Haldane's differential equation [79]:

$$M'(d) = 1 - 2c_0(d)M(d). \tag{1.4}$$

A reasonable mapping function must satisfy the following conditions:

$$C1: \quad M(0) = 0;$$
$$C2: \quad M'(d) > 0, \text{ for any } d;$$
$$C3: \quad M'(0) = 1;$$
$$C4: \quad M''(d) < 0, \text{ for any } d;$$
$$C5: \quad \lim_{d \to \infty} M(d) < \frac{1}{2}.$$

If $c_0(d)$ is taken such that $c_0(d)M(d) < \frac{1}{2}$ then $C2$ and $C4$ are automatically satisfied. $C1$ provides the initial condition for Haldane's differential equation. It is more convenient to consider the inverse $d = M^{-1}(\theta)$ of the mapping function. It is obtained from (1.4) that

$$\frac{\partial M^{-1}(\theta)}{\partial \theta} = \frac{1}{1 - 2c(\theta)\theta},$$

where $c(\theta) = c_0(M^{-1}(\theta))$. We then have the general form of the inverse of the mapping function:

$$d = \int_0^\theta \frac{1}{1 - 2c(r)r} dr. \tag{1.5}$$

1.5.6 Special Mapping Functions

Four widely used mapping functions are discussed in this sub section. All of them can be derived from (1.5) by assuming $c(r)$ as certain particular forms.

Haldane mapping function [79]: By assuming $c(r) \equiv 1$ in (1.5), the inverse of Haldane mapping function is obtained as

$$d = -\frac{1}{2}\ln(1 - 2\theta).$$

It yields the Haldane mapping function:

$$\theta = \frac{1}{2}(1 - e^{-2d}).$$

The Haldane mapping function corresponds to the case of no crossover interference.

Kosambi mapping function [107]: By assuming $c(r) = 2r$ in (1.5), the inverse of Kosambi mapping function is obtained as

$$d = \frac{1}{2}\tanh^{-1}(2\theta) = \frac{1}{4}\ln\left(\frac{1 + 2\theta}{1 - 2\theta}\right).$$

Kosambi mapping function is then given by

$$\theta = \frac{1}{2}\tanh(2d) = \frac{1}{2}\frac{e^{4d} - 1}{e^{4d} + 1}.$$

By condition $C5$, $2r < 1$. Hence Kosambi mapping function captures a certain degree of positive crossover interference. It was found that this mapping function provides good fit to Drosophila data [107].

Carter-Falconer mapping function [25]. By assuming $c(r) = 8r^3$ in (1.5), the inverse of Carter-Falconer mapping function is obtained as

$$d = \frac{1}{4}[\tan^{-1}(2\theta) + \tanh^{-1}(2\theta)].$$

Carter-Falconer mapping function itself cannot be expressed in closed form. But it can be expressed in the following iterative form:

$$\theta = \frac{1}{2}\tanh[4d - \tan^{-1}(2\theta)].$$

Since $8r^3$ is always less than $2r$ and is substantially smaller than $2r$ when r is small, Carter-Falconer mapping function captures a much stronger degree of positive crossover interference than Kosambi mapping function. It was found to be better than other mapping functions for fitting mouse data which exhibit strong positive interference [25].

Felsenstein mapping function [58]: By assuming $c(r) = k - (k-1)2r$ in (1.5) for some $0 \leq k < 2$, the inverse of Felsenstein mapping function is obtained as

$$d = \frac{1}{2(2-k)} \ln \left[1 + \frac{2\theta(2-k)}{1-2\theta} \right].$$

Felsenstein mapping function is then given by

$$\theta = \frac{1}{2} \left[\frac{1 - e^{2(k-2)d}}{1 - (k-1)e^{2(k-2)d}} \right], \ 0 \leq k < 2.$$

The number k measures the degree of crossover interference. If $k = 1$, $c(r) \equiv 1$, Felsenstein mapping function reduces to Haldane mapping function. If $k > 1$, $c(r) > 1$, Felsenstein mapping function captures negative crossover interference. If $k < 1$, $c(r) < 1$, Felsenstein mapping function captures positive crossover interference. Especially, if $k = 0$, Felsenstein mapping function reduces to Kosambi mapping function.

The mapping functions discussed above are depicted in Figure 1.6. In terms of degree of positive crossover interference, the four mapping functions are in the order: Haldane, Felsenstein ($k = 0.5$), Kosambi, Carter-Falconer. As is seen from Figure 1.6, at any fixed recombination fraction, the mapped genetic distance decreases when the positive crossover inference becomes stronger. This is because positive crossover interference depresses crossovers, noting that the genetic distance is a measure on the average number of crossovers.

The occurrence of chiasmata along the bundle of sister chromatids can be described by a stochastic point process. The chiasmata process further induces four identically distributed but correlated crossover processes. Mapping functions can be obtained in the framework of stochastic point processes. Zhao and Speed [221] showed that each of the above mapping functions corresponds to a certain stationary renewal process. Specifically, the Haldane mapping function is related to a Poisson process. Other mapping functions were also considered in the framework of stationary renewal process in [221]. For details of many issues related to mapping functions, the reader is referred to [171].

1.6   Experimental Crosses

Experimental crosses are important approaches to the genetic study of plants, animals and experimental organisms. They are used to produce experimental subjects with certain desirable properties for linkage analysis or QTL mapping. Experimental crosses start with two pure parental strains which are

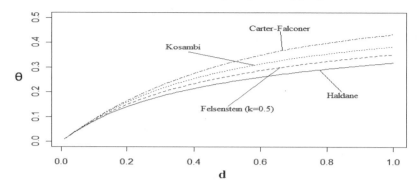

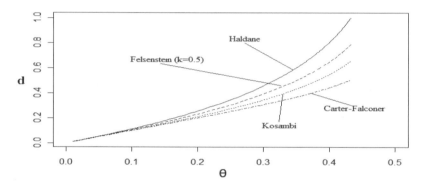

FIGURE 1.6: Comparison of mapping functions.

obtained by either selfing (in plants) or sibling mating (in animals). Then different mating schemes are employed to produce various kinds of generations. In this section, we discuss three most commonly used experimental crosses: backcross, intercross and recombinant inbred lines.

1.6.1 The Mechanism of Selfing and Sibling Mating

A pure parental strain is one that, in Mendel's terminology, always breeds true with respect to the trait under study. For example, on the trait of flower color in Mendel's pea plant experiments, one parental strain always breeds purple flowers and the other always breeds white flowers. For a parental strain to breed true, the genotypes of all individuals in the strain at the gene affecting the trait must all be homozygous with the same allele. The parental stains used in experimental crosses require more. Not only the genotypes at the gene affecting the trait but also the genotypes at all the markers must be made homozygous with the same alleles. Such parental strains are usually obtained by using the scheme of selfing or sibling mating.

To explain the mechanism of selfing and sibling mating, we first introduce the notions of *identity by descent (IBD)*, *inbreeding coefficient* (denoted by β) and *kingship coefficient* (denoted by κ). Two alleles are said to be IBD if they are copies of the same ancestral allele. For example, suppose a parent has a paternal allele A (inherited from its father) and a maternal allele A (inherited from its mother) at a certain locus, if it passes the same paternal (or maternal) A allele to two of its offspring, the two A alleles received by the offspring are the copies of the same allele and hence are IBD; but if it passes the paternal allele to one offspring and the maternal allele to another, the two A alleles received by the offspring are not IBD though they are of the same type. The inbreeding coefficient of an individual is the probability that at a randomly chosen locus the two alleles of the individual are IBD. The kingship coefficient between two individuals is the probability that an allele randomly selected from one individual is IBD with an allele randomly selected at the same locus from the other individual. Suppose individual i has parents j and k, there is a simple relationship between the inbreeding coefficient of i, β_i, and the kingship coefficent of j and k, κ_{jk}; that is, $f_i = \kappa_{jk}$. Also, $\kappa_{ii} = \frac{1}{2}(1 + \beta_i)$, where κ_{ii} is the kingship coefficient of i with itself.

In selfing, a plant is self fertilized. In any generation obtained by selfing, the inbreeding coefficient is the same for all individuals. Let $\beta^{(n)}$ denote the common inbreeding coefficient and $\kappa^{(n)}$ the common kingship coefficient of an individual with itself in the nth generation. Then it is easy to derive that $\beta^{(n+1)} = \kappa^{(n)} = \frac{1}{2}(1 + \beta^{(n)}) = 1 - \frac{1}{2^n}$, noting that $\beta^{(0)} = 0$. This implies that after a few generations, the inbreeding coefficient becomes almost 1. When selfing is conducted with selective breeding with respect to the trait under study, then two parental strains with distinct aspects of the trait can be obtained after a few generations.

In sibling mating of animals, brothers are mated with sisters. In any generation, the inbreeding coefficient is the same for all individuals, and the kingship coefficient is the same for any brother-sister pair. Still denote by $\beta^{(n)}$ and $\kappa^{(n)}$ the common inbreeding and kingship coefficients. A recursive formula in $\kappa^{(n)}$ is easily deduced using the diagram in Figure 1.7. Let the parents have genotypes $P_1 P_2$ and $M_1 M_2$ respectively. Let B and S be two random alleles selected respectively from a brother-sister pair in generation $n + 1$. The allele B is equally likely to be one of the four parental alleles and so is S. Thus there are 16 equally likely pairs which are shown in the middle of Figure 1.7 where each pair is connected by a line segment. If only one of the paternal (or maternal) alleles appears in the pair, the pair is IBD with probability 1. If both alleles are from the same parent but not the same allele, the pair is IBD with probability $\beta^{(n)}$. If one is a paternal allele and the other is a maternal allele, the pair is IBD with probability $\kappa^{(n)}$. Adding up the probabilities for all 16 possible pairs leads to:

$$\kappa^{(n+1)} = \frac{1}{2}\kappa^{(n)} + \frac{1}{4}\beta^{(n)} + \frac{1}{4} = \frac{1}{2}\kappa^{(n)} + \frac{1}{4}\kappa^{(n-1)} + \frac{1}{4}.$$

Father's Genotype   **Mother's Genotype**

P1 P2   **M1 M2**

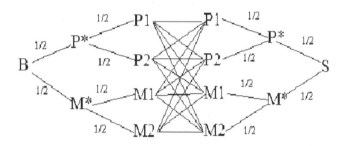

FIGURE 1.7: Illustration of random pairs of alleles selected from brother and sister.

By solving this difference equation,

$$\beta^{(n+1)} = \kappa^{(n)} = 1 - \left(\frac{1}{2} + \frac{1}{\sqrt{5}}\right)\left(\frac{1+\sqrt{5}}{4}\right)^n - \left(\frac{1}{2} - \frac{1}{\sqrt{5}}\right)\left(\frac{1-\sqrt{5}}{4}\right)^n.$$

As in the scheme of selfing, the inbreeding coefficient approaches 1 very quickly, and the genotypes at almost all markers become homozygous after a few generations.

1.6.2 The F_1 Generation

Let P_j, $j = 1, 2$, denote the two parental strains. The F_1 generation is produced by crossing P_1 and P_2 (i.e., letting individuals in P_1 mate with individuals in P_2). Since both parental strains are homozygous at any locus with different alleles, all the individuals in the F_1 generation are heterozygous at any locus. The parental strain crossing is illustrated in Figure 1.8 (a). The alleles in P_1 are denoted by capital letters and those in P_2 by small case letters. Two loci are demonstrated in the figure. The genotypes of P_1 at the two loci are AA and BB, and those of P_2 are aa and bb. The genotypes of F_1 are therefore Aa, Bb.

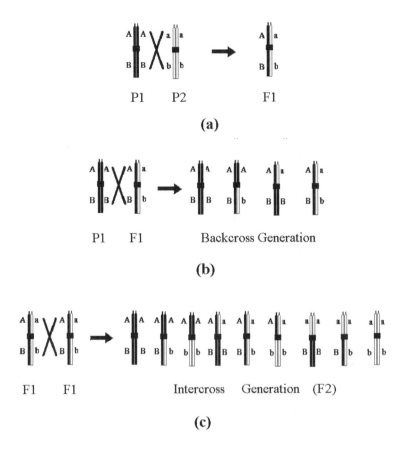

FIGURE 1.8: An illustration of backcross and intercross generations.

1.6.3 Backcross

A backcross generation is obtained by crossing the F_1 generation to one of the parental strains. For the sake of convenience, assume the parental strain involved in the crossing is P_1. The backcross is illustrated in Figure 1.8 (b). In the backcross generation, there are only two different genotypes at any locus, one homozygous and the other heterozygous. At any locus, half of the individuals in the backcross generation have the homozygous genotype and the other half have the heterozygous genotype. For example, at the first locus, half have genotype AA, and the other half have genotype Aa.

A backcross generation has two desirable natures. (i) The recombination status between any two loci can be completely determined. Given two loci, they are either both homozygous, both heterozygous, or one homozygous and one heterozygous. In the first two cases, there is no recombination between the two loci. In the third case, recombination occurs between the two loci. (ii)

The sample from a backcross generation is balanced, i.e., when the individuals are classified according to their genotypes, the sub sample sizes for the sub classes are the same. However, if at a QTL the capital letter allele is dominant to the small letter allele regarding the trait under study then the QTL cannot be detected by a backcross design, since there is no difference between the homozygous genotype and the heterozygous genotype in the trait.

1.6.4   Intercross

An intercross generation (or F_2) is obtained by crossing the F_1 generation with itself. The intercross is illustrated in Figure 1.8 (c). In the intercross generation, there are three different genotypes at any locus, two homozygous ones and a heterozygous one. At any locus, a quarter of the individuals have the homozygous genotype with the capital letter allele, a quarter of the individuals have the homozygous genotype with the small letter allele, and the remaining half of the individuals have the heterozygous genotype. For example, the ratio of the individuals having genotypes AA, Aa and aa at the first locus is 1:2:1.

Contrary to the backcross, in an intercross generation, the recombination status between two loci cannot always be completely determined, and the sample derived from an intercross generation is unbalanced. The recombination status between two loci can be determined only when both loci are homozygous or one is homozygous and the other is heterozygous. When both loci are heterozygous, the recombination status cannot be completely determined. See Table 1.1 for details. But, unlike the backcross, the intercross design is capable of dealing with any genetic modes of the QTL alleles.

TABLE 1.1:   Recombination status between two loci in an intercross generation.

Genotype at locus 1	Genotype at locus 2	Probability of Recombination
AA	BB	0
aa	bb	0
AA	Bb	1
Aa	Bb	0.5

1.6.5   Recombinant Inbred Lines

Recombinant inbred lines (RILs) are obtained from an F_1 generation by either repeated selfing (in plants) or repeated sibling mating (in animals) for many generations. Let RIL(1) denote the F_1 generation and RIL(t) denote the $(t-1)$st generation produced by selfing or sibling mating. Theoretically, when $t \to \infty$, the genotypes in RILs at all loci become homozygous. Practically, when t reaches 7 to 10, almost all the loci become homozygous and the sites with heterozygous genotypes can be ignored. The advantage of RILs is that they can increase the power of detecting QTL, especially, for QTL with small effects. The disadvantage is that RILs are expensive to obtain.

We illustrate the process of generating RILs by selfing with two loci $\mathcal{A}$ and $\mathcal{B}$ having alleles A, a and B, b respectively in the following. In generation RIL(1), i.e., F_1, there is only one joint phase-known genotype, i.e., $AB|ab$. When RIL(1) is self-fertilized, it produces RIL(2) with ten possible phase-known genotypes classified into five classes as follows:

$$
I \quad \frac{A}{B} \bigg| \frac{A}{B} \bigg| \frac{a}{b} \bigg| \frac{a}{b}
$$

$$
II \quad \frac{A}{b} \bigg| \frac{A}{b} \bigg| \frac{a}{B} \bigg| \frac{a}{B}
$$

$$
III \quad \frac{A}{B} \bigg| \frac{A}{b} \bigg| \frac{A}{B} \bigg| \frac{a}{B} \bigg| \frac{A}{b} \bigg| \frac{a}{b} \bigg| \frac{a}{B} \bigg| \frac{a}{b}
$$

$$
IV \quad \frac{A}{B} \bigg| \frac{a}{b}
$$

$$
V \quad \frac{A}{b} \bigg| \frac{a}{B}
$$

When RIL(2) is self-fertilized, it produces RIL(3) with the same possible phase-known genotypes. But the genotype frequencies are changed. Class I and Class II genotypes generate only Class I and Class II genotypes respectively. Class III genotypes generate genotypes in Class I, II and III. Class IV and V genotypes generate genotypes in all five classes. This is the case in general when RIL(t) is self-fertilized to generate RIL($t + 1$) for $t \geq 2$. The process is demonstrated in Figure 1.9.

The genotypes within the same class have the same genotype frequency. Let C_t, D_t, E_t, F_t and G_t denote, respectively, the genotype frequencies for the genotypes in Class I, II, III, IV and V. Since the ten genotypes are all possible genotypes, we have $2C_t + 2D_t + 4E_t + F_t + G_t = 1$ for all $t \geq 1$. Thus, $C_1 = D_1 = E_1 = G_1 = 0$ and $F_1 = 1$. Let r be the recombination fraction between $\mathcal{A}$ and $\mathcal{B}$. It can be easily derived that the relationships between the genotype frequencies in RIL($t + 1$) and those in RIL(t) are as follows:

$$
\begin{aligned}
C_{t+1} &= C_t + \frac{1}{2}E_t + \frac{1}{4}(1 - r)^2 F_t + \frac{1}{4}r^2 G_t, \\
D_{t+1} &= D_t + \frac{1}{2}E_t + \frac{1}{4}r^2 F_t + \frac{1}{4}(1 - r)^2 G_t, \\
E_{t+1} &= \frac{1}{2}E_t + \frac{1}{2}r(1 - r)(F_t + G_t), \qquad (1.6) \\
F_{t+1} &= \frac{1}{2}(1 - r)^2 F_t + \frac{1}{2}r^2 G_t, \\
G_{t+1} &= \frac{1}{2}r^2 F_t + \frac{1}{2}(1 - r)^2 G_t.
\end{aligned}
$$

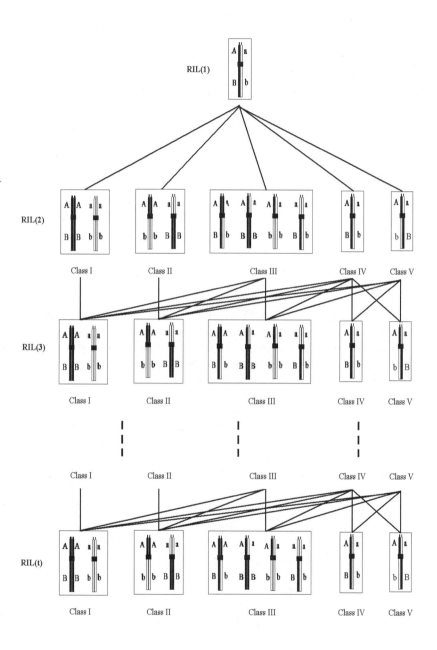

FIGURE 1.9: An illustration of the process of producing RILs by selfing.

We derive C_{t+1} as an illustration. Consider the genotype $AB|AB$. The genotypes which can generate $AB|AB$ when self-fertilized are $AB|AB$ in Class I, $AB|Ab$ and $AB|aB$ in Class III, $AB|ab$ in Class IV and $Ab|aB$ in Class V. $AB|AB$ always generates $AB|AB$, hence the contribution to C_{t+1} of $AB|AB$ in RIL(t) is C_t. The probability for $AB|Ab$ to generate $AB|AB$ is $\frac{1}{4}$ since, at locus $\mathcal{A}$ it always generates $A|A$ and at locus $\mathcal{B}$, each parental genotype transmits B with probability $\frac{1}{2}$ independently. Hence it has a contribution $\frac{1}{4}E_t$ to C_{t+1}. Similarly, $AB|aB$ also has a contribution $\frac{1}{4}E_t$. Thus together they have a contribution $\frac{1}{2}E_t$ to C_{t+1}. The probability for $AB|ab$ to generate $AB|AB$ is derived as follows. At locus $\mathcal{A}$, the probability for having $A|A$ is $\frac{1}{4}$. To have $B|B$ at locus $\mathcal{B}$, the haplotype from each parental genotype should not have a recombination, and hence the probability is $(1-r)^2$. Therefore the probability for $AB|ab$ to generate $AB|AB$ is $\frac{1}{4}(1-r)^2$, and hence the contribution of $AB|ab$ to C_{t+1} is $\frac{1}{4}(1-r)^2F_t$. Similarly, it is derived that the contribution of $Ab|aB$ to C_{t+1} is $\frac{1}{4}r^2G_t$. Adding up all these components gives C_{t+1} in (1.7).

It can be easily seen from (1.7) that $E_t \to 0, F_t \to 0, G_t \to 0$ and that $2C_t + 2D_t \to 1$ as $t \to \infty$. Eventually, the genotypes at both $\mathcal{A}$ and $\mathcal{B}$ in RIL all become homozygous and there are only Class I and Class II genotypes. Class I consists of the genotypes without recombination and Class II consists of genotypes with double recombinations. Unlike the original inbred parental strains P_1 and P_2, which are homogeneous, i.e., all the individuals have the same homozygous genotype at any locus within the same parental strain, RILs are not homogeneous at any locus since half of the individuals have one homozygous genotype and the other half have the other homozygous genotype as shown above.

The frequencies of the genotypes in Class I and II are derived as follows. Let

$$2(C_t - D_t) = c_t, \quad F_t - G_t = d_t.$$

Then

$$c_{t+1} = c_t + \frac{1}{2}(1 - 2r)d_t, \quad d_{t+1} = \frac{1}{2}(1 - 2r)d_t.$$

Find λ such that

$$c_{t+1} + \lambda d_{t+1} = c_t + \lambda d_t \quad \text{for all } t.$$

It turns out that

$$\lambda = \frac{1 - 2r}{1 + 2r}.$$

Let $\lim_{t\to\infty} C_t = C_\infty$, $\lim_{t\to\infty} D_t = D_\infty$ and $\lim_{t\to\infty} d_t = d_\infty$. Denote $c_\infty = 2(C_\infty - D_\infty)$. Note that $d_\infty = 0$ since $F_t \to 0$ and $G_t \to 0$. Thus,

$$c_\infty = c_\infty + \lambda d_\infty = c_1 + \lambda d_1 = \lambda = \frac{1 - 2r}{1 + 2r}.$$

Note that $2(C_\infty + D_\infty) = 1$. It follows that

$$2D_\infty = \frac{2r}{1 + 2r}.$$

Thus the frequency of Class II genotypes, i.e., the genotypes with double recombination, denoted by R, is $R = \frac{2r}{1+2r}$.

The theoretical mechanism behind RILs was first studied by Haldane and Waddington [80]. The above frequency of Class II genotypes was derived in [80]. The theoretical mechanism behind RILs generated by sibling mating is similar but more complicated. Haldane and Waddington also derived the frequency of Class II genotypes in RILs generated by sibling mating which is given by $R = \frac{4r}{1+6r}$. For details of the derivation, see [80].

1.6.6   Other Experimental Crosses

Besides the experimental crosses discussed above, there are other experimental crosses which can be used in QTL mapping. We consider a few in this final sub section.

Multiple Way Recombinant Inbred Lines. The RILs discussed above are originated from two parental strains P_1 and P_2 and are referred to as two-way RILs. Multiple way RILs are originated from 2^n ($n \geq 2$) parental strains with each strain containing different genetic materials. For example, 2^2-way RILs are generated as follows. It starts with four parental strains with different genetic materials: P_1, P_2, P_3, P_4. First, P_1 and P_2 are crossed, and P_3 and P_4 are crossed and two different F_1's are generated. Then these two F_1's are crossed to produce a generation denoted by $F_1^{(2)}$. This $F_1^{(2)}$ plays the role of F_1 in two-way RILs. 2^2-way RILs are obtained by repeated selfing or sibling mating initiated from $F_1^{(2)}$. For 2^4-way RILs, eight parental strains are prepared. Four of them generate a $F_1^{(2)}$ and the other four generate another $F_1^{(2)}$ in the same way as above. Then the two $F_1^{(2)}$'s are crossed to produce a generation denoted by $F_1^{(4)}$. The 2^4-way RILs are then obtained by repeated selfing or sibling mating initiated from $F_1^{(4)}$. For more details and the theoretical mechanism behind multiple way RILs, the reader is referred to [20].

Intermated Recombinant Inbred Lines (IRILs). The process of generating IRILs is as follows. It initiates from an F_2, i.e., an intercross generation. The F_2 population is allowed random mating for a few generations, say, t generations, to produce a generation denoted by $F_2^{(t)}$. Then selfing or sibling mating begins from $F_2^{(t)}$ in the same way as in RILs. For more details, see, e.g., [178] and [198].

Chapter 2

Selected Topics in Statistics

In this chapter, we discuss some selected topics in statistics. These topics include basic statistical concepts, § 2.1 and § 2.2, estimation, § 2.3, hypothesis testing, § 2.4, linear and generalized linear models, § 2.5, mixture models and EM algorithm, § 2.6, Bayesian analysis, § 2.7, samplers for Markov chain Monte Carlo simulations, § 2.8, feature selection in small-n-large-p models, § 2.9, and variable selection criteria for small-n-large-p models, § 2.10. In the discussion of these topics, we choose to elucidate the statistical principles rather than to dwell on the details so that the reader can have a deeper understanding of the essence of statistical analysis.

2.1 Population and Distribution

The essence of statistics is to make inferences on characteristics of a population based on a sample obtained from the population. In this section, we discuss what we mean by a population in a statistical sense and how to characterize a population mathematically.

2.1.1 Natural Population and Statistical Population

First, we need to distinguish between a statistical population and a natural population. A statistical population differs from a natural population, though they might be related. A statistical population is narrower. It focuses on particular aspects of a natural population but ignores all other aspects of the natural population. Statistical populations also differ in different studies even if the studies are on the same natural population.

The plantation species of radiata pine in Southern Australia is an example of a natural population. The following traits have been considered in QTL studies: the annual brown cone number at eight years of age (CN), the diameter of stem at breast height (DBH) and the branch quality score ranging from 1 (poorest) to 6 (best) (BS). When these traits are studied separately, each trait gives rise to a statistical population which is the collection of the trait values in the natural population. Thus, associated with the natural population, we have three statistical populations: a CN population, a DBH population and a BS population. We can also study these three traits together, then the statistical population of concern is the collection of all the triplets of CN, DBH and BS values of the natural population.

2.1.2   Distributions

From now onwards, we refer to a statistical population simply as a population unless otherwise mentioned. In general, a population can be considered as a collection of real numbers such as those of CN or DBH of the radiata pine. A population can be described by a distribution. Roughly speaking, a distribution describes for any set of numbers its proportion in the population. For example, the CN of the radiata pine ranges from 0 to a maximum 45. The distribution of CN specifies for each number its proportion in the population. When random sampling is repeatedly carried out, the frequency of the numbers in a particular set obtained approaches the proportion of the set in the population. Random sampling means that any number in the population has the same chance to be sampled. Thus, the proportion of a set is also the probability of the set in random sampling. These two terms are used interchangeably. A rigorous formulation of a population and its distribution is based on a probability measure space. However, the rigorous formulation, which is beyond the reach of a general reader who is not trained in mathematics, is not necessary for our discussion. We will not go any further on the rigorous formulation besides what is mentioned here.

A distribution can be characterized by a cumulative distribution function (c.d.f.) $F(x)$ which specifies the probability that a randomly sampled number is less than or equal to x. A c.d.f. has the following properties:

(i) $0 \leq F(x) \leq 1$;

(ii) $F(x)$ is non-decreasing;

(iii) $\lim_{x \to -\infty} F(x) = 0$, $\lim_{x \to \infty} F(x) = 1$.

In most practical populations, the c.d.f. is either a step function or a continuous function; see Figure 2.1 and Figure 2.2.

A step c.d.f. describes a discrete population. A discrete population consists of only a finite or a countable number of distinct real numbers and each number has a positive probability. A continuous c.d.f. describes a population consisting of numbers (uncountable) in a continuous range and the probability of any single number is essentially zero. For such populations, the distribution can also be characterized by a probability density function (p.d.f.) $f(x)$. The p.d.f. is related to the c.d.f. as follows:

$$F(x) = \begin{cases} \int_{-\infty}^{x} f(t)dt, & \text{for continuous distribution,} \\ \sum_{x_i \leq x} f(x_i), & \text{for discrete distribution,} \end{cases}$$

where, in the case of discrete distribution, the x_i's are the points with positive

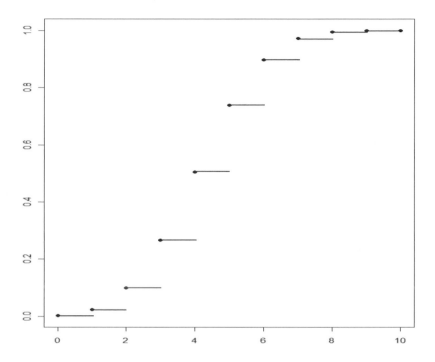

FIGURE 2.1: The cumulative distribution function of a discrete distribution.

probability. In what follows, the integral notation is used for both discrete and continuous distributions since the sum is indeed an integral with respect to a counting measure. A probability density function must satisfy the following properties:

$$f(x) \geq 0, \qquad \int_{-\infty}^{\infty} f(x)dx = 1.$$

The most important characteristics of a distribution are mean, variance, moments and quantiles, denoted by μ, σ^2, m_k and x_α, respectively. They are defined as follows.

$$\mu = \int_{-\infty}^{\infty} x f(x)dx, \qquad \sigma^2 = \int_{-\infty}^{\infty} (x - \mu)^2 f(x)dx,$$

$$m_k = \int_{-\infty}^{\infty} x^k f(x)dx, \qquad x_\alpha = \min\{x : F(x) \geq \alpha\}.$$

For any function $g(x)$, we denote by $Eg(X)$ and $\text{Var}(g(X))$ the two integrals $\int_{-\infty}^{\infty} g(x)f(x)dx$ and $\int_{-\infty}^{\infty} [g(x) - Eg(x)]^2 f(x)dx$, respectively.

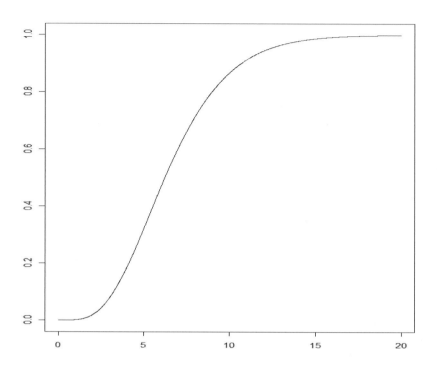

FIGURE 2.2: The cumulative distribution function of a continuous distribution.

2.1.3   Special Distributions

For practical populations, the distribution is rarely known. Furthermore, the actual distribution is usually too complicated to study though, conceptually, it can be characterized by a c.d.f. or a p.d.f.. For any statistical inference to be feasible, the distribution of a practical population is approximated by a simpler and mathematically manageable distribution. In this sub section, we discuss a few such distributions which are commonly used in QTL mapping.

Normal Distribution

The normal distribution, also known as Gaussian distribution, has the following p.d.f.:

$$\phi(x) = \frac{1}{\sqrt{2\pi}\sigma} \exp\{-\frac{1}{2\sigma^2}(x - \mu)^2\}, \quad -\infty < x < \infty,$$

where σ is a positive number. In fact, μ and σ^2 are, respectively, the mean and

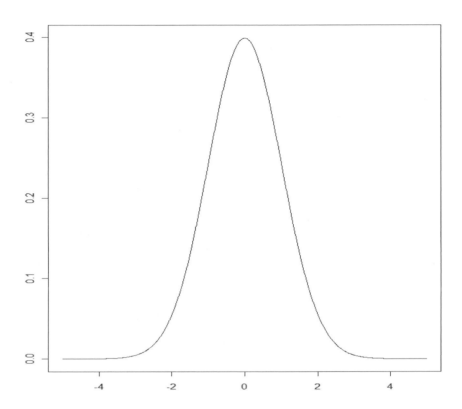

FIGURE 2.3: The probability density function of the standard normal distribution.

variance of the distribution. The normal distribution with mean μ and variance σ^2 is denoted by $N(\mu, \sigma^2)$. The p.d.f. of the normal distribution is symmetric about its mean and has a bell shape. The normal distribution $N(0, 1)$ is called the standard normal distribution. Its p.d.f. and c.d.f. are denoted respectively by $\phi(x)$ and $\Phi(x)$. The p.d.f. $\phi(x)$ is depicted in Figure 2.3.

The normal distribution is usually used to approximate the distribution of the continuous traits whose variation is caused by many unknown factors but each factor has only a tiny effect on the variation. For example, consider the DBH of the radiata pine with the same genetic variants, i.e., the same genotype at the unknown QTL. The factors that cause variation of DBH among individuals might include soil degradation, land topography, sunshine, rainfall, forest canopy, plant density, etc. But none of these factors has a dominating effect on the variation. Thus the distribution of the DBH of radiata pine can well be approximated by a normal distribution.

Bernoulli Distribution and Binomial Distribution

If a population consists of only two distinct objects, the distribution can be described by a Bernoulli distribution. The two distinct objects can be represented by 0 and 1. Let p be the proportion of 1 in the population, which is also conventionally called the probability of success. The Bernoulli distribution has the p.d.f.:

$$f(x) = p^x (1-p)^{1-x}, \ x = 0, 1.$$

The mean and variance of the Bernoulli distribution are p and $p(1-p)$, respectively.

The two distinct objects are usually individuals with or without a certain characteristic. For example, for a natural population of an agricultural plant, if we are concerned with whether or not a plant is drought resistant, then the two distinct objects are the one that is drought resistant and the other one that is not drought resistant.

Let $X_i, i = 1, \ldots, n$, be independent and identically distributed (i.i.d.) Bernoulli random variables with probability of success p. The distribution of $T = \sum_{i=1}^{n} X_i$ is called a binomial distribution with index n and probability of success p and the p.d.f. of T is given by

$$P(T = t) = \binom{n}{t} p^t (1-p)^{n-t}, \ t = 0, 1, \ldots, n.$$

The mean and variance of the binomial distribution are np and $np(1-p)$ respectively.

Another way to describe the binomial distribution is as follows. Suppose there are n independent trials. Each trial has two possible outcomes, success and failure, and the probability of success is p. Let T be the number of successes in these n trials. Then the distribution of T is the binomial distribution with p.d.f. given above. When $p = 0.5$, the binomial distribution is symmetric about its mean np. When $p < 0.5$, it skews to the left. When $p > 0.5$, it skews to the right. The p.d.f.'s of binomial distributions with $n = 10$, $p = 0.25, 0.5$ and 0.75 are depicted in Figure 2.4.

Polytomous Distribution and Multinomial Distribution

A population consisting of $k(> 2)$ distinct categories can be described by a polytomous distribution. An individual in the population can be represented by a $(k-1)$-vector $\boldsymbol{x} = (x_1, \ldots, x_{k-1})$ where $x_j = 0$ or 1 and $\sum_{j=1}^{k-1} x_j \leq 1$. If an individual falls into the jth $(j \neq k)$ category, then the jth component of $\boldsymbol{x}$ is 1 and all the others are 0. The individuals in the kth category are

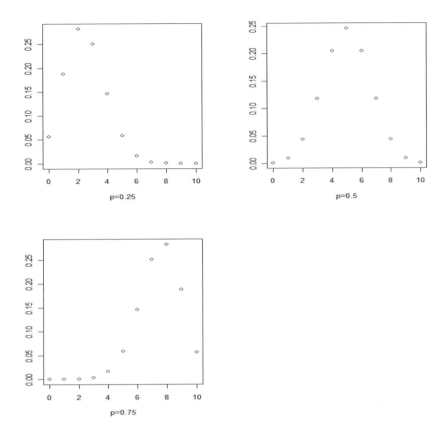

FIGURE 2.4: The probability density functions of Binomial distributions with $n = 10$ and probability of success $p = 0.25, 0.5$ and 0.75.

represented by a zero $(k-1)$-vector $\boldsymbol{x} = (0, \ldots, 0)$. Let p_j be the proportion of the individuals in the jth category. The polytomous distribution has the following p.d.f.:

$$f(\boldsymbol{x}) = p_1^{x_1} \cdots p_{k-1}^{x_{k-1}} \left(1 - \sum_{j=1}^{k-1} p_j\right)^{1 - \sum_{j=1}^{k-1} x_j}, x_j = 0 \text{ or } 1, \sum_{j=1}^{k-1} x_j \leq 1.$$

Let $\boldsymbol{X}_i, i = 1, \ldots, n$, be i.i.d. random vectors with the above polytomous distribution. The distribution of $\boldsymbol{N} = \sum_{i=1}^n \boldsymbol{X}_i = (n_1, \ldots, n_{k-1})^\tau$ is called a multinomial distribution with index n and parameters $\boldsymbol{p} = (p_1, \ldots, p_{k-1})^\tau$. The p.d.f. of $\boldsymbol{N}$ is given by

$$P((n_1, \ldots, n_{k-1})^\tau) = \frac{n!}{n_1! n_2! \cdots n_k!} p_1^{n_1} \cdots p_{k-1}^{n_{k-1}} p_k^{n_k}, \ 0 \leq n_j \leq n, \sum_{j=1}^{k-1} \leq n,$$

where $n_k = n - \sum_{j=1}^{k-1} n_j$ and $p_k = 1 - \sum_{j=1}^{k-1} p_j$.

Like the binomial random variable, the multinomial random variable $\boldsymbol{N}$ can also be considered arising from n independent trials, each having k possible outcomes with probabilities $p_1, p_2, \cdots, p_k$. The component n_j of $\boldsymbol{N}$ is the number of trials with the jth outcome.

Poisson Distribution

The Poisson distribution is a discrete distribution with positive masses on non-negative integers. The p.d.f. of a Poisson distribution is given by

$$f(x) = \frac{e^{-\lambda}\lambda^x}{x!}, \; x = 0, 1, 2, \ldots.$$

The mean and variance of the Poisson distribution are both equal to λ. For small λ, the distribution is skewed to the left. As λ gets larger, the distribution approaches symmetry. The p.d.f. of the Poisson distributions with $\lambda = 2, 4, 6$ and 10 are plotted in Figure 2.5.

A Poisson distribution is usually used to model the number of rare events. It can be used to describe populations consisting of counts of certain events such as the population of CN of the radiata pine.

Exponential Family

The distributions discussed above belong to a general family of distributions known as the exponential family. An exponential family has a general p.d.f. given as follows:

$$f(x) = \exp\{\frac{\boldsymbol{\theta}' T(x) - b(\theta)}{\phi} + c(x, \phi)\} I_A(x),$$

where $\phi > 0$ is called the dispersion parameter, $a(\cdot), b(\cdot)$ and $c(\cdot, \cdot)$ are specific known functions and $I_A(x)$ is the indicator function of the set A. The particular forms of ϕ, A, $b(\cdot)$, $c(\cdot, \cdot)$ and the form of θ for the four distributions discussed above are given in Table 2.1.

In addition to the four distributions above, the exponential families include many other distributions such as exponential, gamma and log-normal distributions which are useful for modelling lifetimes. For more details about exponential families, the reader is referred to, e.g., [132].

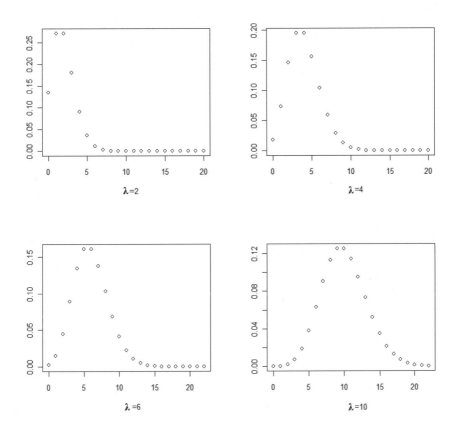

FIGURE 2.5: The probability density functions of Poisson distributions with $\lambda = 2, 4, 6$ and 10.

2.2 Random Variable, Samples, Statistics and Related Distribution

2.2.1 Random Sample and Statistics

In practical problems, the characteristics of a population (or its distribution) are unknown. To make inferences on population characteristics, we can take samples from the population. In a sampling scheme, the probability of each individual to be sampled is specified. If each individual of the population has the same probability to be sampled, the sampling is called simple random sampling. Let X denote the value of a randomly sampled individual. X is called a random variable. If individuals are repeatedly sampled with re-

TABLE 2.1: Characteristics of a few distributions in the exponential family.

	Normal $N(\mu, \sigma^2)$	Bernoulli $B(1, \pi)$	Polytomous $M(1, (\pi_1, \ldots, \pi_{k-1}))$	Poisson $P(\lambda)$
$T(x)$	x	x	$(x_1, \ldots, x_{k-1})$	x
A	$(-\infty, +\infty)$	$\{0, 1\}$	$(x_1, \ldots, x_{k-1})$ $x_j = 0$ or $1,$ $\sum_{j=1}^{k-1} x_j \leq 1$	$\{0, 1, 2, \cdots\}$
ϕ	σ^2	1	1	1
$b(\theta)$	$\theta^2/2$	$\ln(1 + e^\theta)$	$\ln(1 + \sum_{j=1}^{k-1} e^{\theta_j})$	e^θ
$c(x, \phi)$	$-\frac{1}{2}(\frac{x^2}{\phi} + \ln 2\pi\phi)$	1	1	$-\ln x!$
θ	μ	$\ln \frac{\pi}{1-\pi}$	$\theta_j = \ln \frac{\pi_j}{1-\sum_{j=1}^{k-1} \pi_j}$	$\ln \lambda$

placement indefinitely according to the same sampling scheme, the set of all sampled values forms a sampling population. The distribution of the sampling population is called the distribution of X. In the case of simple random sampling, the distribution of X is the same as the distribution of the population from which X is sampled. However, when sampling is not simple random sampling, the distribution of X, in general, is different from the distribution of the population.

We will focus on simple random sampling. A random sample of size n is a set of n independent and identically distributed random variables $(X_1, \ldots, X_n)$ sampled from the population. A data set of size n is the set of n fixed values generated by a random sample. A data set is a realization of the random sample. For convenience, we call both a random sample and its realization simply a sample. A sample mimics the population in all aspects. The larger the sample size, the more alike the sample and the population. It is like a Russian nesting doll. A Russian nesting doll, see Figure 2.6, consists of many layers of carved wooden dolls, each being cut into two halves. The halves fit perfectly top to bottom. The dolls gradually reduce their scales from the outmost one (the largest) until the innermost one (the smallest) is so tiny that it cannot be cut into halves. The features of all the inner dolls mimic the outmost doll. The closer to the outmost doll, the more alike an inner doll and the outmost doll. The population is like the outmost doll. A sample is like an inner doll. The characteristics of the population can be inferred from those of a sample. The larger the sample size, the more accurate the inference.

The characteristics of a sample can be described by various statistics. A statistic is a function of the sample. A statistic is a random variable, but it is different from a general random variable. A statistic is computable when the sample is realized. However, a general random variable might not be computable. The following are a few important statistics:

Empirical distribution: $F_n(x) = \frac{1}{n} \sum_{i=1}^{n} I\{X_i \leq x\}$

Sample mean: $\bar{X} = \frac{1}{n} \sum_{i=1}^{n} X_i$

FIGURE 2.6: The Russian doll.

Sample variance: $S^2 = \frac{1}{n} \sum_{i=1}^{n} (X_i - \bar{X})^2$
Sample moments: $\hat{m}_k = \frac{1}{n} \sum_{i=1}^{n} X_i^k, k = 1, 2, \ldots$
Sample quantile: $\hat{x}_\alpha = \min\{x : F_n(x) \geq \alpha\}$.

2.2.2 t, χ^2 and F Distributions

To make inference about the characteristics of the population from the characteristics of a sample, we need to find the distribution of the statistics of a sample. The most important distributions related to the distribution of statistics are normal, t, χ^2 and F distributions. The normal distribution has been briefly discussed in the previous section. We consider the other three distributions in the following.

Let $Z_1, \cdots, Z_m$ be m i.i.d. standard normal random variables. The distribution of $\chi_m^2 = \sum_{j=1}^{m} Z_j^2$ is referred to as a χ^2-distribution with degrees of freedom m.

Let Z be a standard normal random variable, and χ_m^2 a χ^2 random variable with degrees of freedom m independent of Z. The distribution of $t = \frac{Z}{\sqrt{\chi_m^2/m}}$ is referred to as a t-distribution with degrees of freedom m.

Let χ_m^2 and χ_n^2 be two independent χ^2 random variables with degrees of freedom m and n respectively. The distribution of $F = \frac{\chi_m^2/m}{\chi_n^2/n}$ is referred to as an F-distribution with degrees of freedom m and n.

The three distributions above arise naturally from the distributions of statistics, which we consider next.

2.2.3   Distributions of Sample Mean and Sample Variance

The distributions of the sample mean $\bar{X}$ and sample variance S^2 can be derived either exactly or asymptotically depending on the assumption made for the distribution of the population.

If the distribution of the population is normal with mean μ and variance σ^2, then

$$\bar{X} \sim N(\mu, \frac{\sigma^2}{n}) \text{ and } \frac{nS^2}{\sigma^2} \sim \chi_{n-1}^2,$$

where $\sim$ stands for "distributed as". Furthermore, $\bar{X}$ and S^2 are independent. As an immediate consequence, we have

$$\frac{\sqrt{n}(\bar{X} - \mu)}{\sqrt{nS^2/(n-1)}} = \frac{\sqrt{n-1}(\bar{X} - \mu)}{S} \sim t_{n-1}.$$

If the distribution of the population is not normal, the exact distributions of the sample mean and sample variance usually cannot be derived. The central limit theorems in probability can be used to derive their asymptotic distributions. In particular, if the distribution of the population has a finite second moment, then both $\frac{\sqrt{n}(\bar{X}-\mu)}{\sigma}$ and $\frac{\sqrt{n-1}(\bar{X}-\mu)}{S}$ have an asymptotic standard normal distribution.

2.3   Estimation

The estimation of population characteristics based on a sample is basic in statistical inference. There are various methods for estimation. The most commonly used methods are maximum likelihood estimation (MLE) and the method of moments (MM).

2.3.1   Maximum Likelihood Estimation

Suppose that the distribution of the population belongs to a family with p.d.f. indexed by a parameter θ (θ is either a scaler or a vector), say, $f(x; \theta)$.

Then any characteristic of the population is a function of the parameter θ. The estimation of the population characteristics reduces to the estimation of θ. Let $\boldsymbol{X} = (X_1, \ldots, X_n)$ be a random sample. The joint p.d.f. is then

$$f(\boldsymbol{x}; \theta) = \prod_{i=1}^{n} f(x_i; \theta)$$

where $\boldsymbol{x} = (x_1, \ldots, x_n)$. Mathematically, the joint p.d.f. is a function of both the sample and the parameter. Statistically, we can consider the joint p.d.f. from two different viewpoints. First, if θ is fixed, it is a function of the sample only and specifies how probable the sample will be realized. Second, if we have already observed the sample and fix the sample at its realization, the p.d.f. is then a function of θ only. For each particular θ, the function measures the likelihood that the observed data is produced by the particular θ. The larger the value of the function, the more likely that the corresponding θ could have produced the data. As a function of θ, it is therefore called the likelihood function and is denoted by $L(\theta; \boldsymbol{X})$ or simply by $L(\theta)$. The likelihood function is not defined only for simple random samples. For any random sample whose components are not necessarily independent and identically distributed, the joint p.d.f. of the random sample considered as a function of the parameter is called the likelihood function. In the case of simple random sampling, it is more convenient to deal with the log transformation of the likelihood since it transforms a product to a sum. The log transformation of the likelihood is called the log likelihood and denoted by $l(\theta; \boldsymbol{X})$ or $l(\theta)$, i.e.,

$$l(\theta; \boldsymbol{X}) = \sum_{i=1}^{n} \ln f(X_i, \theta).$$

It is natural and reasonable to take, as an estimate of θ, the value of θ that is most likely to produce the observed data. The maximum likelihood estimator (MLE) of θ is defined as

$$\hat{\theta} = \hat{\theta}(\boldsymbol{X}) = \operatorname{argmin}_\theta l(\theta; \boldsymbol{X}).$$

When $\boldsymbol{X}$ is observed as $\boldsymbol{x}$, the estimate of θ provided by the MLE is $\hat{\theta}(\boldsymbol{x})$. We need to distinguish between the estimator and the estimate. The estimator is a random variable. The estimate is a particular value of the estimator evaluated at the observed sample.

The MLE has several desirable properties. First, it is invariant; that is, if one is interested in $\xi = \xi(\theta)$ then the MLE of ξ is given by $\hat{\xi} = \xi(\hat{\theta})$. Second, more importantly, the asymptotic distribution of $\hat{\theta}$ can be derived. Suppose θ is a k-dimensional vector. Let

$$I(\theta) = -\mathrm{E} \begin{pmatrix} \frac{\partial^2 l(\theta)}{\partial \theta_1^2} & \frac{\partial^2 l(\theta)}{\partial \theta_1 \partial \theta_2} & \cdots & \frac{\partial^2 l(\theta)}{\partial \theta_1 \partial \theta_k} \\ \frac{\partial^2 l(\theta)}{\partial \theta_2 \partial \theta_1} & \frac{\partial^2 l(\theta)}{\partial \theta_2^2} & \cdots & \frac{\partial^2 l(\theta)}{\partial \theta_2 \partial \theta_k} \\ \cdots & \cdots & \cdots & \cdots \\ \frac{\partial^2 l(\theta)}{\partial \theta_k \partial \theta_1} & \frac{\partial^2 l(\theta)}{\partial \theta_k \partial \theta_2} & \cdots & \frac{\partial^2 l(\theta)}{\partial \theta_k^2} \end{pmatrix}.$$

If $k = 1$, $I(\theta)$ is called the Fisher information number. In general, $I(\theta)$ is called the Fisher information matrix. Under some regularity conditions which are satisfied by most practically assumed distributions, especially by distributions of exponential families, the MLE $\hat{\theta}$ of θ has an asymptotic normal distribution with mean θ and variance-covariance matrix given by the inverse of the Fisher information matrix, i.e., $I^{-1}(\theta)$, i.e., $\hat{\theta} \sim N(\theta, I^{-1}(\theta))$. Furthermore, in the asymptotic distribution, the $I^{-1}(\theta)$ can be replaced by its estimator $I^{-1}(\hat{\theta})$.

2.3.2   Method of Moments

For the MLE to be applicable, we must have the information about the form of the density function of the underlying distribution. The method of moments however does not require such information. The essence of the method of moments is to estimate any characteristic of a distribution F by the corresponding characteristic of the empirical distribution $\hat{F}_n$ based on a sample of size n from the distribution. For example, the mean and variance of F, μ and σ^2, are estimated, respectively, by the sample mean and sample variance, $\bar{X}$ and S^2. Note that the sample mean and sample variance are the mean and variance of $\hat{F}_n$ respectively. In general, Suppose ξ is a parameter of interest and can be expressed as a function of the moments of F up to order k, i.e., $\xi = g(m_1, \cdots, m_k)$ where $m_j, j = 1, \ldots, k$ are the population moments. The method of moments (MM) estimator of ξ based on a sample $(X_1, \ldots, X_n)$ is then defined as

$$\hat{\xi} = g(\hat{m}_1, \cdots, \hat{m}_k), \quad \text{where} \quad \hat{m}_j = \frac{1}{n}\sum_{i=1}^{n} X_i^j, \ j = 1, \ldots, k.$$

While the MLE is usually unique, the MM estimator is not necessarily unique. For example, for the parameter λ in the Poisson distribution, we have $\lambda = \mu$, the mean of the distribution, and also $\lambda = \sigma^2 = m_2 - m_1^2$, the variance of the distribution. Thus both $\bar{X}$ and S^2 are MM estimators of λ.

There is a desirable property of the MM estimators: the asymptotic distribution can be easily obtained. Suppose g is differentiable. Let $\boldsymbol{\mu} = (m_1, \ldots, m_k)^\tau$. Denote by ∂g the vector of the first derivatives of g evaluated at $\boldsymbol{\mu}$. If $\partial g \neq 0$ and the underlying distribution has finite moments up to order $2k$, then, asymptotically, $\sqrt{n}[g(\boldsymbol{X}_n) - g(\boldsymbol{\mu})]$ follows a normal distribution with mean zero and variance $(\partial g)^\tau \Sigma \partial g$, where $\Sigma = \text{Var}((X, X^2, \cdots, X^k))$. For example, we can express $S^2 = g(\hat{m}_1, \hat{m}_2) = \hat{m}_2 - \hat{m}_1^2$. It is easy to calculate that $\partial g = (-2m_1, 1)^\tau$ and

$$\Sigma = \begin{pmatrix} m_2 - m_1^2 & m_3 - m_1 m_2 \\ m_3 - m_1 m_2 & m_4 - m_2^2 \end{pmatrix},$$

where m_k is the kth moment of the underlying distribution. Some algebra yields that

$$(\partial g)^\tau \Sigma \partial g = m_4 - 4m_1^4 + 8m_1^2 m_2 - 4m_1 m_3 - m_2^2 = E(X - \mu)^4 - \sigma^4.$$

Note that $\mu = m_1, \sigma^2 = m_2 - m_1^2$. Thus, $\sqrt{n}(S^2 - \sigma^2)$ has an asymptotic normal distribution with mean zero and variance $E(X - \mu)^4 - \sigma^4$.

2.3.3 Unbiasedness, Mean Square Error and Consistency

Any particular estimate of the population parameter is the estimator calculated on a single realization of the sample. It is natural to ask whether or not there is a systematic bias when the estimator is repeatedly used for the same estimation problems. The bias of an estimator is defined as $\text{bias}(\hat{\theta}) = \text{E}\hat{\theta} - \theta$. If the bias is zero, the estimator is said to be unbiased. But unbiasedness alone does not ensure that an estimator is good. The goodness of an estimator is measured by its mean square error (MSE) defined as follows:

$$\text{MSE}(\hat{\theta}) = \text{E}(\hat{\theta} - \theta)^2.$$

The MSE can be decomposed as

$$\text{MSE}(\hat{\theta}) = \text{E}(\hat{\theta} - \text{E}\hat{\theta})^2 + (\text{E}\hat{\theta} - \theta)^2 = \text{Var}(\hat{\theta}) + \text{bias}^2(\hat{\theta}).$$

When an estimator is unbiased, its MSE is the same as its variance. If there are many estimators for a parameter, the one with the minimum MSE should be chosen. However, it is not always possible to have an estimator that achieves the minimum MSE if the estimators are not confined to a certain class. An example is the class of all unbiased estimators. If the underlying distribution belongs to an exponential family, a uniformly minimum variance unbiased estimator usually exists.

The mean square error can be extended to a general *risk* function. Instead of the squared difference $(\hat{\theta} - \theta)^2$, let the error be measured by a general *loss* function $L(\hat{\theta}, \theta)$. The squared difference is referred to as the squared error loss. The expected loss $\text{E}L(\hat{\theta}, \theta)$ is called the risk function. The goodness of estimators can also be measured by the risk function. We then can consider minimum risk estimators within certain classes of estimators.

For any estimator, a least requirement is that the estimator must be consistent in the following sense. An estimator, which is in the form of a function of the sample, depends on the sample size n. Thus we denote by $\hat{\theta}_n$ the estimator of a parameter θ constructed from a sample of size n. The consistency requires that, as n goes to infinity, $\hat{\theta}_n$ must converge to θ in a certain sense. This is a natural requirement. Since, as the sample size gets larger and larger, eventually we have all the information about the population, we should be able to know any characteristic of the population exactly. If an estimator is not consistent, it is not an acceptable estimator. Usually, a better estimator can be found. The MLE and MM estimators are usually consistent.

There are other criteria to assess whether or not an estimator is a desirable one. Those criteria include admissibility, minimaxity, asymptotic efficiency, etc. The interested reader is referred to any standard textbook on advanced statistics.

2.3.4   Interval Estimation

An estimator of a parameter provides a single estimated value of the parameter. The estimator is thus called a point estimator. The point estimator does not provide information on the accuracy of the estimation. In contrast, an interval estimation provides an interval (a range of the parameter) as well as the probability that the interval will cover the true parameter. Such an interval is called the confidence interval. Given a random sample X, a confidence interval of a parameter θ is constructed as a random interval $[\underline{\theta}(X), \bar{\theta}(X)]$ such that $P_\theta(\theta \in [\underline{\theta}(X), \bar{\theta}(X)]) \geq 1 - \alpha$ for any θ, where $0 < \alpha < 1$. The number $1 - \alpha$ is called the confidence level. If $1 - \alpha = \inf_\theta P_\theta(\theta \in [\underline{\theta}(X), \bar{\theta}(X)])$ then $1 - \alpha$ is called the confidence coefficient. Thus if $[\underline{\theta}(X), \bar{\theta}(X)]$ is a confidence interval of θ of confidence coefficient $1 - \alpha$, suppose that we were able to draw the random sample X repeatedly and each time compute the interval $[\underline{\theta}(X), \bar{\theta}(X)]$, then roughly $100(1 - \alpha)\%$ of such computed intervals will cover the true unknown θ. Therefore the confidence coefficient is also called the coverage probability. For a particular parameter θ, many different confidence intervals can be constructed. To compare different confidence intervals, two characteristics must be considered: the coverage probability $1 - \alpha$ and the the expected length $E[\bar{\theta}(X) - \underline{\theta}(X)]$. One confidence interval is better than the other if either it has a shorter expected length and a coverage probability not lower than the other, or it has higher coverage probability and an expected length not longer than the other. Without considering the length, the interval $(-\infty, +\infty)$ provides a trivial interval with the highest coverage probability 1. Without considering the coverage probability, any trivial interval containing a single point has the shortest length 0. However, these intervals are useless. A good confidence interval is the one that has high coverage probability and short expected length.

2.3.4.1   Asymptotic Confidence Intervals

Confidence intervals can be constructed from appropriate estimators. Let $\hat{\theta}$ be an estimator of θ. Let $\hat{\sigma}_{\hat{\theta}}$ be an consistent estimator of the standard deviation of $\hat{\theta}$. The confidence interval of confidence coefficient $1 - \alpha$ obtained from $\hat{\theta}$ is of the form

$$[\,\hat{\theta} - c_{1-\alpha/2}\hat{\sigma}_{\hat{\theta}}, \hat{\theta} + c_{1-\alpha/2}\hat{\sigma}_{\hat{\theta}}\,],$$

where $c_{1-\alpha/2}$ is the quantile of the distribution (or asymptotic distribution) of $(\hat{\theta} - \theta)/\hat{\sigma}_{\hat{\theta}}$. For example, the sample mean $\bar{X}$ provides an estimator for the population mean μ. A consistent estimator of the standard deviation of $\bar{X}$ is given by $S/\sqrt{n-1}$. Thus the confidence interval of μ is

$$[\,\bar{X} - c_{1-\alpha/2}S/\sqrt{n-1},\quad \bar{X} + c_{1-\alpha/2}S/\sqrt{n-1}\,].$$

We know from § 2.2.3 that $\sqrt{n-1}(\bar{X} - \mu)/S$ has a t-distribution with degrees of freedom $n-1$ if the population has a normal distribution, and an asymptotic standard normal distribution otherwise. Thus $c_{1-\alpha/2} = t_{n-1}(1 - \alpha/2)$ if the distribution of the population is normal and $z_{1-\alpha/2}$ otherwise.

We now consider the construction of confidence intervals by using MLE and MM estimators. Let $\hat{\theta}_{\text{MLE}}$ be the MLE of θ. By the properties of MLE given in § 2.3.1, asymptotically, $(\hat{\theta}_{\text{MLE}} - \theta)/\sqrt{I(\hat{\theta}_{\text{MLE}}))^{-1}} \sim N(0, 1)$. Then we obtain a confidence interval of θ with approximate confidence coefficient $1 - \alpha$ given by

$$[\, \hat{\theta}_{\text{MLE}} - z_{1-\alpha/2}\sqrt{I^{-1}(\hat{\theta}_{\text{MLE}})}, \quad \hat{\theta}_{\text{MLE}} + z_{1-\alpha/2}\sqrt{I^{-1}(\hat{\theta}_{\text{MLE}})} \,]$$

Let $\hat{\theta}_{\text{MM}} = g(\hat{m}_1, \ldots, \hat{m}_k)$ be a MM estimator of θ, where $\hat{m}_j$ is the jth sample moment. Let $\sigma_g^2 = (\partial g)^\tau \Sigma \partial g$, as given in § 2.3.2. Denote by $\hat{\sigma}_g^2$ the estimator of σ_g^2 obtained by replacing the population moments involved in σ_g^2 with the corresponding sample moments. Then we have $(\hat{\theta}_{\text{MM}} - \theta)/\hat{\sigma}_g \sim N(0, 1)$. A confidence interval of θ with approximate confidence coefficient $1 - \alpha$ is obtained as

$$[\, \hat{\theta}_{\text{MM}} - z_{1-\alpha/2}\hat{\sigma}_g, \quad \hat{\theta}_{\text{MM}} + z_{1-\alpha/2}\hat{\sigma}_g \,].$$

2.3.4.2   Bootstrap Confidence Intervals

The nominal coverage probability of a confidence interval based on the asymptotic distribution of an estimator is only an approximation to the real coverage probability. The accuracy of the approximation depends on the distributional nature of the estimator and the sample size. If the distribution of the estimator is close to normal, e.g., the distribution is symmetric, the approximation is quite accurate even with a small sample size. But if the distribution of the estimator is far from normal, e.g., the distribution is seriously skewed and heavily tailed, the approximation is inaccurate even with a very large sample size. In the latter case, there usually exists a transformation of the estimator such that the transformed estimator has a distribution more like a normal distribution. The confidence interval based on the asymptotic distribution of the transformed estimator is then more accurate. However, in practical problems, such transformation is not always clear and easy to obtain. The confidence interval based on the untransformed estimator could be misleading, especially, when the sample size is small. There are also situations where even the asymptotic distributions of the estimators are not easy to derive. In these circumstances, bootstrap confidence intervals provide better alternatives to asymptotic confidence intervals. In the following, we briefly discuss bootstrap confidence intervals. We only describe bootstrap percentile confidence intervals because of their desirable properties and simplicity. For other bootstrap confidence intervals and more details, the reader is referred to [50].

Let $X = (X_1, \ldots, X_n)$ be a random sample from the population and $\hat{\theta}(X)$ an estimator of θ based on X. Let $X^* = (X_1^*, \ldots, X_n^*)$ be a random sample from the empirical distribution $F_n(x) = \frac{1}{n}\sum_{i=1}^{n} I\{X_i \leq x\}$; that is, the

$X_1^*, \ldots, X_n^*$ are chosen from $\{X_1, \ldots, X_n\}$ at random with replacement. X^* is called a bootstrap sample. A bootstrap percentile confidence interval of θ is constructed as follows. First, draw bootstrap samples X_b^* for $b = 1, \ldots, B$, B is taken to be a moderately large number. For each bootstrap sample X_b^* compute $\hat{\theta}_b^* = \hat{\theta}(X_b^*)$. Let $\hat{\theta}_{\alpha/2}^*$ and $\hat{\theta}_{1-\alpha/2}^*$ be the $(\alpha/2)$- and $(1 - \alpha/2)$-quantiles of the empirical distribution of $\hat{\theta}_1^*, \ldots, \hat{\theta}_B^*$, i.e.,

$$\hat{\theta}_{\alpha/2}^* = \hat{\theta}_{([B\alpha/2])}^*, \quad \hat{\theta}_{1-\alpha/2}^* = \hat{\theta}_{([B(1-\alpha/2)+1])}^*,$$

where the notation $[x]$ denotes the smallest integer bigger than or equal to x and $\hat{\theta}_{(k)}^*$ denotes the k-th order statistic of $\hat{\theta}_1^*, \ldots, \hat{\theta}_B^*$. The bootstrap percentile confidence interval of θ with confidence coefficient $1 - \alpha$ is approximated by the interval

$$[\, \hat{\theta}_{\alpha/2}^*, \quad \hat{\theta}_{1-\alpha/2}^* \,].$$

Conceptually, the bootstrap interval is the limit of the above interval as B goes to infinity. If B is large, say $B \approx 5,000$, the above approximation is accurate enough.

Unlike the confidence intervals based on the asymptotic distribution of the estimator where the accuracy depends on whether the distribution of the estimator $\hat{\theta}$ is close to normal or whether a transformation can be found such that the transformed estimator has a distribution close to normal, the bootstrap percentile confidence intervals are transformation-respecting. In other words, the accuracy of the bootstrap percentile confidence intervals is not affected by the distributional nature of the estimator and it is as if the desirable transformation is automatically applied when the percentile confidence intervals are constructed. See [50]. Another desirable nature of the percentile confidence interval is its range-preserving property; that is, the percentile interval is always within the range of the parameter. This property is not always possessed by the confidence intervals based on asymptotic distributions. These properties make the bootstrap percentile confidence intervals better than the intervals based on asymptotic distributions especially when the sample size is not large.

2.4 Hypothesis Testing

Many practical problems can be formulated as a problem of hypothesis testing. For example, for a given locus in the genome of a species of plant or animal, we want to answer the question whether or not the locus is a QTL for a certain quantitative trait. Suppose there are three possible genotypes at the locus in the population of the species, say, AA, Aa and aa. Let μ_{AA}, μ_{Aa} and μ_{aa} denote the trait means of all the individuals with genotypes AA, Aa and aa respectively. Then answering the postulated question amounts to testing the hypothesis $\mu_{AA} = \mu_{Aa} = \mu_{aa}$. The topic of hypothesis testing is covered in this section.

2.4.1 Neyman-Pearson Framework

We start with the Neyman-Pearson framework of hypothesis testing. We describe the framework through a parametric family of distributions $\{P(x, \theta) : \theta \in \Theta\}$ where Θ is called the parameter space which is the set of all possible values of θ. By the nature of a practical problem, the parameter space Θ is partitioned into two subspaces Θ_0 and Θ_1. The two subspaces represent two different classes of distributions of concern in the practical problem. Two hypothesis, a null hypothesis denoted by H_0 and an alternative hypothesis denoted by H_1, are formulated as $H_0 : \theta \in \Theta_0$ and $H_1 : \theta \in \Theta_1$. A test statistic T is found to make the decision rule: if $T > c$ for some specified constant c then H_0 is rejected; otherwise H_0 is not rejected. There are two types of errors which could be committed in the decision. The first type is that H_0 is rejected but it is indeed true. This type of error is called Type I error. The second type is that H_0 is not rejected but it is indeed false (H_1 is true). This type of error is called Type II error. The decision rule is determined such that the probability of committing the Type I error (or Type I error rate) is controlled at a specified low level, say α; that is, the constant c is determined as c_α such that $\sup_{\theta \in \Theta_0} P(T > c_\alpha | \theta) \leq \alpha$. The constant c_α is called the critical value of the test statistic at level α. For a particular $\theta \in \Theta_1$, the probability $P(T > c_\alpha | \theta)$ is called the power of the test at θ. As a function of θ, $P(T > c_\alpha | \theta)$ is called the power function. Suppose that at the observed sample the test statistic is computed as T_0. The $\sup_{\theta \in \Theta_0} P(T > T_0 | \theta)$ is called the p-value of the test. The decision rule can be stated in terms of the p-value as: if $p < \alpha$ then H_0 is rejected at level α, otherwise, H_0 is not rejected at level α.

It should be noted that in the above framework the two hypotheses, H_0 and H_1, are not treated on an equal footing. Indeed, the hypotheses are formulated in such a way that the Type I error is a more serious one than the Type II error, since only the probability of Type I error is controlled, not the probability of Type II error. For example, consider the diagnosis of dengue fever, a disease transmitted by mosquitoes. The symptoms of dengue fever and those of an ordinary flu are similar. A patient having the symptoms can be diagnosed either as suffering from dengue fever or flu. If a patient who is suffering from dengue fever is diagnosed as having flu, the consequence could be very serious. In the worst case, it might lead to the death of the patient because of the wrong treatment for flu. When we formulate the diagnosis as a hypothesis test, we should take the null hypothesis as that the patient is suffering from dengue fever. Such formulation justifies the control of Type I error at a low rate. If the hypotheses are not formulated this way, then the emphasis on controlling Type I error is not justifiable.

The standard choice for α is 0.05 or 0.01. Historically the critical values were tabled and these values of α were chosen to effect a reduction in the tables. Unfortunately, these choices become a convention. However, there is no reason why we should stick to these values. The control of Type I error is only one side of the coin. The other side of the coin is the power of the test at certain particular alternatives. For a fixed sample size, the lower the Type I error rate, the lower the power (equivalently, the higher the Type II error rate). While Type I error rate is controlled, a desired power can only be achieved by increasing the sample size, which amounts to increasing the cost of the study. If one is limited by the budget of the study, then a balance between the Type I error rate and power must be stricken, especially if the Type I error is not necessarily more serious when the null hypothesis is formulated simply by convenience.

In QTL mapping, the null hypothesis is always that a locus is not a QTL. The Type I error is committed if a locus is falsely declared to be a QTL. The Type II error is committed if a QTL is falsely declared to be a non-QTL. The hypotheses are indeed formulated by convenience. It is not clear in general which type of error is more serious. If the study is a preliminary study for searching candidates of QTL, the Type II error is probably more serious since it will lead to the loss of chance for discovering real QTLs. If the study is to confirm whether or not a candidate QTL is really a QTL, the Type I error is then more serious because of its possible scientific consequences. Thus in the choice of α value the nature of the study should be taken into account.

2.4.2 t-Test and F-Test

The t-test arises from the comparison of two populations. Let $X_1, \ldots, X_{n_1}$ be a sample from population 1 and $Y_1, \ldots, Y_{n_2}$ a sample from population 2. Let μ_X, μ_Y and σ_X^2, σ_Y^2 denote the means and variances of the two populations respectively. Let $\bar{X}$, $\bar{Y}$ and S_X^2, S_Y^2 be the sample means and sample variances of the two samples respectively. We want to test $H_0 : \mu_X = \mu_Y$. If $\sigma_X^2 = \sigma_Y^2$, the test statistic is given by

$$T_1 = \sqrt{\frac{n_1 n_2}{n_1 + n_2}} \frac{\bar{X} - \bar{Y}}{\sqrt{\frac{n_1 S_X^2 + n_2 S_Y^2}{n_1 + n_2 - 2}}}.$$

If $\sigma_X^2 \neq \sigma_Y^2$, the test statistic is given by

$$T_2 = \frac{\bar{X} - \bar{Y}}{\sqrt{\frac{S_X^2}{n_1 - 1} + \frac{S_Y^2}{n_2 - 1}}}.$$

If the two populations can be assumed to have normal distributions then, under the null hypothesis, T_1 has an exact t-distribution with degrees of freedom $n_1 + n_2 - 2$, giving rise to the name t-test. The null hypothesis is rejected at significance level α if $|T_1| \geq t_{n_1 + n_2 - 2, \alpha/2}$ where $t_{n, \xi}$ denotes the upper

ξ-quantile of the t-distribution with degrees of freedom n. But the exact distribution of T_2 under the null hypothesis is unknown.

If the populations have finite second moments, then both T_1 and T_2 have an asymptotic standard normal distribution under the null hypothesis. The test is also called the z-test to emphasize the asymptotic nature. For the asymptotic test, the null hypothesis is rejected if $|T_1|$ (or $|T_2|$ in the case of unequal population variances) is greater than or equal to $z_{\alpha/2}$, the upper $\alpha/2$ quantile of the standard normal distribution.

In QTL mapping experiments with a backcross design, each locus has only two different genotypes, i.e., a homozygous one and a heterozygous one. To test whether or not a particular locus is a QTL, we partition the mapping population according to the genotypes at the particular locus and then test whether or not the mean trait values of the two sub populations are equal.

The F-test arises from the comparison of more than two populations. Suppose there are K populations with different means but the same population variance. Denote by μ_k the mean of population k, $k = 1, \ldots, K$, and σ^2 denotes the common population variance. The null hypothesis of interest is $H_0 : \mu_1 = \cdots = \mu_K$. To test the null hypothesis, a random sample from each population is drawn. Let $X_{k1}, \ldots, X_{kn_k}$ be the sample from population k, $k = 1, \ldots, K$. Let $\bar{X}_{k\cdot}$ denote the kth sample mean and $\bar{X}_{\cdot\cdot}$ the overall mean of all the K samples. The F-test statistic is derived from a decomposition of the total variation in the data, i.e.,

$$\text{SST} = \sum_{k=1}^{K} \sum_{i=1}^{n_k} (X_{ki} - \bar{X}_{\cdot\cdot})^2.$$

The total variations are caused by two sources: the differences among the means of the populations and the random differences within the populations. The variations due to these two sources are measured respectively by

$$\text{SSB} = \sum_{k=1}^{K} n_k (\bar{X}_{k\cdot} - \bar{X}_{\cdot\cdot})^2 \quad \text{and} \quad \text{SSE} = \sum_{k=1}^{K} \sum_{i=1}^{n_k} (X_{ki} - \bar{X}_{k\cdot})^2.$$

In fact, we have

$$\text{SST} = \text{SSB} + \text{SSE}.$$

The F-test statistic for testing H_0 is given by

$$F = \frac{\text{SSB}/(K-1)}{\text{SSE}/(n-K)} = \frac{\text{MSB}}{\text{MSE}},$$

where $n = \sum_{k=1}^{K} n_k$ and $\text{MSB} = \text{SSB}/(K-1)$ and $\text{MSE} = \text{SSE}/(n-K)$.

If the populations are assumed to have normal distributions then under the null hypothesis SSB/σ^2 and SSE/σ^2 are independent χ^2 random variables with degrees of freedom $K-1$ and $n-K$ respectively; therefore, the F-statistic follows a F-distribution with degrees of freedom $K-1$ and $n-K$. The null hypothesis is rejected at significance level α if $F \geq f_{K-1,n-K,\alpha}$ the upper α quantile of the F-distribution with degrees of freedom $K - 1$ and $n - K$.

If no particular distributions can be assumed for the populations and the populations can only be assumed to have finite second moments, then $(K-1)F$ has an asymptotic χ^2-distribution with degrees of freedom $K - 1$ under the null hypothesis. The null hypothesis is rejected at approximate significance level α if $(K-1)F \geq \chi_{K-1,\alpha}$ the upper α quantile of the χ^2-distribution with degrees of freedom $K - 1$.

In QTL mapping experiments with an intercross design, each locus has three different genotypes. The problem of testing whether or not a particular locus is a QTL reduces to a F-test for the equality of the three means associated with the three different genotypes.

2.4.3   Pearson's χ^2-Test

Pearson's χ^2-test is another common test applied in genetic studies. The Pearson's χ^2-test is related to a contingency tables. Let Y and X be two categorical characteristics of the population with k and m levels respectively. The sampling subjects are classified on these two variables into km categories. The number of sampling subjects that fall into each category are summarized into the following table:

		Factor Y				
		1	2	$\cdots$	m	
Factor X	1	n_{11}	n_{12}	$\cdots$	n_{1m}	$n_{1\cdot}$
	2	n_{21}	n_{22}	$\cdots$	n_{2m}	$n_{2\cdot}$
	$\vdots$	$\vdots$	$\vdots$	$\vdots$	$\vdots$	$\vdots$
	k	n_{k1}	n_{k2}	$\cdots$	n_{km}	$n_{k\cdot}$
		$n_{\cdot1}$	$n_{\cdot2}$	$\cdots$	$n_{\cdot m}$	$n_{\cdot\cdot}$

On the margins of the table are row totals and column totals. The above table is called a $k \times m$ contingency table.

A contingency table can arise from one of two situations. In the first situation, the contingency table is resulted from a simple random sample. Both Y and X are considered as response variables. The joint counts on Y and X, $\{n_{ij} : i = 1, \ldots, k; j = 1, \ldots, m\}$, follow a multinomial distribution with index $n_{..}$ and probabilities $\{\pi_{ij} : i = 1, \ldots, k; j = 1, \ldots, m\}$. The marginal counts on Y, $\{n_{\cdot j} : j = 1, \ldots, m\}$, follow a marginal multinomial distribution with $n_{..}$ and probabilities $\{\pi_{\cdot j} : j = 1, \ldots, m\}$ where $\pi_{\cdot j} = \sum_{i=1}^{k} \pi_{ij}$. Similarly, the marginal counts on X, $\{n_{i\cdot} : i = 1, \ldots, k\}$, follow a marginal multinomial distribution with $n_{..}$ and probabilities $\{\pi_{i\cdot} : i = 1, \ldots, k\}$ where $\pi_{i\cdot} = \sum_{j=1}^{m} \pi_{ij}$. The usual concern is whether or not Y and X are independent, i.e., the null hypothesis is $H_0 : \pi_{ij} = \pi_{i\cdot}.\pi_{\cdot j}$ for all i and j.

In the second situation, the contingency table arises from a conditional sampling conditioning on the levels of X. The variable Y is considered as the response variable and X is considered as a covariate. Unlike the first situation where all the margins of the contingency table are random, in the second situation, the row margins on X are fixed numbers. For $i = 1, \ldots, k$, the counts in row i, $\{n_{ij}, j = 1, \ldots, m\}$, follow a conditional multinomial distribution with index $n_i.$ and probabilities $\{\pi_{j|i} : j = 1, \ldots, m\}$. The null hypothesis of concern is whether or not the conditional distribution of Y depends on the levels of X, i.e., $H_0 : \pi_{j|1} = \cdots = \pi_{j|k}, j = 1, \ldots, m$.

The Pearson χ^2 statistic for the hypothesis testing of a contingency table has a unified form given below:

$$X^2 = \sum_{i=1}^{k} \sum_{j=1}^{m} \frac{(O_{ij} - E_{ij})^2}{E_{ij}},$$

where O_{ij} and E_{ij} stand for, respectively, the observed counts and estimated expected counts under the null hypothesis in cell (i, j). In both situations above, the Pearson χ^2 statistic turns out to have the same form:

$$X^2 = \sum_{i=1}^{k} \sum_{j=1}^{m} \frac{\left(n_{ij} - \frac{n_i. n_{.j}}{n_{..}}\right)^2}{\frac{n_i. n_{.j}}{n_{..}}}.$$

Under the null hypothesis in both situations, the Pearson χ^2 statistic X^2 follows an asymptotic χ^2 distribution with degrees of freedom $(k-1)(m-1)$.

Consider a case-control study for disease gene mapping or QTL mapping. A case group (disease group, or group with extreme quantitative trait values at one end) of size n_1 and a control group of size n_2 (normal group, or group with extreme quantitative trait values at the other end) are obtained. Suppose we want to test whether or not a marker is a disease gene (QTL) or a locus closely linked to a disease gene (QTL). Suppose that the marker under study has three possible genotypes, AA, Aa and aa. The genotype formations of the two groups at the marker, i.e., the number of individuals having each of the genotypes in the two groups, are obtained as follows:

	Genotype			Total
Group	AA	Aa	aa	
Case	n_{11}	n_{12}	n_{13}	$n_1.$
Control	n_{21}	n_{22}	n_{23}	$n_2.$
Total	$n_{.1}$	$n_{.2}$	$n_{.3}$	$n_{..}$

where $n_i. = n_i, i = 1, 2$. If the marker is not related to the disease (or quantitative trait) then the two groups should have similar genotype formation at this locus. The testing can be conducted using the Pearson χ^2 statistic

$$X^2 = \sum_{i=1}^{2} \sum_{j=1}^{3} \frac{\left(n_{ij} - \frac{n_i. n_{.j}}{n_{..}}\right)^2}{\frac{n_i. n_{.j}}{n_{..}}}.$$

2.4.4   Likelihood Ratio Test

The likelihood ratio test is a standard procedure for hypothesis testing with parametric families. Suppose the population has a distribution in a parametric family with probability density function $f(\theta, x)$ where θ belongs to a parameter space Θ. The hypotheses are formulated in the form: $H_0 : \theta \in \Theta_0$ versus $H_1 : \theta \in \Theta_1$, where $\Theta_0 \cap \Theta_1 = \phi$ and $\Theta_0 \cup \Theta_1 = \Theta$. Let $L(\theta, \boldsymbol{X})$ denote the likelihood function based on a sample $\boldsymbol{X} = (X_1, \ldots, X_n)$. In general, the likelihood ratio test statistic is given by

$$\tilde{\lambda}(\boldsymbol{X}) = \frac{\max_{\theta \in \Theta_1} L(\theta, \boldsymbol{X})}{\max_{\theta \in \Theta_0} L(\theta, \boldsymbol{X})}.$$

If the likelihood function is a continuous function of θ, and the null parameter space Θ_0 has a lower dimension than the full parameter space Θ, as is the case in genetic studies, the following form of the likelihood ratio statistic is more convenient to deal with:

$$\lambda(\boldsymbol{X}) = \frac{\max_{\theta \in \Theta} L(\theta, \boldsymbol{X})}{\max_{\theta \in \Theta_0} L(\theta, \boldsymbol{X})} = \max\{\tilde{\lambda}(\boldsymbol{X}), 1\}.$$

The likelihood ratio test rejects the null hypothesis H_0 if $\lambda(\boldsymbol{X}) \geq c_\alpha$ where c_α is the critical value of $\lambda(\boldsymbol{X})$ at level α. The steps to derive the likelihood ratio test are as follows:

1. Calculate the MLE $\hat{\theta}_{\text{FULL}}$ of θ when θ varies over Θ.

2. Calculate the MLE $\hat{\theta}_{\text{NULL}}$ when θ varies only over Θ_0.

3. Form $\lambda(\boldsymbol{X}) = L(\hat{\theta}_{\text{FULL}}, \boldsymbol{X})/L(\hat{\theta}_{\text{NULL}}, \boldsymbol{X})$.

4. Find a function $h(\lambda(\boldsymbol{X}))$ which is strictly increasing in $\lambda(\boldsymbol{X})$ such that $h(\lambda(\boldsymbol{X}))$ has a simple form.

Since h is an strictly increasing function, the likelihood ratio test is equivalent to rejecting H_0 when $h(\lambda(\boldsymbol{X})) \geq h(c_\alpha)$. In certain cases, the exact distribution of $h(\lambda(\boldsymbol{X}))$ under H_0 is given by one of the t-, χ^2-, and F-distributions. The critical value of $h(\lambda(\boldsymbol{X}))$ can be directly obtained while the derivation of c_α is more complicated. For example, the usual t-tests for testing the population mean of a normal population are in fact likelihood ratio tests.

In general, the exact distribution of the likelihood ratio test statistic cannot be derived. In this case, we can resort to its asymptotic distribution. Under mild conditions, which are usually satisfied in QTL mapping problems,

$$2 \ln \lambda(\boldsymbol{X}) = 2[\ln L(\hat{\theta}_{\text{FULL}}, \boldsymbol{X}) - \ln L(\hat{\theta}_{\text{NULL}}, \boldsymbol{X})]$$

has an asymptotic χ^2 distribution with degrees of freedom r, where r is the difference between the number of parameters in the full model and the number of parameters in the null model.

2.5   Linear and Generalized Linear Models

Linear models and generalized linear models are widely used in many applied fields. They are important tools for QTL mapping as well. We provide a general discussion on these models in this section.

2.5.1   Linear Models

A linear model describes the relationship of a response variable y with a number of covariates x_j's. For example, in QTL mapping, the measurement of a quantitative trait is taken as the response variable, and the marker genotypes are taken as covariates. A linear model can be formulated to describe the relationship between the quantitative trait and the markers. Let $\{y_i, (x_{i1}, \ldots, x_{ip}) : i = 1, \ldots, n\}$ be the observations of the response variable and the covariates on n individuals. A linear model is given by the following form:

$$y_i = \beta_0 + \sum_{j=1}^{p} \beta_j x_{ij} + \epsilon_i, \ i = 1, \ldots, n, \tag{2.1}$$

where β_j's are unknown parameters, ϵ_i's are random variables with mean zero and the covariates are treated as non-random. A linear model can be used for (a) detection of causal covariates, (b) prediction of future responses and (c) control of the process underlying the model. In QTL mapping, we are mainly concerned with (a) and (b). The usual assumptions made on model (2.1) are:

(i) The ϵ_i's are independent with variances σ^2/w_i; or

(ii) The ϵ_i's are i.i.d. with variance σ^2; or

(iii) The ϵ_i's are i.i.d. normal random variables with variance σ^2.

Under assumption (iii), exact inference can be made about model (2.1). Under assumptions (i) or (ii), only asymptotic inference can be made on model (2.1).

Model (2.1) can be expressed in a more concise form in matrix notation. Let $\boldsymbol{y} = (y_1, \ldots, y_n)^\tau$, $\boldsymbol{\beta} = (\beta_0, \beta_1, \ldots, \beta_p)^\tau$, $\boldsymbol{\epsilon} = (\epsilon_1, \ldots, \epsilon_n)^\tau$ and $X = (\mathbf{1}, \boldsymbol{x}_1, \ldots, \boldsymbol{x}_p)$, where $\mathbf{1}$ is a column vector with all components 1 and $\boldsymbol{x}_j = (x_{1j}, \ldots, x_{nj})^\tau$. Model (2.1) can then be expressed as

$$\boldsymbol{y} = X\boldsymbol{\beta} + \boldsymbol{\epsilon}.$$

The β parameters in model (2.1) are estimated by the least squares method; that is, the estimator of $\boldsymbol{\beta}$ is obtained by minimizing

$$\sum_{i=1}^{n} (y_i - \beta_0 - \sum_{j=1}^{p} \beta_j x_{ij})^2 = \|\boldsymbol{y} - X\boldsymbol{\beta}\|_2^2,$$

where $\| \cdot \|_2^2$ is the squared L_2-vector norm. The least squares estimator (LSE) of $\boldsymbol{\beta}$ is given by

$$\hat{\boldsymbol{\beta}} = (X^\tau X)^{-1} X^\tau \boldsymbol{y}.$$

Implicitly it is assumed here that X is of full column rank. Under assumption (i), if w_i's are known, a better estimator of $\boldsymbol{\beta}$ can be obtained by the weighted least squares method which minimizes

$$\sum_{i=1}^{n} w_i (y_i - \beta_0 - \sum_{j=1}^{p} \beta_j x_{ij})^2 = (\boldsymbol{y} - X\boldsymbol{\beta})^\tau W (\boldsymbol{y} - X\boldsymbol{\beta}),$$

where W is the diagonal matrix with diagonal elements w_i's. The weighted least squares estimator (WLSE) of $\boldsymbol{\beta}$ is given by

$$\hat{\boldsymbol{\beta}} = (X^\tau W X)^{-1} X^\tau W \boldsymbol{y}.$$

The σ^2 is estimated by

$$\hat{\sigma}^2 = \begin{cases} \frac{1}{n-p-1} \| \boldsymbol{y} - X\hat{\boldsymbol{\beta}} \|_2^2, & \text{in the case of LSE;} \\ \frac{1}{n-p-1} (\boldsymbol{y} - X\hat{\boldsymbol{\beta}})^\tau W (\boldsymbol{y} - X\hat{\boldsymbol{\beta}}), & \text{in the case of WLSE.} \end{cases}$$

The estimator $\hat{\boldsymbol{\beta}}$ has an asymptotic multivariate normal distribution with mean vector $\boldsymbol{\beta}$ and variance-covariance matrix $\Sigma = \sigma^2 (X^\tau X)^{-1}$ ($= \sigma^2 (X^\tau W X)^{-1}$ in the case of WLSE). The estimator $\hat{\Sigma}$ of the variance-covariance matrix is obtained by replacing σ^2 in Σ with $\hat{\sigma}^2$. With the estimated matrix, we still have, asymptotically,

$$\hat{\boldsymbol{\beta}} \sim N(\boldsymbol{\beta}, \hat{\Sigma}). \tag{2.2}$$

The asymptotic distribution (2.2) provides the basis for the inference on model (2.1). Let $\hat{\boldsymbol{\beta}}_0$ be any sub-vector of $\hat{\boldsymbol{\beta}}$. By the properties of multivariate normal distributions, $\hat{\boldsymbol{\beta}}_0$ also has an asymptotic multivariate normal distribution with mean vector $\boldsymbol{\beta}_0$ and the variance-covariance matrix given by the rows and columns of $\hat{\Sigma}$ which correspond to $\hat{\boldsymbol{\beta}}_0$. Specifically, $\hat{\beta}_j \sim N(\beta_j, \hat{\sigma}_{j+1,j+1})$, where $\hat{\sigma}_{j+1,j+1}$ is the $(j+1)$st diagonal element of $\hat{\Sigma}$.

For the hypothesis testing regarding model (2.1), we introduce below the Wald test which is more convenient than and is asymptotically equivalent to the likelihood ratio test. Consider $H_0 : \boldsymbol{\beta}_0 = 0$, where $\boldsymbol{\beta}_0$ is a sub-vector of $\boldsymbol{\beta}$. Let $\hat{\Sigma}_0$ denote the sub matrix of $\hat{\Sigma}$ corresponding to $\hat{\boldsymbol{\beta}}_0$. The Wald statistic for testing H_0 is given by

$$X_W^2 = \hat{\boldsymbol{\beta}}_0^\tau \hat{\Sigma}_0^{-1} \hat{\boldsymbol{\beta}}_0.$$

The Wald statistic has an asymptotic χ^2-distribution with degrees of freedom r, where r is the number of components of $\boldsymbol{\beta}_0$. Thus H_0 is rejected if $X_W^2 \geq \chi_r^2(1 - \alpha)$.

2.5.2 Generalized Linear Models

We now turn to the generalized linear model (GLIM). The essential assumptions of a linear model are (i) the variance of the response variable y_i does not depend on its mean (the typical distribution of this nature is the normal distribution) and (ii) the linear form $\eta_i = \beta_0 + \sum_{i=1}^{n} \beta_j x_{ij}$ is directly related to the expectation of y_i, i.e., $\mu_i = E y_i = \eta_i$. The GLIM extends the linear model in these two essential assumptions. First, the GLIM allows the variance of y_i to be a function of its mean. The distributions of this nature are exponential families mentioned in § 2.1.3. Second, the linear form η_i is related to a function of μ_i, say $g(\mu_i)$, not necessarily μ_i itself. The function g is called the link function (which links the linear form η_i to the mean μ_i).

A GLIM for the observations $\{y_i, (x_{i1}, \ldots, x_{ip}) : i = 1, \ldots, n\}$ is described as follows:

(i) The response y_i follows an exponential family distribution with p.d.f. given by

$$\exp\left\{\frac{y_i \theta_i - b(\theta_i)}{\phi} + c(\phi, y_i)\right\} I_A(y_i),$$

where the dispersion parameter ϕ is fixed.

For the exponential family, $\mu_i = E y_i = b'(\theta_i)$ and $\mathrm{Var}(y_i) = \phi b''(\theta_i)$, where b' and b'' denote the first and second derivatives of b respectively. Note that b' is strictly increasing; hence θ_i can be expressed as a function of μ_i, i.e., $\theta_i = \theta(\mu_i)$ where $\theta(\cdot)$ is the inverse function of $b'(\cdot)$. Thus $\mathrm{Var}(y_i) = \phi b''(\theta(\mu_i)) = \phi V(\mu_i)$ is a function of μ_i. The function $V(\cdot)$ is called the variance function.

(ii) The covariates $x_{i1}, \ldots, x_{ip}$ affect y_i through a linear predictor

$$\eta_i = \beta_0 + \sum_{i=1}^{n} \beta_j x_{ij}.$$

The linear predictor η_i is related to the distribution of y_i through the link function

$$g(\mu_i) = \eta_i,$$

where g is differentiable and strictly monotone.

The estimator of $\boldsymbol{\beta} = (\beta_0, \beta_1, \ldots, \beta_p)^\tau$ is obtained by the maximum likelihood estimation. Unlike the linear models, there is no explicit expression for the MLE of $\boldsymbol{\beta}$ in the GLIM. The MLE is obtained by the Newton method with Fisher scoring through an iterated weighted least squares (IWLS) procedure, which is briefly sketched in the following.

Denote $\boldsymbol{x}_i = (1, x_{i1}, \ldots, x_{ip})^\tau$. Let $h(\cdot)$ be the inverse function of g. The log likelihood function for the GLIM is given by

$$l(\boldsymbol{\beta}) = \frac{1}{\phi} \sum_{i=1}^{n} [y_i \theta(h(\boldsymbol{x}_i^\tau \boldsymbol{\beta})) - b(\theta(h(\boldsymbol{x}_i^\tau \boldsymbol{\beta})))].$$

Denote by $u(\beta)$ the vector of the first derivatives of l, i.e.,

$$u(\beta) = (\frac{\partial l(\beta)}{\partial \beta_0}, \frac{\partial l(\beta)}{\partial \beta_1}, \ldots, \frac{\partial l(\beta)}{\partial \beta_p})^\tau.$$

Let

$$A(\beta) = -E\left[\frac{\partial^2 l(\beta)}{\partial \beta \partial \beta^\tau}\right],$$

where $\frac{\partial^2 l(\beta)}{\partial \beta \partial \beta^\tau}$ is the matrix of the second derivatives of l with (i, j)th entry $\frac{\partial^2 l(\beta)}{\partial \beta_i \partial \beta_j}$.

It can be derived that

$$u(\beta) = \frac{1}{\phi} X^\tau W(\beta) v(\beta),$$

where W is a diagonal matrix with diagonal elements

$$w_i = [V(\mu_i)]^{-1} \left(\frac{\partial \eta_i}{\partial \mu_i}\right)^{-2},$$

and $v(\beta)$ is a vector given by

$$v(\beta) = \begin{pmatrix} \frac{\partial g(\mu_1)}{\partial \mu_1}(y_1 - \mu_1) \\ \vdots \\ \frac{\partial g(\mu_n)}{\partial \mu_n}(y_n - \mu_n) \end{pmatrix}.$$

It can also be derived that

$$A(\beta) = \frac{1}{\phi} X^\tau W(\beta) X.$$

The procedure of Newton method with Fisher scoring for the GLIM amounts to solving iterately

$$X^\tau W(\beta^{\mathrm{OLD}}) X(\beta^{\mathrm{NEW}} - \beta^{\mathrm{OLD}}) = X^\tau W(\beta^{\mathrm{OLD}}) v(\beta^{\mathrm{OLD}}).$$

Rewrite the above iterative equation as

$$\begin{aligned} X^\tau W(\beta^{\mathrm{OLD}}) X \beta^{\mathrm{NEW}} &= X^\tau W(\beta^{\mathrm{OLD}})[X\beta^{\mathrm{OLD}} + v(\beta^{\mathrm{OLD}})] \\ &= X^\tau W(\beta^{\mathrm{OLD}}) z(\beta^{\mathrm{OLD}}). \end{aligned}$$

The above equation is the normal equation of the weighted least square regression with response vector z, the design matrix X and weight matrix W. Thus the Newton method with Fisher scoring is equivalent to an iterated weighted least square procedure. The following is the general algorithm:

Initialization: Set starting values $z(\beta^{\mathrm{OLD}})$ and $W(\beta^{\mathrm{OLD}})$.

Iteration :

    (i) Solve

$$X^T W(\boldsymbol{\beta}^{\mathrm{OLD}}) X \boldsymbol{\beta}^{\mathrm{NEW}} = X^T W(\boldsymbol{\beta}^{\mathrm{OLD}}) \boldsymbol{z}(\boldsymbol{\beta}^{\mathrm{OLD}}).$$

    (ii) Set $\boldsymbol{\beta}^{\mathrm{OLD}} = \boldsymbol{\beta}^{\mathrm{NEW}}$, repeat (i). Iterate until convergence occurs.

The MLE $\hat{\boldsymbol{\beta}}$ has an asymptotic multivariate normal distribution:

$$\hat{\boldsymbol{\beta}} \sim N(\boldsymbol{\beta}, \ \hat{\phi}(X^T W X)^{-1}),$$

where $\hat{\phi}$ is an estimator of ϕ and W is the final version of the weight matrix in the IWLS procedure. The inference on $\boldsymbol{\beta}$ can be carried out in the same way as for the linear models.

For more details on GLIM, the reader is referred to [132].

2.6   Mixture Models and EM Algorithm

In QTL mapping problems, the search for QTL is not confined to the limited number of markers. The investigation is extended to any locus on the genome. However, at loci other than the markers, the genotypes of the experiment subjects are not available. If a locus is indeed a QTL, the genotypes of the locus divide the population into different sub populations, and the quantitative trait has different distributions in different sub populations. When an individual is drawn at random from the population without its identity of sub population, the randomly drawn quantitative trait has a mixture distribution. In general, a mixture distribution has the form

$$F(y; \boldsymbol{\theta}, \boldsymbol{p}) = \sum_{j=1}^{k} p_j F_j(y; \boldsymbol{\theta}_j),$$

where F_j's are c.d.f.'s, $p_j \geq 0$, $\sum_{j=1}^{k} p_j = 1$, and $\boldsymbol{\theta} = (\boldsymbol{\theta}_1, \ldots, \boldsymbol{\theta}_k)$. If F_j has a p.d.f. f_j, the mixture distribution has a p.d.f.

$$f(y; \boldsymbol{\theta}, \boldsymbol{p}) = \sum_{j=1}^{k} p_j f_j(y; \boldsymbol{\theta}_j).$$

Let $(y_1, \ldots, y_n)$ be a random sample from the population. The log likelihood function based on the sample is

$$l(\boldsymbol{\theta}, \boldsymbol{p}) = \sum_{i=1}^{n} \ln \left(\sum_{j=1}^{k} p_j f_j(y_i; \boldsymbol{\theta}_j) \right).$$

The method of maximum likelihood is still applicable for the estimation of the mixture model. In §2.5, the Newton method with Fisher scoring is used for the estimation of GLIM. Due to the special form of exponential families, the computation procedure is reduced to a simple IWLS procedure. However, for the mixture model above, even if the components of the mixture are exponential family distributions, the Newton method does not turn out to be a proper method, though it can be used in principle. The expectation-maximization (EM) algorithm [43] is an effective method for the computation of mixture models . We discuss the EM-algorithm for the mixture model in the following.

In the mixture model, the individual identities of sub populations are unknown. In other words, the identity information is missing. The identity of each individual can be represented by a vector $\boldsymbol{z} = (z_1, \ldots, z_{k-1})^\tau$ defined as

$$
z_j = \begin{cases} 1, & \text{if the individual is in subpopulation } j, \\ 0, & \text{otherwise,} \end{cases}
$$
$$
j = 1, \ldots, k - 1.
$$

The vector $\boldsymbol{z}$ has the following polytomous distribution, see § 2.1.3,

$$
p(\boldsymbol{z}) = \prod_{j=1}^{k} p_j^{z_j}, \; z_k = 1 - \sum_{j=1}^{k-1} z_j, p_k = 1 - \sum_{j=1}^{k-1} p_j.
$$

For individual i, the vector is denoted by $\boldsymbol{z}_i = (z_{i1}, \ldots, z_{i\,k-1})^\tau$. If the $\boldsymbol{z}_i$'s were observed, we have the "complete" data $\{(y_i, \boldsymbol{z}_i) : i = 1, \ldots, n\}$. With the $\boldsymbol{z}_i$'s missing, the data $\{y_i : i = 1, \ldots, n\}$ is referred to as the "incomplete" data. The joint distribution of $(y_i, \boldsymbol{z}_i)$ has the density

$$
\prod_{j=1}^{k} [p_j f_j(y_i; \boldsymbol{\theta}_j)]^{z_{ij}}.
$$

Thus the log likelihood function based on the "complete" data has the following simple form:

$$
l_c(\boldsymbol{\beta}, \boldsymbol{p}) = \sum_{i=1}^{n} \sum_{j=1}^{k} z_{ij} \ln p_j + \sum_{i=1}^{n} \sum_{j=1}^{k} z_{ij} \ln f_j(y_i; \boldsymbol{\theta}_j).
$$

When f_j's are p.d.f's of exponential families, the above likelihood function can be minimized using an IWLS procedure similar to the one discussed in § 2.5, if the z_{ij}'s are given.

A general EM algorithm consists of two steps: an expectation step (E-step) and a maximization step (M-step). In the E-step, the conditional expectation of the log likelihood function based on the complete data given the "incomplete" data and the parameter values is computed. In the M-step, the conditional expectation is maximized with respect to the parameters. The algorithm alternates between these two steps until convergence occurs.

For the mixture model above, in the E-step, the computation of the conditional expectation of the log likelihood function l_c reduces to the computation of the conditional expectations of the z_{ij}'s. Let $\tilde{p}_j$ and $\tilde{\boldsymbol{\theta}}_j$ be the most updated values of p_j and $\boldsymbol{\theta}_j$ respectively. The conditional expectation of z_{ij} is given by

$$z_{ij}^*(\tilde{\boldsymbol{p}}, \tilde{\boldsymbol{\theta}}) = \frac{\tilde{p}_j f_j(y_i; \tilde{\boldsymbol{\theta}}_j)}{\sum_{l=1}^k \tilde{p}_l f_l(y_i; \tilde{\boldsymbol{\theta}}_l)}.$$

In the M-step, the conditional expectation of l_c can be maximized with respect to $\boldsymbol{p}$ and $\boldsymbol{\theta}$ separately. The EM algorithm for the mixture model is described as follows.

Initialization: specify the starting values $\tilde{\boldsymbol{p}}$ and $\tilde{\boldsymbol{\theta}}$ of the parameters. Set FLAG $= 0$.

Iteration: while FLAG $= 0$, repeat

 E-step: Compute $z_{ij}^*(\tilde{\boldsymbol{p}}, \tilde{\boldsymbol{\theta}})$ for $i = 1, \ldots, n; j = 1, \ldots, k$.

 M-step: Maximize

$$\sum_{i=1}^n \sum_{j=1}^k z_{ij}^*(\tilde{\boldsymbol{p}}, \tilde{\boldsymbol{\theta}}) \ln p_j$$

 with respect to $\boldsymbol{p}$. Denote the maximizer by $\hat{\boldsymbol{p}}$.
 Maximize

$$\sum_{i=1}^n \sum_{j=1}^k z_{ij}^*(\tilde{\boldsymbol{p}}, \tilde{\boldsymbol{\theta}}) \ln f_j(y_i, \boldsymbol{\theta}_j)$$

 with respect to $\boldsymbol{\theta}$. Denote the maximizer by $\hat{\boldsymbol{\theta}}$.

 Check convergence: If both $\|\hat{\boldsymbol{p}} - \tilde{\boldsymbol{p}}\|$ and $\|\hat{\boldsymbol{\theta}} - \tilde{\boldsymbol{\theta}}\|$ reach a pre-specified accuracy, set FLAG $= 1$.

In QTL mapping problems considered in latter chapters, the mixture probabilities p_j's are either known or functions of some common parameters. When the p_j's are known, the maximization step with respect to $\boldsymbol{p}$ is omitted. If the p_j's are functions of common parameters, the maximization step is made with respect to those common parameters.

2.7 Bayesian Analysis

Bayesian analysis has become an important approach to statistical inference in general and to QTL mapping in particular. Statistics deals with uncertainty. The uncertainty is relative to the information we have. The less information we have, the more uncertain. As information accumulates, the uncertainty reduces. The basic philosophy of Bayesian analysis is that any uncertainty can be described by a distribution constructed from the information we have, and statistical inference must be based on the distribution constructed from the most updated information. In the usual statistical analysis, we consider the characteristics of a population as fixed parameters. When data are obtained, we formulate a distribution of the data that depends on the unknown parameters. The statistical analysis is then based on the likelihood function of the parameters provided by the data. In Bayesian analysis, a totally different viewpoint is taken and the parameters are considered as random variables. Before data are collected from a population, we might already have some information on the characteristics of the population. The information is summarized in a distribution called the *prior* distribution. When new data are available, the information is updated by the data. The updated information is summarized in a distribution called the *posterior* distribution. Bayesian statistical inference is then based on the posterior distribution. In this section, we briefly discuss the methods of Bayesian analysis. For a full exposure to Bayesian analysis, the reader is referred to [9].

2.7.1   General Framework of Bayesian Analysis

A Bayesian model consists of three basic components: (i) the prior distribution of the population characteristics, (ii) the conditional distribution of the data given the characteristics and (iii) the posterior distribution of the characteristics. Let the population characteristics be represented by a parameter θ (scaler of vector). The prior distribution is a distribution of θ. Denote by $\pi(\theta)$ the p.d.f. of the prior. Let $\boldsymbol{y} = (y_1, \ldots, y_n)$ be the data from the population. Assume that, given θ, $\boldsymbol{y}$ follows a distribution with p.d.f. $f(\boldsymbol{y}|\theta)$. The posterior distribution of θ is a conditional distribution given $\boldsymbol{y}$, which is given by the Bayes formula:

$$\pi_{\theta|\boldsymbol{y}}(\theta) = \frac{\pi(\theta)f(\boldsymbol{y}|\theta)}{\int \pi(\theta)f(\boldsymbol{y}|\theta)d\theta}. \tag{2.3}$$

The posterior distribution in Bayesian analysis is the equivalence of the likelihood function in the usual statistical analysis. All inferences in Bayesian analysis are based on the posterior distribution.

A point estimate of θ can be obtained by the Bayes estimator. Let $\delta(\boldsymbol{y})$ be a statistic. The Bayes risk of δ is defined as

$$r_\pi(\delta) = \int R(\delta, \theta)\pi(\theta)d\theta = \int \int L(\delta(\boldsymbol{y}), \theta)f(\boldsymbol{y}|\theta)\pi(\theta)d\boldsymbol{y}d\theta,$$

where L and R are loss and risk functions respectively, see § 2.3.3. The Bayes estimator is given by $\delta^*(\boldsymbol{y})$ such that δ^* minimizes the Bayes risk; that is,

$$r_\pi(\delta^*) = \inf_\delta r_\pi(\delta).$$

Under the square error loss $L(\delta(\boldsymbol{y}), \theta) = (\delta(\boldsymbol{y}) - \theta)^2$, the Bayes estimator of θ is given by the posterior expectation of θ, i.e.,

$$\hat{\theta} = \int \theta \pi_{\theta|\boldsymbol{y}}(\theta) d\theta.$$

Interval estimates of θ can be provided by *credible* intervals. A credible interval of θ at level $1 - \alpha$ is a fixed interval (depending on $\boldsymbol{y}$) $[\underline{\theta}(\boldsymbol{y}), \ \bar{\theta}(\boldsymbol{y})]$ such that

$$P_{\theta|\boldsymbol{y}}[\ \underline{\theta}(\boldsymbol{y}) \leq \theta \leq \bar{\theta}(\boldsymbol{y})\ ] \geq 1 - \alpha.$$

There is a conceptual difference between a credible interval and a confidence interval. In a credible interval, the limits are fixed (non-random) and the level is a lower bound of the probability that the random θ falls into the interval. In a confidence interval, the limits are random and the level is the lower bound of the probability that the interval covers the fixed θ.

The Bayesian analysis is straightforward in principle. The challenge of Bayesian analysis lies in two points: the specification of the prior and the computation of the posterior. We address these two points in the next two sub sections.

2.7.2 Empirical and Hierarchical Bayesian Analysis

It is not easy to assign a completely known prior $\pi(\theta)$. The prior is usually assigned as a parametric family depending on further unknown parameters, say ξ, called *hyperparameters*. Thus the prior is of the form $\pi_{\theta|\xi}(\theta|\xi)$. However, the Bayesian analysis cannot be carried out with the unknown parameter ξ. There are two ways to get out of this dilemma. One is *empirical Bayesian analysis* and the other is *hierarchical Bayesian analysis*.

In the empirical Bayesian analysis, the hyperparameter is estimated from the data, the estimate is taken as if it is the known value of the hyperparameter, and then the analysis is carried out exactly the same as the original Bayesian analysis. The hyperparamter is estimated as follows. We can consider $\pi_{\theta|\xi}(\theta|\xi) f(\boldsymbol{y}|\theta)$ as the joint p.d.f. of θ and $\boldsymbol{y}$ parameterized by ξ. From the joint p.d.f., the marginal p.d.f. of $\boldsymbol{y}$ is obtained as

$$m(\boldsymbol{y}|\xi) = \int \pi_{\theta|\xi}(\theta|\xi) f(\boldsymbol{y}|\theta) d\theta. \tag{2.4}$$

The marginal p.d.f. $m(\boldsymbol{y}|\xi)$ is treated as the usual parametric density function. Then ξ can be estimated from the data either by MLE or the method of moments based on this density.

In the hierarchical Bayesian analysis, a completely known second layer prior is specified for the hyperparameter ξ. Denote the second layer prior by $\Lambda(\xi)$. We then have a joint p.d.f. of θ and ξ: $\pi_{\theta|\xi}(\theta|\xi)\Lambda(\xi)$. From this joint p.d.f., we obtain the prior of θ free of any parameters:

$$\pi(\theta) = \int \pi_{\theta|\xi}(\theta|\xi)\Lambda(\xi) d\xi.$$

The process can go on further if the prior of ξ still depends on some parameters. Then a third layer prior is specified for those parameters. The parameter-free prior of θ is obtained as the marginal p.d.f. of the joint distribution of the parameters at all layers. The parameter-free prior $\pi(\theta)$ is then used in the Bayesian analysis. The rationale of the hierarchical Bayesian analysis is that the effect of the inaccuracy of a higher layer prior on the analysis is much less than that of the first layer prior. In most practical problems, a second layer prior is usually enough to reduce the inaccuracy in the specification of the first layer prior.

2.7.3 Markov Chain Monte Carlo Simulation

The posterior distribution arising from practical problems is usually complicated. The direct computation for the characteristics of the posterior distribution is not possible in most of the cases. The computation is done through Markov chain Monte Carlo (MCMC) simulation. By using MCMC, random samples are generated from the posterior distribution. Then empirical distribution of the generated samples is used to approximate the posterior distribution. The characteristics of the posterior distribution are approximated by their counterparts of the empirical distribution.

The idea of MCMC simulation is best illustrated by a finite discrete-space Markov chain. Suppose the state space is given by $S = \{s_1, \ldots, s_k\}$. The chain starts with any one of the states at time 0 and moves at unit time interval. At any time t, it moves from state s_i to s_j with probability p_{ij}, where $\sum_{j=1}^{m} p_{ij} = 1$. Let $X_0, X_1, \ldots, X_t, \ldots$, denote the states of the chain at time $0, 1, \ldots, t, \ldots$. Then

$$P(X_t = s_j | X_{t-1} = s_i) = p_{ij}.$$

Let

$$P = \begin{pmatrix} p_{11} & p_{12} & \cdots & p_{1k} \\ p_{21} & p_{22} & \cdots & p_{2k} \\ \cdots & \cdots & \cdots & \cdots \\ p_{k1} & p_{k2} & \cdots & p_{kk} \end{pmatrix}.$$

The matrix P is called the transition matrix. Suppose the starting state is chosen according to a probability density π_0 on S. Then the probability density of X_t is given by

$$\pi_t = \pi_0 P^t = \pi_{t-1} P.$$

Here π_0, π_t denote row vectors of probabilities. If there is a unique probability density π on S satisfying

$$\pi = \pi P, \tag{2.5}$$

then, for any π_0, as $t \to \infty$, $\pi_t \to \pi$. The distribution with density π is called the equilibrium distribution of the Markov chain. This implies that, if we want to generate a random sample X from π, instead of generating

X directly from π, we can run the Markov chain until t is large and take $X = X_t$. This is the core idea of MCMC simulation. Of course, for generating a random sample from a discrete distribution, the MCMC is not needed. But it illustrates how MCMC works for more general distributions including continuous distributions.

Now consider a continuous Markov chain as follows. Let $h(x, y)$ be a transition probability function on the space $\mathcal{R}$, i.e., for any $y \in \mathcal{R}$, $h(x, y)$ is a density function on $\mathcal{R}$. Note that $\mathcal{R}$ could be a Euclidean space of any given dimension. Let X_0 be distributed with density function $\pi_0(x)$. Upon observing $X_0 = x_0$, generate X_1 by the transition density function $h(x, x_0)$, i.e., the conditional density function of X_1 given $X_0 = x_0$ is given by

$$\pi_{X_1|X_0}(x|x_0) = h(x, x_0).$$

The unconditional density function of X_1 is then given by

$$\pi_1(x) = \int_{\mathcal{R}} h(x, y)\pi_0(y)dy.$$

In general, for $t = 2, 3, \ldots$, upon observing $X_{t-1} = x_{t-1}$, generate X_t by the transition density function $h(x, x_{t-1})$. The unconditional density function of X_t is given by

$$\pi_t(x) = \int_{\mathcal{R}} h(x, y)\pi_{t-1}(y)dy.$$

Suppose that there is a density function $\pi(x)$ on $\mathcal{R}$ satisfying

$$\pi(x) = \int_{\mathcal{R}} h(x, y)\pi(y)dy. \tag{2.6}$$

Note that (2.6) is an analog of (2.5). Then, for any $\pi_0(x)$, as $t \to \infty$, $\pi_t(x) \to \pi(x)$. Again, in order to generate a random sample X from π, we can run the Markov chain until π_t becomes equilibrium and take $X = X_t$.

In the Bayesian analysis, the equilibrium distribution is taken as the posterior distribution of the unknowns. In order to apply the MCMC, we need to construct the transition density function $h(x, y)$ such that condition (2.6) is satisfied. Note that condition (2.6) holds if

$$\pi(x)h(y, x) = \pi(y)h(x, y). \tag{2.7}$$

Thus the MCMC can be implemented if we can construct a $h(x, y)$ satisfying either (2.6) or (2.7). Some general methods for constructing $h(x, y)$ will be described in § 2.8.

2.7.4 Bayes Factor

In many practical problems, we encounter the problem of model selection, i.e., we need to select a model with a certain desirable property among a set of models. In Bayesian analysis, the Bayes factor can be used for the model selection. The Bayes factor of two models is defined as the ratio of the marginal densities of y under the two models. The definition is given in more detail as follows. Suppose that, under model j, the parameter is $\boldsymbol{\theta}_j$, the prior is $\pi_j(\boldsymbol{\theta}_j)$ and the conditional probability density of y given the parameter is $f_j(y|\boldsymbol{\theta}_j)$, $j = 1, 2$. The marginal probability density function of y under model j is $\int \pi_j(\boldsymbol{\theta}_j)f_j(y|\boldsymbol{\theta}_j)d\boldsymbol{\theta}_j$. The Bayes factor of model 1 with respect to model 2 is then given by

$$\mathrm{BF}_{12}(y) = \frac{\int \pi_1(\boldsymbol{\theta}_1)f_1(y|\boldsymbol{\theta}_1)d\boldsymbol{\theta}_1}{\int \pi_2(\boldsymbol{\theta}_2)f_2(y|\boldsymbol{\theta}_2)d\boldsymbol{\theta}_2}.$$

Given the observation y, if $\mathrm{BF}_{12}(y) > 1$, model 1 is preferred to model 2; if $\mathrm{BF}_{12}(y) < 1$, model 2 is preferred to model 1; if $\mathrm{BF}_{12}(y) = 1$, model 1 and model 2 are equally preferred. If there are more than two models under consideration, the model selection using the Bayes factor is equivalent to that using the marginal probability density of y. The Bayes factor selects the model that achieves the maximum of the marginal probabilities of y at the observed value. That is, the selected model, say j^*, is given by

$$j^* = \mathrm{argmax}_j \int \pi_j(\boldsymbol{\theta}_j)f_j(y|\boldsymbol{\theta}_j)d\boldsymbol{\theta}_j.$$

2.8 Samplers for Markov Chain Monte Carlo Simulation

In this section, we discuss three particular methods for MCMC simulation: the Gibbs sampler, the Metropolis-Hastings algorithm and the reversible jump MCMC.

2.8.1 The Gibbs Sampler

The Gibbs sampler became a popular method with the study in [69]. A good tutorial exposition of the Gibbs sampler can be found in [26]. We describe the Gibbs sampler without the verification of (2.6) for its transition density function. In fact, the transition density function is not explicitly given in the Gibbs sampler. For a more comprehensive exposure to the sampler, the reader is referred to [26] and the references therein.

We denote the posterior density by the simplified notation $\pi(\boldsymbol{\xi})$ where $\boldsymbol{\xi}$ is the vector of all the unknowns. Let $\boldsymbol{\xi}$ be partitioned as $\boldsymbol{\xi} = (\xi_1, \xi_2, \ldots, \xi_q)$ where the ξ_j, $j = 1, 2, \ldots q$, which are not necessarily scalers, could be vectors of different dimensions. Let $\pi_j(\xi_j|\xi_1, \ldots, \xi_{j-1}, \xi_{j+1}, \ldots, \xi_q)$ be the conditional distribution of ξ_j given the remaining components of $\boldsymbol{\xi}$. Suppose that random samples can be easily generated from these conditional distributions.

The Gibbs sampler generates the Markov chain

$$\boldsymbol{\xi}^{(0)}, \boldsymbol{\xi}^{(2)}, \ldots, \boldsymbol{\xi}^{(t)}, \ldots,$$

where $\boldsymbol{\xi}^{(t)} = (\xi_1^{(t)}, \ldots, \xi_q^{(t)})$, as follows. Let $\boldsymbol{\xi}^{(0)}$ be a specified initial vector. For $t \geq 1$, generate $\boldsymbol{\xi}^{(t)}$ in the following q successive steps:

1. Generate $\xi_1^{(t)}$ from $\pi_1(\xi_1 | \xi_2^{(t-1)}, \ldots, \xi_q^{(t-1)})$.

2. Generate $\xi_2^{(t)}$ from $\pi_2(\xi_2 | \xi_1^{(t)}, \xi_3^{(t-1)}, \ldots, \xi_q^{(t-1)})$.

$\cdots$

j. Generate $\xi_j^{(t)}$ from $\pi_j(\xi_j | \theta_1^{(t)}, \ldots, \xi_{j-1}^{(t)}, \xi_{j+1}^{(t-1)}, \ldots, \xi_q^{(t-1)})$.

$\cdots$

q. Generate $\xi_q^{(t)}$ from $\pi_q(\xi_q | \theta_1^{(t)}, \ldots, \xi_{q-1}^{(t)})$.

2.8.2 Metropolis-Hastings Algorithm

Although the Gibbs sampler is the most efficient, for the Gibbs sampler to be applicable, there must be a partition of $\boldsymbol{\xi}$ such that each part of the partition can be easily sampled from the conditional distributions, which is not always the case for the posterior distribution $\pi(\boldsymbol{\xi})$. Furthermore, $\pi(\boldsymbol{\xi})$ might be only specified up to a normalizing constant which is practically impossible to compute. The Metropolis-Hasting algorithm is designed for the general case. The algorithm was first developed in [135]. It was later generalized in [83]. A simple exposition of the Metropolis-Hastings algorithm is provided in [37].

In the Metropolis-Hastings algorithm, a new state is obtained first by generating a candidate from a proposal density function and then accepting the candidate with a certain probability. Let $q(\boldsymbol{\xi} | \tilde{\boldsymbol{\xi}})$ denote the proposal density function given $\tilde{\boldsymbol{\xi}}$. Define the acceptance probability as

$$\alpha(\boldsymbol{\xi} | \tilde{\boldsymbol{\xi}}) = \begin{cases} \min\left\{ \dfrac{\pi(\boldsymbol{\xi}) q(\tilde{\boldsymbol{\xi}} | \boldsymbol{\xi})}{\pi(\tilde{\boldsymbol{\xi}}) q(\boldsymbol{\xi} | \tilde{\boldsymbol{\xi}})}, 1 \right\}, & \text{if } \pi(\tilde{\boldsymbol{\xi}}) q(\boldsymbol{\xi} | \tilde{\boldsymbol{\xi}}) > 0; \\ 1, & \text{otherwise.} \end{cases}$$

Suppose that $\tilde{\boldsymbol{\xi}}$ is the current state and $\boldsymbol{\xi}$ is the generated candidate state which is accepted as the next state with probability $\alpha(\boldsymbol{\xi} | \tilde{\boldsymbol{\xi}})$. Then the transition density is given by

$$h(\boldsymbol{\xi}, \tilde{\boldsymbol{\xi}}) = q(\boldsymbol{\xi} | \tilde{\boldsymbol{\xi}}) \alpha(\boldsymbol{\xi} | \tilde{\boldsymbol{\xi}}).$$

We can verify condition (2.7) for $h(\boldsymbol{\xi}, \tilde{\boldsymbol{\xi}})$ as follows.

$$
\begin{aligned}
\pi(\tilde{\boldsymbol{\xi}})h(\boldsymbol{\xi}, \tilde{\boldsymbol{\xi}}) &= \pi(\tilde{\boldsymbol{\xi}})q(\boldsymbol{\xi}|\tilde{\boldsymbol{\xi}}) \min\left\{\frac{\pi(\boldsymbol{\xi})q(\tilde{\boldsymbol{\xi}}|\boldsymbol{\xi})}{\pi(\tilde{\boldsymbol{\xi}})q(\boldsymbol{\xi}|\tilde{\boldsymbol{\xi}})}, 1\right\} \\
&= \min\{\pi(\boldsymbol{\xi})q(\tilde{\boldsymbol{\xi}}|\boldsymbol{\xi}), \pi(\tilde{\boldsymbol{\xi}})q(\boldsymbol{\xi}|\tilde{\boldsymbol{\xi}})\} \\
&= \pi(\boldsymbol{\xi})q(\tilde{\boldsymbol{\xi}}|\boldsymbol{\xi}) \min\left\{\frac{\pi(\tilde{\boldsymbol{\xi}})q(\boldsymbol{\xi}|\tilde{\boldsymbol{\xi}})}{\pi(\boldsymbol{\xi})q(\tilde{\boldsymbol{\xi}}|\boldsymbol{\xi})}, 1\right\} \\
&= \pi(\boldsymbol{\xi})h(\tilde{\boldsymbol{\xi}}, \boldsymbol{\xi}).
\end{aligned}
$$

Thus, $h(\boldsymbol{\xi}, \tilde{\boldsymbol{\xi}})$ is the desired transition density function for the Markov chain with $\pi(\boldsymbol{\xi})$ as its unique equilibrium distribution.

The Metropolis-Hastings algorithm is described as follows. Let $\boldsymbol{\xi}^{(0)}$ be an arbitrary initial state, then

For $t = 1, 2, \ldots$,

Generate $\boldsymbol{\xi}$ from $q(\boldsymbol{\xi}|\boldsymbol{\xi}^{(t-1)})$ and U from the uniform distribution $\mathcal{U}[0, 1]$.

If $U < \alpha(\boldsymbol{\xi}|\boldsymbol{\xi}^{(t-1)})$, let $\boldsymbol{\xi}^{(t)} = \boldsymbol{\xi}$; otherwise, let $\boldsymbol{\xi}^{(t)} = \boldsymbol{\xi}^{(t-1)}$.

Return the sequence $\{\boldsymbol{\xi}^{(1)}, \boldsymbol{\xi}^{(2)}, \ldots\}$.

2.8.3 Reversible Jump MCMC

In the description of the Gibbs sampler and the Metropolis-Hastings algorithm, it is implicitly assumed that the dimension of $\boldsymbol{\xi}$ is fixed. In the Bayesian methods for QTL mapping, this is the case only when the number of QTLs is pre-specified. When the number of QTLs is unknown, the dimension of $\boldsymbol{\xi}$ is variable. The Gibbs sampler is no longer applicable and the Metropolis-Hastings algorithm needs to be modified. The modification of the Metropolis-Hastings algorithm so that the variable dimension of $\boldsymbol{\xi}$ can be handled was done in [77]. The resultant algorithm is referred to as reversible jump MCMC. The general principle of the reversible jump MCMC is discussed below.

In the case of variable dimension of $\boldsymbol{\xi}$, the sample space of $\boldsymbol{\xi}$ can be decomposed as $\mathcal{R} = \cup_k \mathcal{C}_k$ where $\mathcal{C}_k$ is a subspace with a fixed dimension. In the reversible jump Markov chain, there are different types of moves. It could move within the same subspace where the current state belongs, jump to another subspace, and so on. Denote by m the move type, $q_m(\boldsymbol{\xi}|\tilde{\boldsymbol{\xi}})$ the proposal density function to move from $\tilde{\boldsymbol{\xi}}$ with move type m and $\alpha_m(\boldsymbol{\xi}|\tilde{\boldsymbol{\xi}})$ the acceptance probability for the proposed move. These proposal densities and acceptance probabilities determine the transition density of the chain. For condition (2.7) of the transition density to hold, it is required that

$$
\pi(\tilde{\boldsymbol{\xi}}) \sum_m q_m(\boldsymbol{\xi}|\tilde{\boldsymbol{\xi}})\alpha_m(\boldsymbol{\xi}|\tilde{\boldsymbol{\xi}}) = \pi(\boldsymbol{\xi}) \sum_m q_m(\tilde{\boldsymbol{\xi}}|\boldsymbol{\xi})\alpha_m(\tilde{\boldsymbol{\xi}}|\boldsymbol{\xi}).
$$

It is sufficient to require that, for each m,

$$\pi(\tilde{\boldsymbol{\xi}})q_m(\boldsymbol{\xi}|\tilde{\boldsymbol{\xi}})\alpha_m(\boldsymbol{\xi}|\tilde{\boldsymbol{\xi}}) = \pi(\boldsymbol{\xi})q_m(\tilde{\boldsymbol{\xi}}|\boldsymbol{\xi})\alpha_m(\tilde{\boldsymbol{\xi}}|\boldsymbol{\xi}).$$

This is achieved by taking

$$\alpha_m(\boldsymbol{\xi}|\tilde{\boldsymbol{\xi}}) = \min\left\{\frac{\pi(\boldsymbol{\xi})q_m(\tilde{\boldsymbol{\xi}}|\boldsymbol{\xi})}{\pi(\tilde{\boldsymbol{\xi}})q_m(\boldsymbol{\xi}|\tilde{\boldsymbol{\xi}})}, 1\right\}. \tag{2.8}$$

The acceptance probability looks the same as that in the Metropolis-Hastings algorithm. However, when the move type is to jump from one subspace to another, $\boldsymbol{\xi}$ and $\tilde{\boldsymbol{\xi}}$ have different dimensions. Therefore, in defining $\alpha_m(\boldsymbol{\xi}|\tilde{\boldsymbol{\xi}})$, an issue of dimension matching must be addressed. To ensure the dimension matching requirement, the following particular but still quite general approach is considered in [77]. Suppose that the move type is from sub space $\mathcal{C}_{k_1}$ to $\mathcal{C}_{k_2}$ where $\mathcal{C}_{k_l}$ is a Euclidean space of dimension n_l, $l = 1, 2$. Without loss of generality, assume $n_2 > n_1$. Suppose that the posterior distribution π has proper densities $p(\boldsymbol{\xi}^{(k_1)}|k_1)$ and $p(\boldsymbol{\xi}^{(k_2)}|k_2)$, respectively, on $\mathcal{C}_{k_1}$ and $\mathcal{C}_{k_2}$. To move from $\boldsymbol{\xi}^{(k_1)} \in \mathcal{C}_{k_1}$ to $\mathcal{C}_{k_2}$, first a continuous random vector $\boldsymbol{u}$ of dimension $m_1 = n_2 - n_1$ is generated by a density $q(\boldsymbol{u})$. Then set $\boldsymbol{\xi}^{(k_2)} = \boldsymbol{g}(\boldsymbol{\xi}^{(k_1)}, \boldsymbol{u})$, a one-to-one deterministic function of $\boldsymbol{\xi}^{(k_1)}$ and $\boldsymbol{u}$. Denote by $J(\boldsymbol{\xi}^{(k_1)}, \boldsymbol{u})$ the Jacobian

$$\left|\left(\begin{array}{cc} \frac{\partial \boldsymbol{g}^\tau}{\partial \boldsymbol{\xi}^{(k_1)}} & \frac{\partial \boldsymbol{g}^\tau}{\partial \boldsymbol{u}} \end{array}\right)\right|.$$

Let $j(\boldsymbol{\xi}^{(k_1)})$ be the probability to choose the move from $\boldsymbol{\xi}^{(k_1)}$ to $\mathcal{C}_{k_2}$ and $j(\boldsymbol{\xi}^{(k_2)})$ the probability to choose the move from $\boldsymbol{\xi}^{(k_2)}$ to $\mathcal{C}_{k_1}$. The dimension matcing acceptance probability is given by

$$\alpha_m(\boldsymbol{\xi}^{(k_2)}|\boldsymbol{\xi}^{(k_1)}) = \min\left\{\frac{p(\boldsymbol{\xi}^{(k_2)}|k_2)j(\boldsymbol{\xi}^{(k_2)})}{p(\boldsymbol{\xi}^{(k_1)}|k_1)j(\boldsymbol{\xi}^{(k_1)})q(\boldsymbol{u})}J(\boldsymbol{\xi}^{(k_1)}, \boldsymbol{u}), 1\right\}. \tag{2.9}$$

If the move is from $\boldsymbol{\xi}^{(k_2)}$ to $\mathcal{C}_{k_1}$, solve $\boldsymbol{\xi}^{(k_1)}$ and $\boldsymbol{u}$ from $\boldsymbol{\xi}^{(k_2)} = \boldsymbol{g}(\boldsymbol{\xi}^{(k_1)}, \boldsymbol{u})$ and the acceptance probability is given by

$$\alpha_m(\boldsymbol{\xi}^{(k_1)}|\boldsymbol{\xi}^{(k_2)}) = \min\left\{\frac{p(\boldsymbol{\xi}^{(k_1)}|k_1)j(\boldsymbol{\xi}^{(k_1)})q(\boldsymbol{u})}{p(\boldsymbol{\xi}^{(k_2)}|k_2)j(\boldsymbol{\xi}^{(k_2)})}[J(\boldsymbol{\xi}^{(k_1)}, \boldsymbol{u})]^{-1}, 1\right\}. \tag{2.10}$$

The reversible jump MCMC is summarized in the following algorithm:

Let $\boldsymbol{\xi}^{(0)}$ be an initial state.

For $t = 1, 2, \ldots,$

Determine the move type m by the probability density $j(\boldsymbol{\xi}^{(t-1)})$.

- If $\boldsymbol{\xi}^{(t-1)} \in \mathcal{C}_k$ and the move type m is to move within $\mathcal{C}_k$, generate $\boldsymbol{\xi}$ by $q_m(\boldsymbol{\xi}|\boldsymbol{\xi}^{(t-1)})$ (with this move type, $q_m(\boldsymbol{\xi}|\boldsymbol{\xi}^{(t-1)})$ is a proper density on $\mathcal{C}_k$), and compute $\alpha(\boldsymbol{\xi}|\boldsymbol{\xi}^{(t-1)})$ by (2.8);

- If the move type m is to jump from $\mathcal{C}_k$ to $\mathcal{C}_l$ and the dimension of $\mathcal{C}_k$ and $\mathcal{C}_l$ are n_1 and n_2 respectively, then

  – if $n_1 < n_2$, generate $\boldsymbol{u} \in R^{n_2 - n_1}$ by $q(\boldsymbol{u})$, set $\boldsymbol{\xi} = g(\boldsymbol{\xi}^{(t-1)}, \boldsymbol{u})$ and compute $\alpha(\boldsymbol{\xi}|\boldsymbol{\xi}^{(t-1)})$ by (2.9);

  – if $n_1 > n_2$, solve $\boldsymbol{\xi}$ and $\boldsymbol{u}$ from $\boldsymbol{\xi}^{(t-1)} = g(\boldsymbol{\xi}, \boldsymbol{u})$ and compute $\alpha(\boldsymbol{\xi}|\boldsymbol{\xi}^{(t-1)})$ by (2.10) with $\boldsymbol{\xi}^{(k_2)}$ and $\boldsymbol{\xi}^{(k_1)}$ replaced by $\boldsymbol{\xi}^{(t-1)}$ and $\boldsymbol{\xi}$ respectively.

- Generate U from the uniform distribution $\mathcal{U}[0, 1]$.

  – If $U < \alpha(\boldsymbol{\xi}|\boldsymbol{\xi}^{(t-1)})$, let $\boldsymbol{\xi}^{(t)} = \boldsymbol{\xi}$;

  – otherwise, let $\boldsymbol{\xi}^{(t)} = \boldsymbol{\xi}^{(t-1)}$.

Return the sequence $\{\boldsymbol{\xi}^{(1)}, \boldsymbol{\xi}^{(2)}, \dots\}$.

2.8.4 Extracting Information from the MCMC Sequence

By the theory of Markov chain, the MCMC sequence generated in § 2.8.1 to § 2.8.3 converges, i.e., the distribution of $\boldsymbol{\xi}^{(t)}$ converges to the posterior distribution $\pi(\boldsymbol{\xi})$. If we are only concerned with the estimation of the posterior means $E\boldsymbol{\xi}$, we can take $\frac{1}{N} \sum_{t=1}^{N} \boldsymbol{\xi}^{(t)}$ as an estimate of $E\boldsymbol{\xi}$, since, by the Markov convergence theorem, $\frac{1}{N} \sum_{t=1}^{N} \boldsymbol{\xi}^{(t)} \to E\boldsymbol{\xi}$ almost surely. For the inference on some other aspects of $\pi(\boldsymbol{\xi})$, e.g., its density, we need to have an i.i.d. sample from $\pi(\boldsymbol{\xi})$. To extract an i.i.d. sample from the MCMC sequence, we need to ensure that (i) the $\boldsymbol{\xi}^{(t)}$'s to be extracted have reached equilibrium and (ii) the extracted $\boldsymbol{\xi}^{(t)}$'s should be at least approximately independent. A practical way to achieve these two requirements is through *burning* and *thinning*. *Burning* means discarding the generated $\boldsymbol{\xi}^{(t)}$'s in the early part of the sequence. Usually, around 5,000 of them are discarded. Burning is to ensure (i). *Thinning* means extracting, after burning, every rth of the generated $\boldsymbol{\xi}^{(t)}$'s for an appropriate r. Usually, r can be taken somewhere between 20 and 50. Thinning is to ensure (ii). For more rigorous but more time-consuming and costly approaches, see, e.g., [66], [67]and [177].

2.9 An Overview on Feature Selection with Small-n-Large-p Models

Without loss of generality, we focus on small-n-large-p linear models in our discussion. A small-n-large-p linear model is formulated as follows:

$$y_i = \sum_{j=1}^{p} \beta_j x_{ij} + \epsilon_j, \ i = 1, \ldots, n, \tag{2.11}$$

where p is large, usually even much larger than n. A typical nature of the above model is that the number of covariates which are truly associated with the response variable is relatively small; that is, $p_0 = \sum_{j=1}^{p} I\{\beta_j \neq 0\}$ is small compared with p. Model (2.11) is therefore also called a sparse high-dimensional linear model.

The sparse high-dimensional linear model causes two difficulties: (i) the parameters cannot be estimated by the classical least squares method or the maximum likelihood method since the $n \times p$ design matrix, which has a rank at most n, does not have a full column rank; (ii) because $n << p$, the columns of the design matrix are highly correlated, these artifacts cause the phenomenon of spurious correlation. There are two related types of spurious correlation. The first type is the spurious correlation among the covariates: covariates which are not statistically correlated appear correlated in the data. The second type is the spurious correlation between covariates and the response variable: covariates which are independent of the response variable appear correlated with the response variable due to their spurious correlation with some causal covariates. The phenomenon of spurious correlation is particularly serious in QTL mapping. In addition, markers close to each other usually have a high correlation in their genotypes, which is another source of spurious correlation of the markers with the trait values.

A major task in the analysis of sparse high-dimensional linear models is feature selection. A feature is either a covariate or a function of covariates, i.e., the product of two covariates. In general, there are two purposes of feature selection: (i) to build a model with the selected feature for prediction and (ii) to detect causal or truly correlated features. For convenience, causal or truly correlated features are called relevant features. For the first purpose, the emphasis is on the prediction accuracy and whether or not the selected features are all relevant ones does not matter as long as the features have a certain prediction power. For the second purpose, the emphasis is on whether or not the selected features are relevant ones and whether or not all relevant features are selected.

Feature selection is usually done through a certain model framework and hence is also referred to as model selection. The ideal of model selection is to consider all possible models which can be formed by the features under study, and then select the model by a certain model selection criterion. This ideal method is called all-subset method. However, the all-subset method can only be realized when there are only a very few features under study. It can never be realized when the number of features p is large, since the number of all possible models is 2^p which is far beyond any computational capacity.

In recent years, various penalized likelihood methods are developed for sparse high-dimensional models. In a penalized likelihood method, a penalty function is added to the likelihood function. The penalty function, which imposes penalties on the parameters β_j's, is regularized by a penalty parameter. By properly choosing the value of the penalty parameter, the penalized likelihood method can select features and estimate the β_j's simultaneously. Different penalized likelihood methods differ in the penalty function. The following are the major penalty functions studied in the literature: L_1 penalty (also referred to as the Lasso penalty) [181], smoothly clipped absolute deviation (SCAD) penalty [55], bridge penalty [63] and minimax concave penalty (MCP) [220]. Under different conditions, some mild and some restricted, the penalized likelihood method with each of the above penalty functions has a so-called oracle property, meaning that, in the asymptotic sense, it performs as well as if the analyst had known in advance which coefficients were zero and which were nonzero.

In addition to the penalized likelihood methods, sequential methods have also been used for feature selection in sparse high-dimensional models. Sequential methods have a common so-called greedy nature; that is, whenever a feature is selected due to its significance to the local model at any step, the feature is retained in the subsequent steps. This greedy nature might be detrimental if the purpose is for prediction. However, it is beneficial if the purpose is to detect relevant features, see [184]. The traditional forward regression selection has been re-considered in [193]. Other sequential methods such as least angle regression (LAR) [49] and orthogonal matching pursuit (OMP) [187] have been proposed and studied recently. The OMP is related to a sequential penalized likelihood procedure called sequential Lasso [124] which selects features step by step through partially penalized likelihoods where the coefficient parameters of the features already selected are not penalized. A property desired of a sequential method is that all the relevant features should be selected before any irrelevant feature can be selected in the sequential steps. It is shown under the framework of sequential Lasso in [124] that the OMP possesses the desired property.

Strictly speaking, neither the penalized likelihood methods nor the sequential methods mentioned above fulfill the task of model selection, if the value of the penalty parameter is not specified in the penalized likelihood methods and it is not told at which step to stop in the sequential methods. Essentially, these approaches only generate a sequence of candidate models. These candidate models need to be assessed by a certain model selection criterion to yield the final selected model. Model selection criteria are discussed in the next section.

2.10   Model Selection Criteria for Small-n-Large-p Models

There are a number of classical model selection criteria in statistical analysis. Those include cross validation (CV) score [174], Akaike information criterion (AIC) [1] and the Bayes information criterion (BIC) [159], etc..

The CV score is an approximation to the prediction error. Let $\{(y_i, \boldsymbol{x}_i) : i = 1, \ldots, n\}$ be the data. For a given model, the original CV score is obtained as follows. The data points are left out one-at-a-time, the remaining $n-1$ data points are used to fit the model, the fitted model is used to predict the response variable which is left out, and the prediction error is computed. The CV score is then the average of the prediction error over all the data points. In the context of linear model, the CV score is given by

$$\text{CV} = \frac{1}{n} \sum_{i=1}^{n} [y_i - \boldsymbol{x}_i^\tau \hat{\boldsymbol{\beta}}^{(-i)}]^2,$$

where $\hat{\boldsymbol{\beta}}^{(-i)}$ is the estimate of $\boldsymbol{\beta}$ based on the data with $(y_i, \boldsymbol{x}_i)$ being left out. The computation of the CV score above is time-consuming, especially when the method used to estimate $\boldsymbol{\beta}$ is sophisticated. Instead of the original CV, a d-fold CV is usually applied in many problems. In the d-fold CV, the data set is divided into d portions. The portions are left out one-at-a-time and the remaining $d - 1$ portions are used to fit the model. The fitted model is then used to predict the portion which is left out. The prediction errors over all the portions are averaged to give the CV score. The d-fold CV score in the context of linear model is given by

$$\text{CV}_{d\text{-fold}} = \frac{1}{n} \sum_{j=1}^{d} \sum_{(y_i, \boldsymbol{x}_i) \in \text{portion } j} [y_i - \boldsymbol{x}_i^\tau \hat{\boldsymbol{\beta}}^{(-j)}]^2,$$

where $\hat{\boldsymbol{\beta}}^{(-j)}$ is the estimate of $\boldsymbol{\beta}$ based on the data with the j-th portion being left out.

The AIC is an approximation to the Kullback-Leibler information. Suppose that $f(y)$ is the density function of y under the true model, and $g(y|\boldsymbol{\beta})$ the density function of y based on a model indexed by $\boldsymbol{\beta}$ which is used to approximate the true model. The Kullback-Leibler information is given by

$$I(f, g) = \int f(y) \ln \left(\frac{f(y)}{g(y|\boldsymbol{\beta})} \right) dy.$$

However, the Kullback-Leibler information cannot be computed. As an approximation to the Kullback-Leibler information, the AIC, which can be computed, is given as follows:

$$\text{AIC} = -2 \ln L_n \{\hat{\boldsymbol{\beta}}(s)\} + 2\nu(s),$$

where s stands for the model used to approximate the true model, $\hat{\boldsymbol{\beta}}(s)$ is the MLE of $\boldsymbol{\beta}(s)$ under model s, L_n is the likelihood function and $\nu(s)$ is the number of components of $\boldsymbol{\beta}(s)$.

The BIC of a model s is an approximation to a transform of the posterior probability of model s under the framework of Bayesian analysis. Assume that $p(s)$ is the prior probability of model s and that $\pi\{\boldsymbol{\beta}(s)\}$ is the prior density of $\boldsymbol{\beta}(s)$ given s. The posterior probability of s given $\boldsymbol{y} = (y_1, \ldots, y_n)^\tau$ is

$$p(s|\boldsymbol{y}) = \frac{m(\boldsymbol{y}|s)p(s)}{\sum_s p(s)m(\boldsymbol{y}|s)},$$

where $m(\boldsymbol{y}|s)$ is the likelihood of model s given by

$$m(\boldsymbol{y}|s) = \int e^{L_n\{\boldsymbol{\beta}(s)\}} \pi\{\boldsymbol{\beta}(s)\}d\boldsymbol{\beta}(s).$$

Under some regularity conditions, $m(\boldsymbol{y}|s)$ is approximated by the Laplace approximation $e^{L_n\{\hat{\boldsymbol{\beta}}(s)\}} n^{-\nu(s)/2}$. Thus,

$$-2\ln p(s|\boldsymbol{y}) = -2\ln L_n\{\hat{\boldsymbol{\beta}}(s)\} + \nu(s)\ln n - 2\ln p(s),$$

up to a constant $\sum_s p(s)m(\boldsymbol{y}|s)$. Taking $p(s)$ as a constant over all s gives rise to BIC.

It is obvious that the criterion of CV puts more emphasis on prediction error and that both AIC and BIC put more emphasis on the contribution of selected features to the likelihood. These criteria have been widely used for model selection in various model contexts. However, these criteria are not appropriate for model selection in sparse high-dimensional models, especially when the purpose is to identify true features. These criteria all tend to select too many irrelevant features. This phenomenon has been observed in QTL mapping studies, see [14] [22] [161] .

Model selection criteria for sparse high-dimensional models have been studied only recently. A version of modified BIC dubbed mBIC is proposed and studied in [14] [15]. A family of extended BIC referred to as EBIC is developed in [27]. The EBIC includes the original BIC and the mBIC as its special cases. If a criterion is able to select the true model (the model consisting of exactly all the relevant features) among all possible models or among a set of models that contains the true model almost surely in an asymptotic sense, the criterion is said to be selection consistent. It has been shown that the EBIC is selection consistent for feature selection in various sparse high-dimensional models, see [28], [62], [28], [125] and [123].

We have seen that the BIC is developed under a Bayesian framework by taking the prior probability of models as a constant. This constant prior causes the problem mentioned above. Let S_j denote the class of models consisting of j features. With the constant prior, the prior probability of S_j is proportional to the size of S_j. To be more specific, let $p = 1000$. The class of models containing a single covariate, S_1, has size 1000, while the class of models containing two covariates, S_2, has size $1000 \times 999/2$. Thus the prior probability of S_2 is $999/2$ times that of S_1. The size of S_j increases at an exponential rate as j increases to $j = p/2 = 500$, so does the prior probability of S_j. Thus models with a larger number of features have much higher prior probabilities than models with a smaller number of features, which makes the BIC favoring models with larger number of features.

The EBIC extends BIC by taking the prior distribution such that the prior probability of S_j is proportional to a power (lower than 1) of its size. Let $\tau(S_j)$ denote the size of S_j. The specified prior probability of S_j is proportional to $\tau^\xi(S_j)$ for $0 \leq \xi \leq 1$. The prior probability $p(s)$ for $s \in S_j$ is then proportional to $\tau^{-\gamma}(S_j)$ where $\gamma = 1 - \xi$. This gives rise to the EBIC below:

$$\text{EBIC}_\gamma(s) = -2 \ln L_n(\hat{\theta}(s)) + \nu(s) \ln n + 2\gamma \ln \tau(S_j), \quad 0 \leq \gamma \leq 1.$$

As mentioned earlier, the selection consistency of EBIC has been established. Specifically, the selection consistency is the following property. Let s_0 be the true model. Under certain mild conditions,

$$P(\text{EBIC}_\gamma(s_0) < \min_{s: s \neq s_0, \nu(s) \leq k\nu(s_0)} \text{EBIC}_\gamma(s)) \to 1, \quad \text{as } n \to \infty,$$

when $\gamma > 1 - \ln n/(2 \ln p)$, p being the total number of features, where $k \geq 1$ is any fixed number. The result implies that when we consider a set of models of size less than or comparable with that of the true model the EBIC can select the true model almost surely in the asymptotic sense if the set contains the true model. This well serves the purpose of practical feature selection, especially QTL mapping, since practically only models having size less than or comparable with that of the true model are considered.

The selection consistency of EBIC is an asymptotic property. However, in practical problems, we only have a finite sample size. Though, for any $\gamma > 1 - \ln n/(2 \ln p)$, the selection consistency holds, we need to make a proper choice of γ in this consistency range for a particular practical problem. The larger the γ, the fewer features which can be selected. A larger γ bars more strictly the false features from being selected, but at the same time it also makes the true features with less effect more difficult to be selected. In other words, with a larger γ, we have less false selection but also we have less positive selection, and vice versa. We cannot have less false selection and more positive selection at the same time by choosing any value of γ. This is similar to the dilemma which we encounter in hypothesis testing: we cannot have a lower Type I error rate and a higher power at the same time by using any critical value. The same strategy in hypothesis testing can be adopted for the choice of γ, that is, we choose the γ value such that the positive selection is maximized while the false selection is reasonably controlled. If we take a value of γ as close to the lower bound of its consistency range as possible, we have the maximum positive selection while the false selection is guaranteed to converge to zero.

Thus, a value of γ which is slightly bigger than the lower bound is a proper choice. Practically, we can well choose the lower bound $1 - \ln n/(2\ln p)$.

In the development above, the linear model is implicitly assumed as a main-effect model with p covariates. The features and covariates are the same. If we consider an interaction model (main effects plus two-way interactions) with p covariates, we then have p main-effect features, which are the same as the covariates, and $p(p-1)/2$ interaction features, which are functions of the covariates. For a model s with ν_M main-effect features and ν_I interaction features, the EBIC becomes

$$
\begin{aligned}
\text{EBIC}_\gamma(s) &= -2\ln L(\hat{\boldsymbol{\theta}}) + (\nu_M + \nu_I)\ln(n) + 2\gamma\ln\binom{p}{\nu_M}\binom{\frac{p(p-1)}{2}}{\nu_I} \\
&= -2\log L(\hat{\boldsymbol{\theta}}) + (\nu_M + \nu_I)\ln(n) \\
&\quad + 2\gamma\ln\binom{p}{\nu_M} + 2\gamma\ln\binom{\frac{p(p-1)}{2}}{\nu_I}.
\end{aligned}
\tag{2.12}
$$

The selection consistency of EBIC for main-effect models extends to the EBIC for interaction models when γ is taken such that $\gamma > 1 - \ln n/(4\ln p)$. The quantity $2\gamma\ln p$ is roughly the penalty imposed on adding one feature (main-effect or interaction) to the model. By using the same γ, the main-effect features and interaction features are treated on an equal footing. However, in sparse high-dimensional interaction models, there are many more interaction features than main-effect features. Therefore there are many more possible artifacts in the interaction features than in the main-effect features. It has been observed in some studies that, if the main-effect features and the interaction features are treated on the same footing, the true main-effect features could easily be masked by interaction artifacts, see § 6.2.1. It suggests that a less severe penalty should be imposed for adding a main-effect feature than adding an interaction feature. This gives rise to the following version of EBIC for interaction models:

$$
\text{EBIC}_{\gamma_M\gamma_I}(s) \tag{2.13}
$$
$$
= -2\ln L(\hat{\boldsymbol{\theta}}) + (\nu_M + \nu_I)\ln(n) + 2\gamma_M\ln\binom{p}{\nu_M} + 2\gamma_I\ln\binom{\frac{p(p-1)}{2}}{\nu_I}.
$$

It is shown in [84] that the $\text{EBIC}_{\gamma_M\gamma_I}$ is selection consistent for interaction models (linear or generalized linear) when $\gamma_M > 1 - \frac{\ln n}{2\ln p}$ and $\gamma_I > 1 - \frac{\ln n}{4\ln p}$. By the same consideration as that for main-effect models, in the application of EBIC in practical problems, the two γ's in $\text{EBIC}_{\gamma_M\gamma_I}$ can be taken as their lower bounds, i.e., $\gamma_M = 1 - \frac{\ln n}{2\ln p}$ and $\gamma_I = 1 - \frac{\ln n}{4\ln p}$.

Chapter 3

Quantitative Genetics and General Issues on QTL Mapping

Quantitative genetics is concerned with the inheritance of genetic quantitative traits. Stretches of DNA containing or linked to the genes that underlie a quantitative trait are called quantitative trait loci (QTL) (locus for singular form). The abbreviation QTL is used for both the plural and singular forms. QTL mapping is the study for locating QTL in the genome by using trait and marker genotype data obtained from certain populations. The understanding of the principles of quantitative genetics is crucial for QTL mapping. Unlike qualitative traits which are less influenced by environmental factors and have a deterministic relationship with the affecting genes, quantitative traits are influenced by environmental factors to such an extent that their relationships with the affecting genes are usually masked by the environmental effects. Genetic quantitative traits can only be studied from the aspects of their distributions in certain populations. Therefore, statistical methodology is more crucial in the study of quantitative traits. This chapter is devoted to the principles of quantitative genetics and some general statistical issues on QTL mapping. Section 3.1 covers the distributional nature of genetic quantitative traits. Section 3.2 deals with genetic models of QTL effects and genetic variances of quantitative traits. Section 3.3 tackles the problem of inference on unknown QTL genotypes from marker genotypes. Section 3.4 provides a general discussion on statistical models of quantitative traits. Section 3.5 discusses issues related to QTL mapping data.

3.1 Distributional Features of Genetic Quantitative Traits

We concentrate on quantitative traits that are measurable in continuous scales, though there are discrete quantitative traits which we will also deal with later. There are genetic and non-genetic quantitative traits. The distributional features of non-genetic and genetic quantitative traits are different. For genetic quantitative traits, difference in the number of QTL, the magnitude of

QTL effects and environment effects also give rise to different distributional features.

3.1.1   Features of Genetic and Non-genetic Quantitative Traits

Non-genetic quantitative traits are not affected by any genetic factors. The variation of a non-genetic quantitative trait in a population is caused by environmental factors only. A non-genetic quantitative trait in a population can be described by

$$y_i = \mu + \epsilon_i$$

for any individual i, where ϵ_i is a random variable with expectation zero. This ϵ_i, the deviation from μ, is caused solely by environmental effect. The environmental effect is in general an overall effect of many identifiable and unidentifiable factors, each causing a small fluctuation. For example, environmental factors affecting a plant could include identifiable ones such as amount of sun light, rainfall, degradation of soil, temperature, humidity and many other unidentifiable sources. The distribution of a random variable of this nature can be well approximated by a normal distribution, which is justifiable by the central limit theorem in probability theory. Thus, a non-genetic quantitative trait must have the features of a normal distribution. Specifically, the distribution is unimodal, symmetric and bell-shaped.

Genetic quantitative traits are affected by both genetic and environmental factors. The variation of a genetic trait in a population is partially due to the variation in genetic formation and partially due to random environmental effect. The population can be conceptually divided into groups with different genetic formations corresponding to the genes affecting the trait. The trait values in a particular group k can be described as

$$y_{ki} = \mu_k + \epsilon_{ki},$$

where μ_k, which varies from group to group, is conferred by the genetic formation of group k, and ϵ_{ki} is a random deviation caused by environmental effect. The distributional feature of each group is the same as that of a non-genetic quantitative trait. However, the grouping of the population is unknown since the genes are unknown. The distribution of the trait in the population is a mixture of those in the unknown groups. The distribution of a genetic quantitative trait departs from a normal distribution and usually exhibits bumps or multiple modals. Two histograms of samples from distributions of genetic and non-genetic quantitative traits are illustrated in Figure 3.1.

3.1.2   Features of Genetic Traits Affected by Major and Minor QTL

Major QTL are the ones that account for a large proportion of the variation of the trait in a population. The distribution of a genetic quantitative

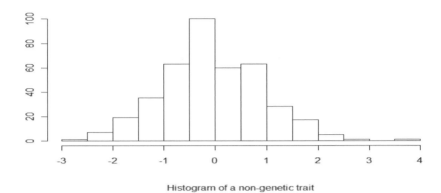

Histogram of a non-genetic trait

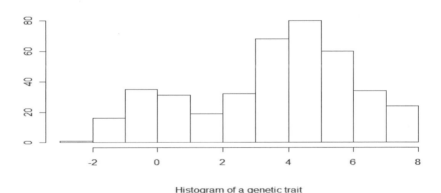

Histogram of a genetic trait

FIGURE 3.1: Illustration of the distributions of genetic and non-genetic quantitative traits.

trait affected by major QTL exhibits explicit multi-modality. In the extreme case, the trait values are completely separated between groups of different genetic formations, like the plant height in Mendel's experiments. In this case, the quantitative trait can be treated as qualitative trait. Except for a few rare examples, there are overlapping of trait values among different groups. The typical features of genetic quantitative traits affected by major QTL are demonstrated in Figure 3.2

A minor QTL is one that accounts for a small proportion of the total variation of the trait values. If a trait is affected by many minor QTL, the trait values become extremely overlapped between different groups and the distribution of the trait usually does not display multi-modality. The typical features of genetic quantitative traits affected by minor QTL are demonstrated in Figure 3.3.

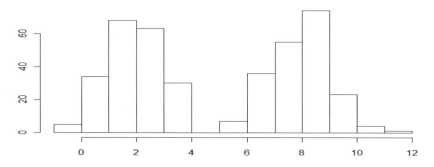

Trait values from two groups are completely separated

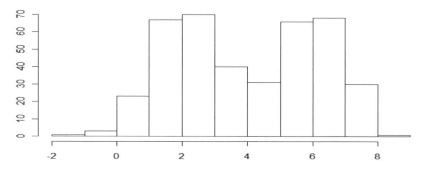

Trait values from two groups are overlapping

FIGURE 3.2: Illustration of distributions of traits affected by major QTL.

3.2 Genetic Values and Variances

Because of the overlapping of trait values between groups of different genetic formations, genetic quantitative traits cannot be studied at individual level. Two individuals with different genetic formations might have similar trait values due to environmental effect. To unmask the differences caused by genetic formations from environmental effect, the population aspects of QTL must be considered. In this section, we deal with these aspects.

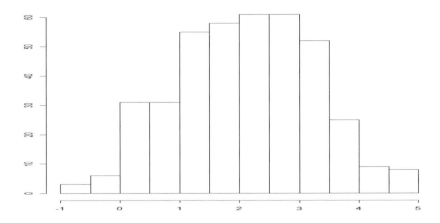

FIGURE 3.3: Illustration of distributions of traits affected by minor QTL.

3.2.1 Genotype and Allele Frequencies

The genetic constitution of a population is characterised by genotype and allele frequencies. Suppose that we are concerned with a single locus with two alleles: A, a, and three genotypes: AA, Aa, aa, in the population. A genotype frequency is the proportion of a particular genotype in the population. For example, if 16% of the individuals in the population have genotype AA then the genotype frequency of AA is 0.16. An allele frequency is the proportion of a particular allele in the population. For example, if the proportion of the A allele is 40% then the allele frequency of A is 0.4.

The allele frequencies can be obtained from genotype frequencies. Denote the genotype frequencies of AA, Aa and aa, respectively, by P, Q and R. Let N be the population size. Then there are N genotypes and $2N$ alleles in the population. PN individuals have genotype AA, each of them has two A alleles and together they contribute $2PN$ A-alleles. QN individuals have genotype Aa, each of them has one A-allele and together they contribute QN A-alleles. Thus, in the population, there are $2PN + QN$ A-alleles and the frequency of the A-allele is given by $p = (2PN + QN)/(2N) = (2P + Q)/2$. Similarly, the frequency of the a-allele is given by $q = (Q + 2R)/2$.

For example, suppose that the genotype frequencies of of AA, Aa and aa are, respectively, 0.16, 0.5 and 0.34, then the allele frequencies of A and a are $p = (2 \times 0.16 + 0.5)/2 = 0.41$, $q = (0.5 + 2 \times 0.34)/2 = 0.59 = 1 - p$.

If random mating has been made within a population then, under conditions such as no selection, mutation or migration, the genotype frequencies are constant from generation to generation, and the genotype frequencies are simply the products of the allele frequencies of their constituent alleles. A population with these properties is said to have reached *Hardy-Weinberg equilibrium*

(HWE). If the allele frequencies of A and a are, respectively, p and q then, under HWE, the genotype frequencies of AA, Aa and aa are, respectively, p^2, $2pq$ and q^2. For instance, suppose that the allele frequencies of A and a are, respectively, 0.41 and 0.59 when the population has reached HWE. Then the genotype frequencies of of AA, Aa and aa are given by

$$p_{AA} = p^2 = 0.1681, \ p_{Aa} = 2pq = 0.4838, \ p_{aa} = q^2 = 0.3481.$$

If the parental sources of the alleles in a genotype are known, it is said that the *phase* of the genotype is known. Note that the genotype Aa consists of two phase-known genotypes, i.e., $A|a$ and $a|A$. By convention, the left allele denotes the paternal one and the right the maternal one. When more than one locus is studied, in addition to genotypes, haplotypes are also of interest. A *haplotype* of two or more than two loci is a combination of the alleles on the same parental chromosome at the loci. For example, suppose at two loci $\mathcal{A}$ and $\mathcal{B}$ the phase-known genotypes are $A|a$ and $b|B$ respectively, then the two haplotypes are Ab and aB. Haplotype frequencies are also characteristics of a population. If the haplotype frequencies of two loci are the products of the allele frequencies of the consituent alleles, the two loci are said to be in *linkage equilibrium*, otherwise, they are said to be in *linkage disequilibrium* (LD).

If we consider the alleles in the genotypes or haplotypes of a randomly chosen individual as random objects, the HWE implies that the two random alleles at a genotype are statistically independent, and the linkage equilibrium implies that the alleles at different loci are statistically independent. Throughout this section, the HWE and linkage equilibrium are assumed.

3.2.2 Genetic Values

Genetic quantitative trait values are referred to as phenotypic values from now onwards. A phenotypic value Y can be decomposed into a component attributable to the effect of genotypes, denoted by G, and a deviation caused by environment, denoted by E; that is,

$$Y = G + E.$$

The case of one QTL for a quantitative trait

Suppose that there are only two alleles of the QTL in the population. The situation of more than two alleles can be treated similarly, but it does not concern QTL mapping with experimental crosses. Let A and a denote the two alleles and p and q be their allele frequencies. Each genotype confers a different value to the trait. These values are called genotypic values. Without loss of generality, the genotypic values can be assigned as follows:

$$G = \begin{cases} c, & \text{for } AA, \\ d, & \text{for } Aa, \\ -c, & \text{for } aa. \end{cases}$$

We can consider genotypic value G as a function of the two alleles of the genotype. Let x_1 and x_2 represent the paternal and maternal alleles of an individual randomly chosen from the population. We can write the function as

$$G(x_1, x_2) = \begin{cases} c, & \text{if } (x_1, x_2) = (A, A), \\ d, & \text{if } (x_1, x_2) = (A, a) \text{ or } (a, A), \\ -c, & \text{if } (x_1, x_2) = (a, a). \end{cases}$$

We now consider a decomposition of $G(x_1, x_2)$. Let $H(x_1, x_2) = G(x_1, x_2) - \mu$, where $\mu = E[G(x_1, x_2)] = c(p - q) + 2pqd$. First, we seek functions $h_1(x_1)$ and $h_2(x_2)$ such that they minimize

$$E[H(x_1, x_2) - \tilde{h}_1(x_1) - \tilde{h}_2(x_2)]^2$$

among all $\tilde{h}_1(x_1)$ and $\tilde{h}_2(x_2)$ satisfying $E\tilde{h}_1(x_1) = E\tilde{h}_2(x_2) = 0$. Let

$$h_1(x_1) = E[H(x_1, x_2)|x_1], \quad h_2(x_2) = E[H(x_1, x_2)|x_2].$$

Noting that x_1 and x_2 are independent, it can be easily verified that

$$\begin{aligned} & E[H(x_1, x_2) - \tilde{h}_1(x_1) - \tilde{h}_2(x_2)]^2 \\ = \ & E[H(x_1, x_2) - h_1(x_1) - h_2(x_2)]^2 \\ & + E[h_1(x_1) - \tilde{h}_1(x_1)]^2 + E[h_2(x_2) - \tilde{h}_2(x_2)]^2. \end{aligned}$$

Thus $h_1(x_1)$ and $h_2(x_2)$ are the minimizers. Because of the symmetry of $H(x_1, x_2)$ and the i.i.d. property of x_1 and x_2, we obtain

$$h_1(x) = h_2(x) \equiv h(x)$$

and

$$\begin{aligned} h(A) &= pc + qd - \mu = q[c + d(q - p)] \equiv \alpha_1, \\ h(a) &= pd - qc - \mu = -p[c + d(q - p)] \equiv \alpha_2. \end{aligned}$$

In the quantitative genetics literature, α_1 is called the *average effect of allele A*, α_2 the *average effect of allele a*, and the difference $\alpha = \alpha_1 - \alpha_2$ the *average effect of the gene substitution*. From the equalities above, we have $\alpha = c + d(q - p)$. Let

$$D(x_1, x_2) = H(x_1, x_2) - h(x_1) - h(x_2).$$

Thus we have the decomposition of $g(x_1, x_2)$ as

$$g(x_1, x_2) = \mu + h(x_1) + h(x_2) + D(x_1, x_2).$$

The component $h(x_1) + h(x_2)$ is called the *additive effect* (or *main effects*). It is also referred to as the *breeding value* of the genotype x_1x_2. Thus, the breeding values of the genotypes AA, Aa and aa are, respectively, $2q\alpha$, $(q-p)d$ and $-2p\alpha$. The component $D(x_1, x_2)$ is called the *dominance deviation* (or interaction effect, in statistical terminology). The reader is referred to [53] for a genetic interpretation of the concepts: *average gene effect, average effect of the gene substitution* and *breeding value.*

Explicitly

$$D(x_1, x_2) = \begin{cases} -2q^2 d, & \text{if } (x_1, x_2) = (A, A), \\ 2pqd, & \text{if } (x_1, x_2) = (A, a) \text{ or } (a, A), \\ -2p^2 d, & \text{if } (x_1, x_2) = (a, a); \end{cases}$$

and

$$\begin{aligned} G(x_1, x_2) &= \begin{cases} \mu + 2q\alpha - 2q^2 d, & \text{if } (x_1, x_2) = (A, A), \\ \mu + (q - p)\alpha + 2pqd, & \text{if } (x_1, x_2) = (A, a) \text{ or } (a, A), \\ \mu - 2p\alpha - 2p^2 d, & \text{if } (x_1, x_2) = (a, a), \end{cases} \\ &\equiv \mu + A(x_1, x_2) + D(x_1, x_2), \end{aligned} \tag{3.1}$$

where $A(x_1, x_2)$ denotes the additive effect. Since $D(x_1, x_2) = 0$ if and only if $d = 0$, d is referred to as dominant effect.

The case of two or more than two QTL for a quantitative trait

First, consider the case of two QTL. Let $\boldsymbol{x}_k = (x_{k1}, x_{k2})$, $k = 1, 2$, denote the random genotypes of the two QTL. The genotypic value can be expressed as a function $G(\boldsymbol{x}_1, \boldsymbol{x}_2)$. We first seek functions $G_1(\boldsymbol{x}_1)$ and $G_2(\boldsymbol{x}_2)$ satisfying $E[G_1(\boldsymbol{x}_1)] = 0$ and $E[G_2(\boldsymbol{x}_2)] = 0$ to approximate $H(\boldsymbol{x}_1, \boldsymbol{x}_2) = G(\boldsymbol{x}_1, \boldsymbol{x}_2) - \mu$ in the sense that they minimize

$$E[H(\boldsymbol{x}_1, \boldsymbol{x}_2) - \tilde{G}_1(\boldsymbol{x}_1) - \tilde{G}_2(\boldsymbol{x}_2)]^2.$$

The same argument as in the decomposition of genotypic effect into allele effects yields that the minimizers are given by

$$G_1(\boldsymbol{x}_1) = E[H(\boldsymbol{x}_1, \boldsymbol{x}_2)|\boldsymbol{x}_1], \quad G_2(\boldsymbol{x}_2) = E[H(\boldsymbol{x}_1, \boldsymbol{x}_2)|\boldsymbol{x}_2].$$

Denote

$$I(\boldsymbol{x}_1, \boldsymbol{x}_2) = H(\boldsymbol{x}_1, \boldsymbol{x}_2) - G_1(\boldsymbol{x}_1) - G_2(\boldsymbol{x}_2).$$

Then we have the decomposition

$$G(\boldsymbol{x}_1, \boldsymbol{x}_2) = \mu + G_1(\boldsymbol{x}_1) + G_2(\boldsymbol{x}_2) + I(\boldsymbol{x}_1, \boldsymbol{x}_2).$$

The components $G_1(\boldsymbol{x}_1)$, $G_2(\boldsymbol{x}_2)$ are referred to as the main effects of the two QTL, and $I(\boldsymbol{x}_1, \boldsymbol{x}_2)$ is referred to as their interaction effect.

As in the case of one QTL, the main effects $G_k(\boldsymbol{x}_k), k = 1, 2$, are further decomposed as

$$\begin{aligned} G_k(\boldsymbol{x}_k) &= \begin{cases} 2q_k\alpha_k - 2q_k^2 d_k, & \text{if } \boldsymbol{x}_k = (A_k, A_k), \\ (q_k - p_k)\alpha_k + 2p_k q_k d_k, & \text{if } \boldsymbol{x}_k = (A_k, a_k) \text{ or } (a_k, A_k), \\ -2p_k\alpha_k - 2p_k^2 d_k, & \text{if } \boldsymbol{x}_k = (a_k, a_k) \end{cases} \\ &\equiv A_k(\boldsymbol{x}_k) + D_k(\boldsymbol{x}_k), \end{aligned}$$

where A_k, a_k are the alleles, p_k, q_k are the allele frequencies, α_k is the average effect of the gene substitution, and d_k is the dominant effect at QTL k.

Thus, we have the decomposition

$$
\begin{aligned}
G(\boldsymbol{x}_1, \boldsymbol{x}_2) &= \mu + [A_1(\boldsymbol{x}_1) + A_2(\boldsymbol{x}_2)] + [D_1(\boldsymbol{x}_1) + D_2(\boldsymbol{x}_2)] + I(\boldsymbol{x}_1, \boldsymbol{x}_2) \\
&\equiv \mu + A + D + I.
\end{aligned}
$$

In general, if there are K QTL, we have the decomposition

$$
\begin{aligned}
G(\boldsymbol{x}_1, \ldots, \boldsymbol{x}_K) &= \mu + \sum_{k=1}^{K} A_k(\boldsymbol{x}_k) + \sum_{k=1}^{K} D_k(\boldsymbol{x}_k) + I(\boldsymbol{x}_1, \ldots, \boldsymbol{x}_K) \\
&\equiv \mu + A + D + I;
\end{aligned}
$$

that is, the joint genotypic value is decomposed into three components: the additive component A, the dominant component D and the interaction component I. By the way of the decomposition, these three components are statistically independent.

3.2.3   Genetic Variance and Heritability

Phenotypic values vary from individual to individual in a population. The variation of phenotypic values is caused by the variation in genotypic values and the variation in environmental effects. As discussed in the previous sub section, the phenotypic value is decomposed as

$$
\begin{aligned}
Y &= G + E \\
&= \mu + A + D + I + E.
\end{aligned}
$$

The variation of the phenotypic value can be measured by the variances of its components. Since the components are independent, we have

$$
\begin{aligned}
\mathrm{Var}(Y) &= \mathrm{Var}(G) + \mathrm{Var}(E) \\
&= \mathrm{Var}(A) + \mathrm{Var}(D) + \mathrm{Var}(I) + \mathrm{Var}(E) \\
&\equiv V_A + V_D + V_I + V_E.
\end{aligned}
$$

The variance $\mathrm{Var}(G)$ of the genotypic value is referred to as the *genetic variance*. The genetic variance consists of three components: V_A, V_D and V_I. The component V_A is called the *additive genetic variance*. The component V_D is called the *dominance genetic variance*. The sum $V_D + V_I$ is also called non-additive genetic variance. The component V_E is called the *environmental variance*.

The additive and dominance genetic variances can be derived in explicit forms in terms of allele frequencies and genetic values. For a single QTL, the variances are easily derived through the decomposition:

$$
G_k(\boldsymbol{x}_k) = \mu + h(x_{k1}) + h(x_{k2}) + D(x_{k1}, x_{k2}).
$$

The additive and dominance variances due to a single QTL are given by

$$V_{A_k} = \text{Var}(h(x_{k1}) + h(x_{k2})) = 2\text{Var}(h(x_{k1})) = 2p_kq_k\alpha_k^2.$$
$$V_{D_k} = \text{Var}(D(x_{k1}, x_{k2})) = (2p_kq_kd_k)^2.$$

The final expression of the variances can be derived either by direct calculation using (3.1) or by using a statistic model which will be introduced in the next section. Thus the overall additive and dominance genetic variances are given by

$$V_A = \sum_k 2p_kq_k\alpha_k^2.$$

$$V_D = \sum_k (2p_kq_kd_k)^2.$$

The ratio of the genetic variance V_G and the overall variance $\text{Var}(Y)$, i.e, $V_G/\text{Var}(Y)$, is called the *heritability in broad sense*. It measures the proportion of the variation in the phenotypic value attributable to genetic factors. The ratio of the additive genetic variance V_A and $\text{Var}(Y)$, i.e., $V_A/\text{Var}(Y)$, is called the *heritability in narrow sense*.

3.3   Statistical Models with Known QTL Genotypes

In this section we consider models with known QTL genotypes. Though these models cannot be directly used for statistical inference since QTL genotypes are unknown in practical problems, they are building blocks for the QTL mapping models to be developed in the subsequent chapters.

3.3.1   Models with Single QTL

Let A and a be the two alleles at the QTL of concern, and p and q their allele frequencies. Assume the allele frequencies are known. Let y denote the quantitative trait value (genotypic value), and x_1, x_2 the paternal and maternal alleles at the QTL. First we define the following function for a random allele:

$$I(x) = \begin{cases} q & \text{if } x = A, \\ -p & \text{if } x = a. \end{cases}$$

Note that $EI(x) = 0$ and $\text{Var}(I(x)) = pq$. Then the quantitative trait value y can be modeled as

$$y = \mu + \alpha[I(x_1) + I(x_2)] - 2dI(x_1)I(x_2) + \epsilon, \tag{3.2}$$

where μ is the populaton mean, α is the average effect of the gene substitution, d is the dominant effect, and ϵ is the random error caused by environmental effect. It is easy to check that model (3.2) is equivalent to model (3.1). By using this model, the additive and dominant genetic variances can be immediately derived as

$$V_A = \text{Var}(\alpha[I(x_1) + I(x_2)]) = 2\alpha^2 \text{Var}(I(x_1)) = 2pq\alpha^2,$$
$$V_D = \text{Var}(-2dI(x_1)I(x_2)) = 4d^2 E[I^2(x_1)I^2(x_2)] = 4(pq)^2 d^2.$$

In populations generated by experimental crosses, the two allele frequencies are equal, i.e., $p = q = \frac{1}{2}$. In this case, model (3.2) reduces to

$$y = \begin{cases} \mu + \alpha - \frac{1}{2}d + \epsilon, & \text{if } (x_1, x_2) = (A, A) \\ \mu + \frac{1}{2}d + \epsilon, & \text{if } (x_1, x_2) = (A, a) \text{ or } (a, A) \\ \mu - \alpha - \frac{1}{2}d + \epsilon, & \text{if } (x_1, x_2) = (a, a). \end{cases} \quad (3.3)$$

Note that, when $p = q$, $\alpha = c + (q - p)d = c$. Let $\beta_0 = \mu - c - \frac{1}{2}d$. Model (3.3) can be expressed as

$$y = \begin{cases} \beta_0 + 2c + \epsilon, & \text{if } (x_1, x_2) = (A, A) \\ \beta_0 + c + d + \epsilon, & \text{if } (x_1, x_2) = (A, a) \text{ or } (a, A) \\ \beta_0 + \epsilon, & \text{if } (x_1, x_2) = (a, a). \end{cases} \quad (3.4)$$

We now introduce two dummy variables for the genotype as follows:

$$\delta_1 = \begin{cases} 2, & \text{if genotype is } AA, \\ 1, & \text{if genotype is } Aa, \\ 0 & \text{if genotype is } aa; \end{cases} \qquad \delta_2 = \begin{cases} 1, & \text{if genotype is } Aa, \\ 0 & \text{otherwise.} \end{cases}$$

Finally, we arrive at the following model:

$$y = \beta_0 + \beta\delta_1 + \gamma\delta_2 + \epsilon, \quad (3.5)$$

where $\beta = c$ and $\gamma = d$.

3.3.2 Models with Multiple QTL

Suppose there are K QTL. For each QTL, define two dummy variables as in the last sub section. Let δ_{k1} and δ_{k2} be the dummy variables corresponding to QTL k. First, consider the case that there are no epistasis effects. The quantitative trait value y can be modelled as

$$y = \beta_0 + \sum_{k=1}^{K} (\beta_k \delta_{k1} + \gamma_k \delta_{k2}) + \epsilon, \quad (3.6)$$

where β_0 is a compound parameter common to all individuals and β_k and γ_k measure, respectively, the additive and dominance effect of QTL k.

If epistasis effects among the QTL are of concern, model (3.6) can be extended as

$$
\begin{aligned}
y \;=\; & \beta_0 + \sum_{k=1}^{K}(\beta_k \delta_{k1} + \gamma_k \delta_{k2}) \\
& + \sum_{k=1}^{K}\sum_{l=1}^{K}(\xi_{kl}^{AA}\delta_{k1}\delta_{l1} + \xi_{kl}^{AD}\delta_{k1}\delta_{l2} + \xi_{kl}^{DA}\delta_{k2}\delta_{l1} + \xi_{kl}^{DD}\delta_{k2}\delta_{l2}) \\
& + \epsilon.
\end{aligned}
\tag{3.7}
$$

If dominance effects can be ignored, model (3.7) reduces to a simpler form:

$$
y = \beta_0 + \sum_{k=1}^{K}\beta_k \delta_k + \sum_{k=1}^{K}\sum_{l=1}^{K}\xi_{kl}\delta_k \delta_l + \epsilon,
\tag{3.8}
$$

where $\delta_k = \delta_{k1}, k = 1, \ldots, K$.

3.4   Conditional Probabilities of Putative QTL Genotypes Given Markers

In the study of QTL mapping, the locations and genotypes of QTL are unknown. We can investigate any loci to see whether or not they are QTL. Such loci are referred to as putative QTL. In QTL mapping experiments, only markers are genotyped and the genotypes of putative QTL are generally unknown. However, with experimental crosses, the conditional probabilities of the genotypes of putative QTL given markers can be obtained. These conditional probabilities play a crucial role in interval mapping methodologies. In this section, the conditional probabilities of the genotypes of putative QTL flanked by two markers in populations of various experimental crosses are considered.

3.4.1   Conditional Probabilities of Putative QTL Genotypes in Backcross and Intercross Populations

Let $\mathcal{Q}$ denote the putative QTL, $\mathcal{A}$ and $\mathcal{B}$ denote the left and right flanking markers. Denote the alleles of $\mathcal{Q}$, $\mathcal{A}$ and $\mathcal{B}$, respectively, by Q, q, A, a and B, b. The genotypes are indexed as follows. A homozygous genotype with two capital letter alleles is indexed by 2, a heterozygous genotype is indexed by 1, and a homozygous genotype with two small letter alleles is indexed by 0. Let $p_{ij}, i, j = 0, 1$, denote the joint probability that there is i recombination between $\mathcal{A}$ and $\mathcal{Q}$ and j recombination between $\mathcal{Q}$ and $\mathcal{B}$ on a single chromo-

TABLE 3.1: Probabilities of joint genotypes of a putative QTL and its two flanking markers in backcross population.

Joint genotype	Probability
$AAQQBB$	p_{00}
$AAQqBB$	p_{11}
$AAQQBb$	p_{01}
$AAQqBb$	p_{10}
$AaQQBB$	p_{10}
$AaQqBB$	p_{01}
$AaQQBb$	p_{11}
$AaQqBb$	p_{00}

some. Denote by P_1 the probability of recombination between the $\mathcal{A}$ and $\mathcal{B}$ on a single chromosome. Let $P_0 = 1 - P_1$.

In a backcross population, recombination only concerns one chromosome. There are eight possible joint genotypes of the three loci in a backcross population. These eight genotypes and their probabilities (frequencies in the population) are given in Table 3.1.

Dividing these joint probabilities by the corresponding probabilities of the marker genotypes, we obtain the conditional probabilities of the genotypes of the putative QTL which are given in Table 3.2.

The conditional probabilities of the genotypes of the putative QTL in an intercross population can be derived in a similar way. But the derivation is more complicated since recombination concerns both chromosomes in an intercross population. The conditional probabilities of the genotypes of the putative QTL in an intercross population are given in Table 3.3. As an illustration, the entries of the first row of Table 3.3 are derived as follows. For convenience, we use the indices introduced at the beginning of this section to denote the joint genotypes. Thus, 222, 212 and 202 stand, respectively, for the joint genotypes

TABLE 3.2: Conditional probabilities of putative QTL genotypes given two flanking markers in backcross population.

Marker genotype	Genotype frequency	QTL genotype 2	1
22	$\frac{P_0}{2}$	$\frac{p_{00}}{P_0}$	$\frac{p_{11}}{P_0}$
21	$\frac{P_1}{2}$	$\frac{p_{01}}{P_1}$	$\frac{p_{10}}{P_1}$
12	$\frac{P_1}{2}$	$\frac{p_{10}}{P_1}$	$\frac{p_{01}}{P_1}$
11	$\frac{P_0}{2}$	$\frac{p_{11}}{P_0}$	$\frac{p_{00}}{P_0}$

TABLE 3.3: Conditional probabilities of putative QTL genotypes given two flanking markers in intercross population.

Marker genotype	Genotype frequency	QTL genotype		
		2	1	0
22	$\frac{p_0^2}{4}$	$\frac{p_{00}^2}{p_0^2}$	$\frac{2p_{00}p_{11}}{p_0^2}$	$\frac{p_{11}^2}{p_0^2}$
21	$\frac{p_0 p_1}{2}$	$\frac{p_{00}p_{01}}{p_0 p_1}$	$\frac{p_{00}p_{10}+p_{11}p_{01}}{p_0 p_1}$	$\frac{p_{11}p_{10}}{p_0 p_1}$
20	$\frac{p_1^2}{4}$	$\frac{p_{01}^2}{p_1^2}$	$\frac{2p_{01}p_{10}}{p_1^2}$	$\frac{p_{10}^2}{p_1^2}$
12	$\frac{p_0 p_1}{2}$	$\frac{p_{00}p_{10}}{p_0 p_1}$	$\frac{p_{00}p_{01}+p_{11}p_{10}}{p_0 p_1}$	$\frac{p_{11}p_{01}}{p_0 p_1}$
11	$\frac{p_0^2+p_1^2}{2}$	$\frac{p_{00}p_{11}+p_{10}p_{01}}{p_0^2+p_1^2}$	$\frac{p_{00}^2+p_{10}^2+p_{01}^2+p_{11}^2}{p_0^2+p_1^2}$	$\frac{p_{00}p_{11}+p_{10}p_{01}}{p_0^2+p_1^2}$
10	$\frac{p_0 p_1}{2}$	$\frac{p_{01}p_{11}}{p_0 p_1}$	$\frac{p_{00}p_{01}+p_{10}p_{11}}{p_0 p_1}$	$\frac{p_{10}p_{00}}{p_0 p_1}$
02	$\frac{p_1^2}{4}$	$\frac{p_{10}^2}{p_1^2}$	$\frac{2p_{01}p_{10}}{p_1^2}$	$\frac{p_{01}^2}{p_1^2}$
01	$\frac{p_0 p_1}{2}$	$\frac{p_{10}p_{11}}{p_0 p_1}$	$\frac{p_{10}p_{00}+p_{01}p_{11}}{p_0 p_1}$	$\frac{p_{01}p_{00}}{p_0 p_1}$
00	$\frac{p_0^2}{4}$	$\frac{p_{11}^2}{p_0^2}$	$\frac{2p_{00}p_{11}}{p_0^2}$	$\frac{p_{00}^2}{p_0^2}$

$AAQQBB$, $AAQqBB$ and $AAqqBB$. First, we derive the joint probabilities of 222, 212 and 202. For 222, the probability of having the homozygous genotype of $\mathcal{A}$ is $\frac{1}{4}$. On both chromosomes, there is no recombination both between $\mathcal{A}$ and $\mathcal{Q}$ and between $\mathcal{Q}$ and $\mathcal{B}$. The probability for this to happen is p_{00}^2 since the two chromosomes are independent. Thus the joint probability of 222 is $\frac{1}{4}p_{00}^2$. For 212, there are two possible phases because of the heterozygous putative QTL. The probabilities of the two phases are the same. For each phase, the probability is derived in the same way as for 222. First, the probability for $\mathcal{A}$ to be homozygous is $\frac{1}{4}$. On one of the chromosomes, there is no recombination both between $\mathcal{A}$ and $\mathcal{Q}$ and between $\mathcal{Q}$ and $\mathcal{B}$. Its probability is p_{00}. On the other chromosome, there is recombination both between $\mathcal{A}$ and $\mathcal{Q}$ and between $\mathcal{Q}$ and $\mathcal{B}$. Its probability is p_{11}. Therefore, the joint probability of 212 is $\frac{1}{2}p_{00}p_{11}$. The joint probability of 202, which is similarly derived, is $\frac{1}{4}p_{11}^2$. The derivation of the joint probability of 212 is demonstrated in Figure 3.4. Dividing these joint probabilities by the joint probability of the two markers given in the second column of Table 3.3 yields the conditional probabilities.

We now relate the joint probabilities p_{11}, p_{10}, p_{01} and p_{00} to recombination fractions. Let r, s and γ denote, respectively, the recombination fractions between $\mathcal{A}$ and $\mathcal{Q}$, between $\mathcal{Q}$ and $\mathcal{B}$ and between $\mathcal{A}$ and $\mathcal{B}$. Note that a recombination fraction is the marginal probability of recombination event occurring between two loci. Let E_1, E_2 and E_3 denote, respectively, the events that there is recombination between $\mathcal{A}$ and $\mathcal{Q}$, between $\mathcal{Q}$ and $\mathcal{B}$ and between $\mathcal{A}$ and $\mathcal{B}$. In fact, $r = P(E_1)$, $s = P(E_2)$ and $\gamma = P(E_3)$. Let E_j^c denote the

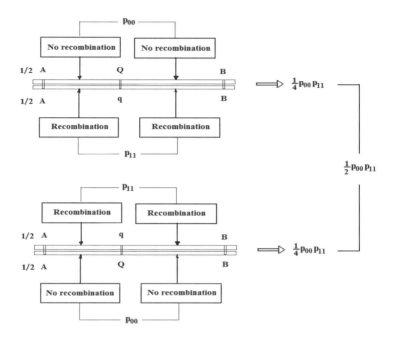

FIGURE 3.4: Illustration of the derivation of joint genotype probabilities of a putative QTL and its two flanking markers.

complement of E_j, $j = 1, 2, 3$. Note that

$$p_{11} = P(E_1 \cap E_2), \quad p_{00} = P(E_1^c \cap E_2^c),$$
$$p_{10} = P(E_1 \cap E_2^c), \quad p_{01} = P(E_1^c \cap E_2),$$
$$p_{10} + p_{01} = P(E_3), \quad p_{11} + p_{00} = P(E_3^c).$$

Furthermore,

$$
\begin{aligned}
p_{00} &= P(E_1^c \cap E_2^c) \\
&= 1 - P(E_1 \cup E_2) \\
&= 1 - P(E_1) - P(E_2) + P(E_1 \cap E_2), \\
&= 1 - P(E_1) - P(E_2) + p_{11}, \\
p_{10} &= P(E_1 \cap E_2^c) \\
&= 1 - P(E_1^c \cup E_2) \\
&= 1 - P(E_1^c) - P(E_2) + P(E_1^c \cap E_2) \\
&= P(E_1) - P(E_2) + p_{01}.
\end{aligned}
$$

Therefore,

$$\begin{aligned}
p_{11} &= 1 - P(E_3) - p_{00} \\
&= 1 - P(E_3) - [1 - P(E_1) - P(E_2) + p_{11}] \\
&= P(E_1) + P(E_2) - P(E_3) - p_{11}, \\
p_{01} &= P(E_3) - p_{10} \\
&= P(E_3) - P(E_1) + P(E_2) - p_{01}.
\end{aligned}$$

Thus, it is obtained that

$$\begin{aligned}
p_{11} &= \frac{1}{2}[P(E_1) + P(E_2) - P(E_3)] = \frac{1}{2}(r + s - \gamma), \\
p_{01} &= \frac{1}{2}[P(E_2) - P(E_1) + P(E_3)] = \frac{1}{2}(s - r + \gamma), \\
p_{10} &= \frac{1}{2}[P(E_1) - P(E_2) + P(E_3)] = \frac{1}{2}(r - s + \gamma), \\
p_{00} &= 1 - \frac{1}{2}[P(E_1) + P(E_2) + P(E_3)] = 1 - \frac{1}{2}(r + s + \gamma).
\end{aligned}$$

In the data of QTL mapping expriments, the genetic linkage map is given, i.e., the genetic distance between any two markers is known. Suppose the genetic distance between A and B is D cM. Let the genetic distance between A and Q be d cM. The recombination fractions r, s and γ are obtained from the genetic distances by a certain mapping function according to the assumption on crossover interference. Let $M(\cdot)$ be the mapping function. Then

$$r = M(d/100), \quad s = M((D - d)/100), \quad \gamma = M(D/100).$$

When D is known, the recombination fractions are functions of d as d varies. Once d is specified, Table 3.2 and Table 3.3 can be calculated.

If the assumption of no crossover interference is made, the Haldane mapping function is then used to compute the recombination fractions. Under the assumption of no crossover interference, the recombinations over non-overlapping intervals are independent and the recombination fractions satisfy

TABLE 3.4: Conditional probabilities of putative QTL genotypes given two flanking markers in backcross population with no crossover interference.

Marker genotype	Genotype frequency	QTL genotype 2	QTL genotype 1
22	$\frac{1-\gamma}{2}$	$\frac{(1-r)(1-s)}{1-\gamma}$	$\frac{rs}{1-\gamma}$
21	$\frac{\gamma}{2}$	$\frac{(1-r)s}{\gamma}$	$\frac{(1-s)r}{\gamma}$
12	$\frac{\gamma}{2}$	$\frac{(1-s)r}{\gamma}$	$\frac{(1-r)s}{\gamma}$
11	$\frac{1-\gamma}{2}$	$\frac{rs}{1-\gamma}$	$\frac{(1-r)(1-s)}{1-\gamma}$

TABLE 3.5: Conditional probabilities of putative QTL genotypes given two flanking markers in intercross population with no crossover interference.

Marker genotype	Genotype frequency	QTL genotype		
		2	1	0
22	$\frac{(1-\gamma)^2}{4}$	$\frac{(1-r)^2(1-s)^2}{(1-\gamma)^2}$	$\frac{2rs(1-r)(1-s)}{(1-\gamma)^2}$	$\frac{r^2s^2}{(1-\gamma)^2}$
21	$\frac{\gamma(1-\gamma)}{2}$	$\frac{(1-r)^2s(1-s)}{\gamma(1-\gamma)}$	$\frac{r(1-r)(1-2s+2s^2)}{\gamma(1-\gamma)}$	$\frac{r^2s(1-s)}{\gamma(1-\gamma)}$
20	$\frac{\gamma^2}{4}$	$\frac{(1-r)^2s^2}{\gamma^2}$	$\frac{2rs(1-r)(1-s)}{\gamma^2}$	$\frac{r^2(1-s)^2}{\gamma^2}$
12	$\frac{\gamma(1-\gamma)}{2}$	$\frac{(1-s)^2r(1-r)}{\gamma(1-\gamma)}$	$\frac{s(1-s)(1-2r+2r^2)}{\gamma(1-\gamma)}$	$\frac{s^2r(1-r)}{\gamma(1-\gamma)}$
11	$\frac{(1-\gamma)^2+\gamma^2}{2}$	$\frac{2rs(1-r)(1-s)}{\gamma^2+(1-\gamma)^2}$	$\frac{(1-2r+2r^2)(1-2s+2s^2)}{\gamma^2+(1-\gamma)^2}$	$\frac{2rs(1-r)(1-s)}{\gamma^2+(1-\gamma)^2}$
10	$\frac{\gamma(1-\gamma)}{2}$	$\frac{s^2r(1-r)}{\gamma(1-\gamma)}$	$\frac{s(1-s)(1-2r+2r^2)}{\gamma(1-\gamma)}$	$\frac{(1-s)^2r(1-r)}{\gamma(1-\gamma)}$
02	$\frac{\gamma^2}{4}$	$\frac{(1-s)^2r^2}{\gamma^2}$	$\frac{2rs(1-r)(1-s)}{\gamma^2}$	$\frac{s^2(1-r)^2}{\gamma^2}$
01	$\frac{\gamma(1-\gamma)}{2}$	$\frac{r^2s(1-s)}{\gamma(1-\gamma)}$	$\frac{r(1-r)(1-2s+2s^2)}{\gamma(1-\gamma)}$	$\frac{(1-r)^2s(1-s)}{\gamma(1-\gamma)}$
00	$\frac{(1-\gamma)^2}{4}$	$\frac{r^2s^2}{(1-\gamma)^2}$	$\frac{2rs(1-r)(1-s)}{(1-\gamma)^2}$	$\frac{(1-r)^2(1-s)^2}{(1-\gamma)^2}$

$\gamma = r + s - 2rs$. Hence, the joint probabilities reduce to

$$
\begin{aligned}
p_{11} &= rs, \\
p_{10} &= r(1-s), \\
p_{01} &= (1-r)s, \\
p_{00} &= (1-r)(1-s).
\end{aligned}
$$

Table 3.2 and Table 3.3 reduce to Table 3.4 and Table 3.5 respectively.

3.4.2 Conditional Probabilities of Putative QTL Genotypes in RILs

In RILs generated either by selfing or sibling mating, the genotypes of all individual experimental subjects at any locus are homozygous. There are eight possible joint genotypes of $\mathcal{A}$, $\mathcal{Q}$ and $\mathcal{B}$. They are: $AAQQBB$, $AAqqBB$, $AAQQbb$, $AAqqbb$, $aaQQBB$, $aaqqBB$, $aaQQbb$ and $aaqqbb$. If, at two loci, one locus has capital letter alleles and the other has small letter alleles, it is said that there is a zygote recombination between these two loci. The frequency of recombinant zygotes is defined as zygote recombination fraction. The zygote recombination fraction differs from the recombination fraction which has been considered so far. To distinguish, we refer to the original recombination fraction as gamete recombination fraction. Gamete recombination fraction is the frequency of recombination on a single chromosome during one meiosis. Zygote recombination fraction is the frequency of recombination of homozygous genotypes on two chromosomes cumulated over many meioses.

We consider the joint probabilities of the above three-locus genotypes in a similar way to the case of gamete recombinations. Let R, S and Γ denote, respectively, the zygote recombination fractions between $\mathcal{A}$ and $\mathcal{Q}$, between $\mathcal{Q}$ and $\mathcal{B}$, and between $\mathcal{A}$ and $\mathcal{B}$. Let g_{ij}, $i,j = 0$ or 1, denote the probability that there is i zygote recombination between $\mathcal{A}$ and $\mathcal{Q}$ and j zygote recombination between $\mathcal{Q}$ and $\mathcal{B}$. Then the same argument leading to the relationship between p_{ij}'s and r, s, γ yields the following relationship between g_{ij}'s and R, S, Γ:

$$g_{11} = \frac{1}{2}(R + S - \Gamma),$$

$$g_{01} = \frac{1}{2}(S - R + \Gamma),$$

$$g_{10} = \frac{1}{2}(R - S + \Gamma),$$

$$g_{00} = 1 - \frac{1}{2}(R + S + \Gamma).$$

The conditional probabilities of putative QTL genotypes given the genotypes of the flanking markers can be easily derived as in the case of backcross. These conditional probabilities are presented in Table 3.6.

The formulae of the conditional probabilities in Table 3.6 are general for all RILs no matter whether the RILs are generated by selfing or sibling mating. These formulae also apply for a general 2^n-way RILs as well as for IRI. However, the derivation of the zygote recombination fractions for different RILs or IRI is different. The zygote recombination fraction is determined by its corresponding gamete recombination fraction. The relationship between zygote recombination fraction and gamete recombination fraction in two-way RILs was derived by Haldane and Waddington [80]:

For selfing:　　　$R = \dfrac{2r}{1 + 2r}$,

For sibling mating:　　　$R = \dfrac{4r}{1 + 6r}$,

where r is the gamete recombination fraction and R is the zygote recombination fraction between the same two loci. The relationship between R and r in

TABLE 3.6: Conditional probabilities of putative QTL genotypes given two flanking markers in RILs.

Marker genotype	Genotype frequency	QTL genotype 2	QTL genotype 0
22	$\frac{1-\Gamma}{2}$	$\frac{2-R-S-\Gamma}{2(1-\Gamma)}$	$\frac{R+S-\Gamma}{2(1-\Gamma)}$
20	$\frac{\Gamma}{2}$	$\frac{S-R+\Gamma}{2\Gamma}$	$\frac{R-S+\Gamma}{2\Gamma}$
02	$\frac{\Gamma}{2}$	$\frac{R-S+\Gamma}{2\Gamma}$	$\frac{S-R+\Gamma}{2\Gamma}$
00	$\frac{1-\Gamma}{2}$	$\frac{R+S-\Gamma}{2(1-\Gamma)}$	$\frac{2-R-S-\Gamma}{2(1-\Gamma)}$

TABLE 3.7: The zygote recombination fraction R as function of gamete recombination fraction r in different types of RILs.

	Selfing	Sibling mating Autosome	Sibling mating X chromosome
2-way	$\frac{2r}{1+2r}$	$\frac{4r}{1+6r}$	$\frac{8r}{3(1+4r)}$
4-way	$\frac{3r}{1+2r}$	$\frac{6r}{1+6r}$	$\frac{4r}{1+4r}$
8-way	$\frac{r(4-r)}{1+2r}$	$\frac{7r}{1+6r}$	$\frac{14r}{3(1+4r)}$
16-way	$\frac{r(5-3r+r^2)}{1+2r}$	$\frac{r(8-r)}{1+6r}$	$\frac{r(16-r)}{3(1+4r)}$
2^n-way	$1-\frac{(1-r)^{n-1}}{1+2r}$	$1-\frac{(1-r)^{n-2}}{1+6r}$	given in (3.9)

IRILs having t generations of random mating was derived in [198] as follows:

For selfing:
$$R = \frac{1}{2}\left[1 - \frac{1-2r}{1+2r}(1-r)^t\right],$$

For sibling mating:
$$R = \frac{1}{2}\left[1 - \frac{1-2r}{1+6r}(1-r)^t\right].$$

Note that when $t = 0$ the above formulae reduce to Haldane and Waddington's formulae. Broman [20] considered a general procedure for the derivation of the relationships for general 2^n-way RILs. He derived the relationships for both autosomes and sex chromosomes. The relationships between R and r in general 2^n-way RILs are summarized in Table 3.7.

The relationship between R and r on X chromosome of a 2^n-way RIL by sibling mating is given by

$$R = 1 - \frac{1}{3(1+4r)}\sum_{k=1}^{n}\chi_{nk}\left(\frac{1-r}{2}\right)^{k-1}, \tag{3.9}$$

where the coefficients χ_{nk} are given as $\chi_{21} = 3$, $\chi_{2k} = 0$ for $k \geq 2$, $\chi_{31} = 1$, $\chi_{32} = 4$, $\chi_{3k} = 0$, for $k \geq 3$, $\chi_{n1} = 0$, and, for $n \geq 4, k \geq 2$, χ_{nk} is computed recursively as

$$\chi_{nk} = \chi_{n-1,k-1} + \chi_{n-2,k-1}.$$

There is another important difference between zygote recombination and gamete recombination. Gamete recombinations over non-overlapping intervals are indepedent if there is no crossover interference. But zygote recombinations are never independent even if there is no crossover interference. In other words, $\Gamma \neq R + S - 2RS$ in any case. Under the assumption of no crossover interference, Martin and Hospital [130] derived that

$$\Gamma = \frac{R + S - 3RS}{1 - 2RS}, \text{ for selfing,}$$

$$\Gamma = \frac{2R + 2S - 7RS}{2(1 - 3RS)}, \text{ for sibling mating.}$$

3.4.3   Comparison of Conditional Probabilities under Different Assumptions on Crossover Interference

In Section 1.5, we discussed four mapping functions: Haldane, Kosambi, Carter-Falconer and Felsenstein. Haldane mapping function corresponds to the assumption of no crossover interference. The other three mapping functions correspond to different kinds and degrees of crossover interference. In the following, we make a comparison of the conditional probabilities obtained by Haldane mapping function with those obtained by the other three mapping functions in intercross populations. For an interval of D cM, consider $d = 1, 2, \ldots, D - 1$. For each d, the matrix of the conditional probabilities is computed, and the differences between the matrix corresponding to Haldane mapping function and those corresponding to the other three mapping functions are obtained. Then the average and standard deviation of differences over the $D - 1$ difference matrices are computed. For $D = 10$ and 20, these averages and standard deviations are given in Table 3.8, the standard deviations are within parentheses. The differences of the magnitude as shown in Table 3.8 have little influence on QTL mapping. This justifies the common practice that the Haldane mapping function is used in QTL mapping no matter whether or not there is crossover interference. There are two advantages of using Haldane mapping function in QTL mapping. First, it is easier for the computation. Second, more importantly, the conditional probabilities of putative QTL genotypes only depend on the two flanking markers, which makes the statistical modeling for QTL mapping much simpler and easier than in the case of crossover interference.

TABLE 3.8: Comparison of the conditional probabilities obtained by Haldane mapping function with those obtained by Kosambi, Carter-Falconer and Felsenstein mapping functions in intercross populations.

Comparison with Kosambi					
$D = 10$			$D = 20$		
-.003(.001)	.003(.001)	.000(.000)	-.007(.003)	.007(.003)	.000(.000)
-.001(.001)	.000 (.001)	.001 (.000)	-.002 (.003)	.000 (.004)	.002 (.001)
.000 (.001)	.001 (.000)	.000 (.001)	-.001 (.003)	.003 (.002)	-.001 (.003)
-.001 (.001)	.000 (.001)	.001 (.000)	-.002 (.003)	.000 (.004)	.002 (.001)
.001 (.000)	-.002 (.001)	.001 (.000)	.001 (.000)	-.002 (.001)	.001 (.000)
.001(.000)	.000 (.001)	-.001 (.001)	.002 (.001)	.000 (.004)	-.002 (.003)
.000(.001)	.001 (.000)	.000 (.001)	-.001 (.003)	.003 (.002)	-.001 (.003)
.001 (.000)	.000 (.001)	-.001 (.001)	.002 (.001)	.000 (.004)	-.002 (.003)
.000 (.000)	.003 (.001)	-.003 (.001)	.000 (.000)	.007 (.003)	-.007 (.003)
Comparison with Felsenstein (k=0.5)					
$D = 10$			$D = 20$		
-.001 (.000)	.001 (.000)	.000 (.000)	-.003 (.001)	.003 (.001)	.000 (.000)
.000 (.001)	.000 (.001)	.000 (.000)	-.001 (.002)	.000 (.002)	.001 (.000)
.000 (.000)	.000 (.000)	.000 (.000)	-.001 (.001)	.001 (.001)	-.001 (.001)
.000 (.001)	.000 (.001)	.000 (.000)	-.001 (.002)	.000 (.002)	.001 (.000)
.001 (.000)	-.001 (.000)	.001 (.000)	.001(.000)	-.001 (.000)	.001 (.000)
.000 (.000)	.000 (.001)	.000 (.001)	.001 (.000)	.000 (.002)	-.001 (.002)
.000 (.000)	.000 (.000)	.000 (.000)	-.001 (.001)	.001 (.001)	-.001 (.001)
.000 (.000)	.000 (.001)	.000 (.001)	.001 (.000)	.000 (.002)	-.001 (.002)
.000 (.000)	.001 (.000)	-.001 (.000)	.000 (.000)	.003 (.001)	-.003 (.001)
Comparison with Carter-Falconer					
$D = 10$			$D = 20$		
-.004 (.001)	.004 (.001)	.000 (.000)	-.012 (.005)	.012 (.005)	.000 (.000)
-.001 (.001)	.000 (.001)	.001 (.000)	-.003 (.003)	.000 (.004)	.003 (.002)
.000 (.000)	.000 (.000)	.000 (.000)	-.001 (.002)	.002 (.001)	-.001 (.002)
-.001 (.001)	.000 (.001)	.001 (.000)	-.003 (.003)	.000 (.004)	.003 (.002)
.001 (.000)	-.003 (.001)	.001 (.000)	.002 (.001)	-.005 (.002)	.002 (.001)
.001 (.000)	.000 (.001)	-.001 (.001)	.003 (.002)	.000 (.004)	-.003 (.003)
.000 (.000)	.000 (.000)	.000 (.000)	-.001 (.002))	.002 (.001)	-.001 (.002)
.001 (.000)	.000 (.001)	-.001 (.001)	.003 (.002)	.000 (.004)	-.003 (.003)
.000 (.000)	.004 (.001)	-.004 (.001)	.000 (.000)	.012 (.005)	-.012 (.005)

3.5 QTL Mapping Data

In this section, we give a general description of QTL mapping data. The data of a QTL mapping study consists of two elements: data on the genetic map and data on trait values and marker genotypes. The data on the genetic map provides the markers on each chromosome of the genome and the genetic distances between adjacent markers. The data on trait values and marker genotypes provides for each individual experimental subject its trait value and genotypes of the markers on the genetic map.

As an example, we describe the radiata pine data as follows. The data was analyzed by in [100] and [33]. Radiata pine is a widely used plantation species in southern Australia. A genetic map of radiata pine was constructed using MapMaker/EXP [120]. In a QTL mapping study, a sample of 134 experimental subjects is obtained from a backcross population. The genotypes of the experimental subjects at 120 markers in 12 chromosomes on the genetic map are obtained. Three quantitative traits of radiata pine are of concern in the study: the annual brown cone number at eight years of age (CN), diameter of stem at breast height (DBH) and branch quality score ranging from 1 (poorest) to 6 (best) (BS). The map data are provided in Table 3.9. A portion of the data on trait values and marker genotypes is provided in Table 3.10. It provides the trait values and genotypes at the markers on the first chromosome for 30 experimental subjects. The digits 1 and 2 stand for heterozygous and homozygous genotypes respectively. The dash sign "-" indicates missing value.

TABLE 3.9: The genetic distances (in units of cM) between adjacent markers on the genetic map of radiata pine tree (the first marker of each chromosome is taken as the left end of the chromosome, the numbers are the genetic distances of the markers from theirs left adjacent markers).

	Chromosomes											
	1	2	3	4	5	6	7	8	9	10	11	12
2	10.8	14.6	10.8	27.1	15.6	9.9	19.9	16.7	3.1	17.7	10.8	15.6
3	7.2	45.4	4.7	12.7	14.6	19.9	11.7	13.6	27.1	18.8	22.2	14.6
4	74.8	2.3	16.7	40.2	14.6	8.1	7.2		16.7	5.5	14.6	6.4
5	14.6	13.6	16.7	12.7	35.4	53.5	19.9		2.3	47.3	24.6	0.8
6	10.8	9.0	8.1	12.7	21.0	16.7	14.6		12.7	14.6	15.6	16.7
7	28.4	9.9	28.4	35.4	16.7	24.6	2.3		9.0	6.4		6.4
8		8.1	23.4		23.4	6.4	14.6			6.4		12.7
9		15.6	14.6		8.1	3.1	11.7			7.2		14.6
10		24.6	3.9		3.1	9.0				9.9		5.5
11		3.9	4.7		19.9	5.5				15.6		21.0
12		3.9	1.5			12.7				12.7		12.7
13		15.6	12.7			63.0						14.6
14						12.7						32.5
15						4.7						5.5
16						16.7						
17						4.7						

TABLE 3.10: A portion of the trait and genotype data of radiata pine trees for 30 experimental subjects from a backcross population.

Index	CN	DBH	BS	Marker Genotypes							
1	13	179.801	4.7199	2	2	2	1	1	2	2	⋯
2	4	207.800	2.7195	1	1	1	2	2	2	2	⋯
3	5	240.801	2.7195	2	2	2	2	2	2	2	⋯
4	14	212.171	2.7242	1	1	1	2	2	1	1	⋯
5	6	202.173	3.7242	2	2	2	1	1	1	2	⋯
6	5	197.173	3.7242	2	2	2	2	1	1	1	⋯
7	13	212.801	1.7199	1	1	1	1	2	2	2	⋯
8	16	201.173	3.7242	1	1	1	1	1	2	2	⋯
9	9	200.318	3.9708	1	1	2	1	1	1	1	⋯
10	10	207.318	3.9708	1	1	1	1	2	2	2	⋯
11	8	209.321	2.9708	2	2	2	1	1	1	2	⋯
12	1	190.318	2.9708	2	2	2	2	2	2	2	⋯
13	-	106.041	5.4619	2	2	2	2	2	2	1	⋯
14	26	209.360	2.8579	1	1	1	1	1	1	1	⋯
15	4	221.360	1.8583	2	2	2	1	1	1	1	⋯
16	-	183.360	3.8579	1	1	1	1	1	1	2	⋯
17	10	179.360	4.8583	2	1	1	1	1	1	1	⋯
18	10	179.360	3.8148	2	2	2	2	2	2	1	⋯
19	8	190.059	3.8148	1	1	1	1	2	2	2	⋯
20	5	226.060	2.8148	1	1	1	1	1	1	1	⋯
21	-	210.060	3.8148	2	2	2	2	2	2	2	⋯
22	5	170.100	3.8449	1	1	1	1	1	1	1	⋯
23	11	195.095	3.8449	2	2	2	2	2	2	1	⋯
24	7	201.095	2.8449	2	2	1	1	1	1	1	⋯
25	-	173.100	4.8453	2	2	2	2	2	2	2	⋯
26	4	184.020	5.0710	1	1	1	1	1	1	1	⋯
27	15	217.020	5.0710	1	1	1	1	1	1	1	⋯
28	11	221.020	4.0714	2	2	2	1	1	1	2	⋯
29	3	109.020	5.0710	1	1	1	2	2	2	2	⋯
30	-	175.490	4.4101	2	2	2	2	2	2	1	⋯
⋯	⋯	⋯	⋯	⋯	⋯	⋯	⋯	⋯	⋯	⋯	⋯

Chapter 4

One-dimensional Mapping Approaches

We start our journey with the discussion on one-dimensional QTL mapping approaches. Though more advanced methods are preferred, the one-dimensional approaches are still commonly used in the genetics discipline because of their simplicity. Besides their own interests, the one-dimensional approaches can also serve as tools for preliminary analysis in more sophisticated methods, which will be elaborated later. By one-dimensional mapping approaches, we refer to the methods based on single-QTL models discussed in § 3.3.1, since these methods search for QTL over the genome in a linear manner. Three one-dimensional mapping approaches: single-marker analysis, single interval mapping and composite interval mapping, are discussed in § 4.1 to § 4.3. In the latter sections of this chapter, we discuss some common issues on one-dimensional mapping approaches. In § 4.4, we consider the determination of threshold values. In § 4.5, we consider the determination of sample size. In § 4.6, we discuss selective genotyping, a strategy for increasing the efficiency of QTL mapping.

4.1 QTL Mapping by Single-marker Analysis

The approach of single marker analysis was used in earlier QTL mapping studies, see [47], [154], [170] and [176]. Single-marker analysis deals with the markers in a QTL study one-at-a-time. For each marker, it carries out a test to see whether or not the marker is a QTL or linked with a QTL. We first describe the models for such tests in the following. Let y denote the trait value and M, m the two alleles of the marker. Treat the marker as a putative QTL. It is obtained in § 3.3 that, for an intercross design, we have the following model:

$$y = \begin{cases} \beta_0 + 2\beta + \epsilon, & \text{if the marker genotype is } MM, \\ \beta_0 + \beta + \gamma + \epsilon, & \text{if the marker genotype is } Mm, \\ \beta_0 + \epsilon, & \text{otherwise.} \end{cases} \qquad (4.1)$$

For a backcross design, model (4.1) reduces to

$$y = \begin{cases} \beta_0 + 2\beta + \epsilon, & \text{if the marker genotype is } MM, \\ \beta_0 + \beta + \gamma + \epsilon, & \text{otherwise.} \end{cases}$$

Let $b_0 = \beta_0 + \beta + \gamma$ and $b_1 = \beta - \gamma$. Then the above model becomes

$$y = \begin{cases} b_0 + b_1 + \epsilon, & \text{if the marker genotype is } MM, \\ b_0 + \epsilon, & \text{otherwise.} \end{cases} \quad (4.2)$$

For a RIL design, model (4.1) reduces to

$$y = \begin{cases} \beta_0 + 2\beta + \epsilon, & \text{if the marker genotype is } MM, \\ \beta_0 + \epsilon, & \text{otherwise.} \end{cases}$$

If we let $b_0 = \beta_0$ and $b_1 = 2\beta$, the model for the RIL design has the same form as (4.2).

For the intercross design, define dummy variables z_1 and z_2 as follows

$$z_1 = \begin{cases} 2, & \text{if the marker genotype is } MM, \\ 1, & \text{if the marker genotype is } Mm, \\ 0 & \text{if the marker genotype is } mm; \end{cases}$$

$$z_2 = \begin{cases} 1, & \text{if the marker genotype is } Mm, \\ 0 & \text{otherwise.} \end{cases}$$

By using the dummy variables, model (4.1) is expressed as

$$y = \beta_0 + \beta z_1 + \gamma z_2 + \epsilon. \quad (4.3)$$

For the backcross design and RIL design, define dummy variable z as follows

$$z = \begin{cases} 1, & \text{if the marker genotype is } MM, \\ 0 & \text{otherwise.} \end{cases}$$

Then the model for backcross design and RIL design can be expressed as

$$y = b_0 + b_1 z + \epsilon, \quad (4.4)$$

where b_1 is the difference between the effects of the two different genotypes.

The single-marker analysis amounts to testing for each marker the hypothesis $H_0 : \beta = \gamma = 0$, if we have an intercross design, and $H_0 : b_1 = 0$, if we have a backcross or RIL design. With a sample of n experiment subjects, models (4.3) and (4.4) can be expressed in matrix form as

$$\boldsymbol{y} = X\boldsymbol{\beta} + \boldsymbol{\epsilon},$$

where $\boldsymbol{y} = (y_1, \ldots, y_n)^\tau$ is the vector of trait values, $X = (\mathbf{1}, \boldsymbol{z}_1, \boldsymbol{z}_2)$, $\boldsymbol{\beta} = (\beta_0, \beta, \gamma)^\tau$, in the case of intercross design, and $X = (\mathbf{1}, \boldsymbol{z})$, $\boldsymbol{\beta} = (b_0, b_1)^\tau$, in the case of backcross or RIL design. Here $\boldsymbol{z}_1, \boldsymbol{z}_2$ and $\boldsymbol{z}$ are vectors of dummy variables. The tests are typical significant tests in the above linear regression model.

Traditionally, a LOD score is used as the test statistic. The LOD score is the $\log_{10}$ of the ratio of the maximum likelihoods under the alternative and null hypotheses. Assume that $\epsilon \sim N(0, \sigma^2 I)$. Let $\hat{\sigma}_k^2$, $k = 0, 1$, denote the MLE of σ^2 under the null and alternative hypothesis. We have

$$\hat{\sigma}_0^2 = \|y - \bar{Y}\mathbf{1}\|_2^2 \text{ and } \hat{\sigma}_1^2 = \|y - X\hat{\beta}\|_2^2,$$

where $\bar{Y} = \frac{1}{n}\sum_{i=1}^n y_i$, $\mathbf{1}$ is a vector of all elements 1, $\hat{\beta} = (X^\tau X)^{-1}X^\tau y$, which is the MLE of β under the alternative hypothesis and $\| \cdot \|_2$ is the L_2 vector norm. The maximum likelihoods are given by

$$\hat{L}_k = \left[\frac{1}{\sqrt{2\pi\hat{\sigma}_k^2}}\right]^n \exp(-n/2), \ k = 0, 1.$$

Thus the LOD score is given by

$$\text{LOD} = \frac{n}{2} \log_{10}\left(\frac{\hat{\sigma}_0^2}{\hat{\sigma}_1^2}\right).$$

The use of $\log_{10}$ instead of the natural logarithm is a long standing practice in human genetics. The LOD score is equivalent to the likelihood ratio test statistic introduced in § 2.4.4. A large LOD score indicates that the alternative hypothesis is more likely to be true. If the LOD score exceeds a threshold value, the marker is declared as a QTL or linked with a QTL. Other test statistics rather than the LOD score such as the Wald statistic introduced in § 2.5.1 can also be used as the test statistic.

Suppose that there are a total of M markers under study. Then the test must be carried out for all the M markers. To reject the null hypothesis for any particular marker (i.e., to claim the marker is a QTL or linked with QTL), a common threshold value must be determined for the M tests to control the overall Type I error rate. We delay the discussion of the determination of such threshold values to § 4.4.

If the markers under study are not dense over the genome, the single marker analysis is not very useful. Among other reasons, the following two are specific for the single marker analysis. The first reason is the lack of power in detecting QTL. The apparent difference among different genotypes of a marker does not adequately reflect the true difference among the effects of different genotypes of a QTL if the marker is not within a short distance from the QTL. For example, consider the case of backcross design. Let the recombination fraction between the QTL and the marker be θ. Denote by μ_{QQ} and μ_{Qq} the effects of the QTL genotype QQ and Qq respectively. Then the effect reflected by the marker genotype MM is

$$\mu_{MM} = (1 - \theta)\mu_{QQ} + \theta\mu_{Qq},$$

and that by Mm is

$$\mu_{Mm} = \theta\mu_{QQ} + (1 - \theta)\mu_{Qq}.$$

Hence

$$\mu_{MM} - \mu_{Mm} = (1 - 2\theta)(\mu_{QQ} - \mu_{Qq}).$$

That is, the difference of the QTL effects is reduced by a factor $1 - 2\theta$. When the markers are not dense, they seldom fall close to a QTL, i.e., θ will be large. Then, even if $\mu_{QQ} - \mu_{Qq}$ is statistically significantly large, $\mu_{MM} - \mu_{Mm}$ will be small and undetectable. The second reason is the inaccuracy of QTL position even if a QTL is detected. A QTL is detected due to the statistical significance of $\mu_{MM} - \mu_{Mm}$. But from the significant difference $\mu_{MM} - \mu_{Mm}$ it cannot be deduced whether or not θ is small. In other words, the position of the QTL relative to the marker cannot be well determined.

The single marker analysis will be mainly used in the case of dense markers as a preliminary procedure for more sophisticated methods in Chapter 6.

4.2   Single Interval Mapping

A typical QTL mapping data set consists of not only the trait values and the marker genotypes but also the genetic linkage map. In the single marker analysis discussed in the previous section, the information of genetic distances in the linkage map is not used. Without using this information, loci other than the available markers cannot be investigated, thus, limiting the power and efficiency of the mapping. Interval mapping makes use of the genetic distances to explore any loci between markers. As a consequence, it increases both the power and accuracy of the QTL mapping. The brilliant idea of interval mapping was first proposed by Lander and Botstein [112]. They developed a normal mixture model approach to tackle the problem. A regression version of interval mapping referred to as Haley-Knott regression was later considered to simplify computation; see [81] and [131]. In this section, we first present the normal mixture model approach to interval mapping and then the Haley-Knott regression approach.

4.2.1   Interval Mapping by Normal Mixture Models

In a QTL mapping experiment, the markers under study divide the whole genome into intervals, each flanked by two markers. At any locus of an interval, the genotype of the locus, though unobserved, can be inferred from the genotypes of the flanking markers, see § 3.4, thus making the evaluation of the locus feasible. We refer to the locus under investigation as a putative QTL.

In order to derive the normal mixture models, we introduce a coding variable x for the flanking marker genotypes and dummy variables for unobserved putative QTL genotypes as follows. Let Q, q denote the alleles of the putative QTL and L, l and U, u the alleles of the left and right flanking marker respectively. Let x be the code of the genotype combination of the two flanking markers. For backcross and RIL designs, the genotype combinations is coded as

$$x = \begin{cases} 1, & \text{for } LL/UU, \\ 2, & \text{for } LL/Uu \text{ or } LL/uu, \\ 3, & \text{for } Ll/UU, \text{ or } ll/UU \\ 4, & \text{for } Ll/Uu \text{ or } ll/uu. \end{cases}$$

For intercross designs, the genotype combinations is coded as

$$x = \begin{cases} 1, & \text{for } LL/UU, \\ 2, & \text{for } LL/Uu, \\ 3, & \text{for } LL/uu, \\ 4, & \text{for } Ll/UU, \\ 5, & \text{for } Ll/Uu, \\ 6, & \text{for } Ll/uu, \\ 7, & \text{for } ll/UU, \\ 8, & \text{for } ll/Uu, \\ 9, & \text{for } ll/uu. \end{cases}$$

For the backcross and RIL designs, we introduce a single dummy variable δ as follows:

$$\delta = \begin{cases} 1, & \text{for } QQ, \\ 0, & \text{otherwise.} \end{cases}$$

For the intercross design, we introduce two dummy variables δ_1 and δ_2 as follows:

$$\delta_1 = \begin{cases} 1, & \text{for } Qq, \\ 0 & \text{otherwise;} \end{cases}$$

$$\delta_2 = \begin{cases} 1, & \text{for } QQ, \\ 0 & \text{otherwise.} \end{cases}$$

Note that these dummy variables are different from that in § 3.3.1. The dummy variables defined above are more convenient in deriving the mixture models.

In the following, we derive the normal mixture model for the intercross design. The mixture models for the backcross and RIL design, which are simpler, can be derived in the same way and will be given later. Assume that, given the genotype of the putative QTL, the quantitative trait follows a normal distribution; that is,

$$y|(\delta_1, \delta_2) \sim \phi(y; \beta_0 + \beta_1\delta_1 + \beta_2\delta_2, \sigma^2), \tag{4.5}$$

where $\phi(y; \mu, \sigma^2)$ denote the probability density function of the normal distribution with mean μ and variance σ^2. By the definition of δ_1 and δ_2, β_0 is the mean trait value of the subjects with genotype qq, $\beta_0 + \beta_1$ and $\beta_0 + \beta_2$ are, respectively, the mean values of the subjects with genotype QQ and Qq. The pair (δ_1, δ_2), which is unobserved, is a random vector having a trinomial distribution. Let r be the recombination fraction between the left flanking marker and the putative QTL, and γ the recombination fraction between the two flanking markers. These recombination fractions are converted from the corresponding genetic distances using Haldane mapping function under the assumption of no crossover interference. Denote by $p_1(x, r, \gamma)$ and $p_2(x, r, \gamma)$, respectively, the conditional probabilities of Qq and QQ given the marker genotype code x, since they are completely determined by x, r and γ. These probabilities are given in Table 3.5. The probability density of (δ_1, δ_2) is then given by

$$(\delta_1, \delta_2) \sim [p_1(x, r, \gamma)]^{\delta_1} [p_2(x, r, \gamma)]^{\delta_2} [p_0(x, r, \gamma)]^{1-\delta_1-\delta_2}, \qquad (4.6)$$

where $p_0(x, r, \gamma) = 1 - p_1(x, r, \gamma) - p_2(x, r, \gamma)$. From (4.5) and (4.6) we arrive at the joint distribution of y and (δ_1, δ_2) as follows:

$$
\begin{aligned}
&f(y, \delta_1, \delta_2) \qquad\qquad\qquad\qquad\qquad\qquad\qquad\qquad\qquad (4.7)\\
&= [p_1(x, r, \gamma)]^{\delta_1} [p_2(x, r, \gamma)]^{\delta_2} [p_0(x, r, \gamma)]^{1-\delta_1-\delta_2} \phi(y; \beta_0 + \beta_1\delta_1 + \beta_2\delta_2, \sigma^2).
\end{aligned}
$$

The marginal probability density function of y is thus obtained as

$$
\begin{aligned}
f(y; x, \boldsymbol{\beta}, \sigma^2) &\qquad\qquad\qquad\qquad\qquad\qquad\qquad\qquad (4.8)\\
&= \sum_{(\delta_1, \delta_2)} [p_1(x, r, \gamma)]^{\delta_1} [p_2(x, r, \gamma)]^{\delta_2} [p_0(x, r, \gamma)]^{1-\delta_1-\delta_2}\\
&\quad \times \phi(y; \beta_0 + \beta_1\delta_1 + \beta_2\delta_2, \sigma^2)\\
&= p_1(x, r, \gamma)\phi(y; \beta_0 + \beta_1, \sigma^2) + p_2(x, r, \gamma)\phi(y; \beta_0 + \beta_2, \sigma^2)\\
&\quad + p_0(x, r, \gamma)\phi(y; \beta_0, \sigma^2),
\end{aligned}
$$

where $\boldsymbol{\beta} = (\beta_0, \beta_1, \beta_2)^\tau$. This is a mixture of normal distributions.

For each interval, present the mapping data in the form $(y_i, x_i) : i = 1, \dots, n$. At any location on the interval, with r determined by the location, we have the following likelihood function

$$L(\boldsymbol{\beta}, \sigma^2) = \prod_{i=1}^{n} f(y_i, x_i, \boldsymbol{\beta}, \sigma^2). \qquad (4.9)$$

A LOD score is computed for this location as

$$\text{LOD} = \log_{10} \frac{L(\hat{\boldsymbol{\beta}}, \hat{\sigma}^2)}{L(\hat{\boldsymbol{\beta}}_0, \hat{\sigma}_0^2)},$$

where $\hat{\boldsymbol{\beta}}, \hat{\sigma}^2$ are the MLEs obtained by maximizing (4.9), and $\hat{\boldsymbol{\beta}}_0, \hat{\sigma}_0^2$ the MLEs under the hypothesis that $\beta_1 = \beta_2 = 0$. In fact, $L(\hat{\boldsymbol{\beta}}_0, \hat{\sigma}_0^2) = \left[\frac{1}{\sqrt{2\pi\hat{\sigma}_0^2}} \right]^n \exp(-n/2)$ where $\hat{\sigma}_0^2 = \frac{1}{n}\sum_{i=1}^{n}(y_i - \bar{y})^2$. The LOD score is related to the likelihood ratio test (LRT) statistic as follows:

$$\text{LOD} = \text{LRT}/(2\ln 10).$$

The MLEs, $\hat{\boldsymbol{\beta}}$ and $\hat{\sigma}^2$, with the likelihood function (4.9) can be computed by the EM algorithm discussed in § 2.6. Here, the "complete" data consists of $(y_i, \delta_{1i}, \delta_{2i}) : i = 1, \ldots, n$. From (4.7), the "complete" data likelihood is given by

$$L_{\text{COM}}(\boldsymbol{\beta}, \sigma^2) = \prod_{i=1}^{n} [p_1(x_i, r, \gamma)]^{\delta_{1i}} [p_2(x_i, r, \gamma)]^{\delta_{2i}} [p_0(x_i, r, \gamma)]^{1 - \delta_{1i} - \delta_{2i}}$$

$$\times \prod_{i=1}^{n} \phi(y_i; \beta_0 + \beta_1 \delta_{1i} + \beta_2 \delta_{2i}, \sigma^2). \qquad (4.10)$$

The log of the likelihood can be expressed as

$$\ell(\boldsymbol{\beta}, \sigma^2; \boldsymbol{\delta}_1, \boldsymbol{\delta}_2) = \ell_1(\boldsymbol{\beta}, \sigma^2; \boldsymbol{\delta}_1, \boldsymbol{\delta}_2) + \ell_2(\boldsymbol{\beta}, \sigma^2; \boldsymbol{\delta}_1, \boldsymbol{\delta}_2),$$

where $\boldsymbol{\delta}_k = (\delta_{k1}, \ldots, \delta_{kn})^\tau, k = 1, 2$, and, up to a constant that does not depend on $\boldsymbol{\beta}, \sigma^2$ and the $\boldsymbol{\delta}_k$'s,

$$\ell_1(\boldsymbol{\beta}, \sigma^2; \boldsymbol{\delta}_1, \boldsymbol{\delta}_2) = \sum_{i=1}^{n} \left[\delta_{1i} \ln \frac{p_1(x_i, r, \gamma)}{p_0(x_i, r, \gamma)} + \delta_{2i} \ln \frac{p_2(x_i, r, \gamma)}{p_0(x_i, r, \gamma)} \right],$$

and

$$\ell_2(\boldsymbol{\beta}, \sigma^2; \boldsymbol{\delta}_1, \boldsymbol{\delta}_2) = -\frac{1}{2\sigma^2} \sum_{i=1}^{n} (y_i - \beta_0 - \beta_1 \delta_{1i} - \beta_2 \delta_{2i})^2 - \frac{n}{2} \ln \sigma^2$$

$$= -\frac{1}{2\sigma^2} \sum_{i=1}^{n} [(y_i - \beta_0)^2 - 2(y_i - \beta_0)(\beta_1 \delta_{1i} + \beta_2 \delta_{2i}) + \beta_1^2 \delta_{1i}$$

$$+ \beta_2^2 \delta_{2i}] - \frac{n}{2} \ln \sigma^2.$$

The last equality holds since $\delta_{1i} \delta_{2i} = 0$ and $\delta_{ki}^2 = \delta_{ki}$.

Since $\ell(\boldsymbol{\beta}, \sigma^2; \boldsymbol{\delta}_1, \boldsymbol{\delta}_2)$ is linear in δ_{ki}'s, the E-step of the EM algorithm reduces to the computation of the conditional expectations of the δ_{ki}'s which are given by

$$\tilde{\delta}_{ki} = \frac{p_k(x_i, r, \gamma) \phi(y_i; \mu_k^{(0)}, \sigma^{2(0)})}{\sum_{k=0}^{2} p_0(x_i, r, \gamma) \phi(y_i; \mu_k^{(0)} \sigma^{2(0)})},$$

where $\mu_0^{(0)}$, $\mu_k^{(0)}, k = 1, 2$, and $\sigma^{2(0)}$ are updated parameter values, $\mu_0^{(0)} = \beta_0^{(0)}, \mu_k^{(0)} = \beta_0^{(0)} + \beta_k^{(0)}$.

Since $\ell_1(\boldsymbol{\beta}, \sigma^2; \boldsymbol{\delta}_1, \boldsymbol{\delta}_2)$, in fact, does not involve the parameters, at the M-step, only $\ell_2(\boldsymbol{\beta}, \sigma^2; \boldsymbol{\delta}_1, \boldsymbol{\delta}_2)$ is maximized with $\boldsymbol{\delta}_k$'s replaced by $\tilde{\boldsymbol{\delta}}_k$'s. The solution at the M-step has the following explicit forms:

$$\tilde{\beta}_0 = \frac{\bar{y} - \bar{\delta}_1 \bar{y}_{\delta_1} - \bar{\delta}_2 \bar{y}_{\delta_2}}{1 - \bar{\delta}_1 - \bar{\delta}_2},$$

$$\tilde{\beta}_1 = \bar{y}_{\delta_1} - \tilde{\beta}_0,$$

$$\tilde{\beta}_2 = \bar{y}_{\delta_2} - \tilde{\beta}_0,$$

$$\tilde{\sigma}^2 = \frac{1}{n} \sum_{i=1}^{n} (y_i - \tilde{\beta}_0 - \tilde{\beta}_1 \tilde{\delta}_{1i} - \tilde{\beta}_2 \tilde{\delta}_{2i})^2,$$

where $\bar{\delta}_k = \frac{1}{n} \sum_{i=1}^{n} \delta_{ki}$, $\bar{y}_{\delta_k} = \frac{\sum_{i=1}^{n} y_i \delta_{ki}}{\sum_{i=1}^{n} \delta_{ki}}$, $k = 1, 2$.

For the backcross and RIL designs, the likelihood function is given by

$$L(\boldsymbol{\beta}, \sigma^2) = \prod_{i=1}^{n} [p(x_i, r, \gamma) \phi(y_i; \beta_0 + \beta_1, \sigma^2) + (1 - p(x_i, r, \gamma)) \phi(y_i; \beta_0, \sigma^2)],$$

where $p(x_i, r, \gamma)$ is the probability of QQ given flanking marker genotype x_i, which is given in Table 3.4, for backcross design, and in Table 3.6, for RIL designs. Note that, for the quantities R, S and Γ in Table 3.6, R is determined by r according to the formulae in § 3.4.2, S is determined by $s = (\gamma - r)/(1 - 2r)$ according to the same formulae since the relationship between R and r and that between S and s are the same, the Γ is determined using the formulae at the end of § 3.4.2. In the EM algorithm, at the E-step,

$$\tilde{\delta}_i = \frac{p(x_i, r, \gamma) \phi(y_i; \beta_0^{(0)} + \beta_1^{(0)}, \sigma^{2(0)})}{p(x_i, r, \gamma) \phi(y_i; \beta_0^{(0)} + \beta_1^{(0)}, \sigma^{2(0)}) + (1 - p(x_i, r, \gamma)) \phi(y_i; \beta_0^{(0)}, \sigma^{2(0)})}$$

are calculated; at the M-step, the parameters are updated as

$$\tilde{\beta}_0 = \frac{\bar{y} - \bar{\delta} \bar{y}_{\delta}}{1 - \bar{\delta}},$$

$$\tilde{\beta}_1 = \bar{y}_{\delta} - \tilde{\beta}_0,$$

$$\tilde{\sigma}^2 = \frac{1}{n} \sum_{i=1}^{n} (y_i - \tilde{\beta}_0 - \tilde{\beta}_1 \tilde{\delta}_i)^2.$$

The strategy of the interval mapping is as follows. First, divide the whole genome by equally spaced grid points, say, 1 cM apart. At each grid point (which falls into a unique interval), compute the LOD score. A LOD curve is then drawn over each chromosome of the genome. The peaks of the curve that exceed a predetermined threshold value are declared as QTL. For each declared QTL, a one-log interval is taken as the supporting interval for the position of the QTL. The one-log interval is determined by the two points on both sides of the peak at which the LOD scores are 1 less than the LOD score at the peak point. The determination of the threshold value will be discussed in § 4.4.

The normal mixture model assumes that the quantitative trait is normally distributed. When a trait is not normally distributed, but a transformation on the trait value can make the transformed trait values approximately normally distributed, which is the case in certain situations, then the interval mapping approach should be applied to the transformed trait values. If no such transformation is available, the mixture model approach is still applicable, but we have to consider other mixture models. The mixture model approach for traits with exponential family distributions will be considered in Chapter 5.

4.2.2 Interval Mapping by Haley-Knott Regression

As mentioned in the last paragraph, the normal mixture model approach requires the assumption of normality for the distribution of the quantitative trait. In the approach of Haley-Knott regression, the distributional assumption is relaxed. In addition, the computation amount needed for Haley-Knott regression is less than the normal mixture model approach. We use model (3.4) in § 3.3.1 to derive the Haley-Knott regression model as follows.

In the designed experimental crosses, the quantitative trait with a single QTL follows model (3.4) as given below.

$$y = \begin{cases} \beta_0 + 2c + \epsilon, & \text{for genotype } QQ \\ \beta_0 + c + d + \epsilon, & \text{for genotype } Qq \\ \beta_0 + \epsilon, & \text{for genotype } qq, \end{cases}$$

where c and d are, respectively, the additive and dominant effects of the QTL, and ϵ has a distribution with mean 0 and variance σ^2. The distribution of ϵ is not necessarily normal.

In the case of intercross design, for an experiment subject with flanking marker genotype x_i, the expectation of y_i is given by

$$\begin{aligned} Ey_i &= \beta_0 p_0(x_i, r, \gamma) + (\beta_0 + c + d)p_1(x_i, r, \gamma) + (\beta_0 + 2c)p_2(x_i, r, \gamma) \\ &= \beta_0 + c[p_1(x_i, r, \gamma) + 2p_2(x_i, r, \gamma)] + dp_1(x_i, r, \gamma). \end{aligned}$$

Let $z_{i1} = p_1(x_i, r, \gamma) + 2p_2(x_i, r, \gamma)$ and $z_{i2} = p_1(x_i, r, \gamma)$. Denote c by β_1 and d by β_2. Then we have the linear model

$$y_i = \beta_0 + \beta_1 z_{i1} + \beta_2 z_{i2} + e_i, i = 1, \ldots, n. \tag{4.11}$$

In the case of backcross design, the expectation of y_i is given by

$$\begin{aligned} Ey_i &= (\beta_0 + c + d)(1 - p(x_i, r, \gamma)) + (\beta_0 + 2c)p(x_i, r, \gamma) \\ &= (\beta_0 + c + d) + (c - d)p(x_i, r, \gamma), \end{aligned}$$

which leads to the linear model

$$y_i = b_0 + b_1 z_i + e_i, i = 1, \ldots, n, \tag{4.12}$$

where $b_0 = \beta_0 + c + d$, $b_1 = c - d$ and $z_i = p(x_i, r, \gamma)$.

In the case of RIL design, the expectation of y_i is given by

$$
\begin{aligned}
Ey_i &= \beta_0(1 - p(x_i, r, \gamma)) + (\beta_0 + 2c)p(x_i, r, \gamma) \\
&= \beta_0 + 2cp(x_i, r, \gamma),
\end{aligned}
$$

which leads to the same model as (4.12) with the different definition $b_0 = \beta_0$, $b_1 = 2c$.

Note that the covariates in models (4.11) and (4.12) implicitly depend on r, the recombination fraction determined from the genetic distance between the left marker and the putative QTL. In the implementation of the Haley-Knott regression approach, the first step is to compute the covariates z_{i1}, z_{i2} from the conditional probabilities given in Table 3.5 for intercross design (or z_i using the Table 3.4 for backcross and Table 3.6 for RIL design). The Haley-Knott regression approach treats, for each position on the interval, the models (4.11) and (4.12) as ordinary regression models, and tests the null hypothesis $H_0 : \beta_1 = \beta_2 = 0$ (or $b_1 = 0$). The test statistic is given by

$$
F = \frac{(\text{RSS}_0 - \text{RSS})/(df_0 - df)}{\text{RSS}/df},
$$

where RSS, RSS_0 and df, df_0 are the residual sums of squares and the corresponding degrees of freedom under the full and the null model, which are given as follows:

$$
\text{RSS} = \|\boldsymbol{y} - Z\hat{\boldsymbol{\beta}}\|_2^2, \quad \text{RSS}_0 = \|\boldsymbol{y} - \bar{y}\boldsymbol{1}\|_2^2,
$$

where $\hat{\boldsymbol{\beta}} = (Z^\tau Z)^{-1} Z^\tau \boldsymbol{y}$, $Z = (\boldsymbol{1}, \boldsymbol{z}_1, \boldsymbol{z}_2)$, for intercross design, $(\boldsymbol{1}, \boldsymbol{z})$, for backcross and RIL design, $df_0 = n - 1$, and $df = n - 3$, for intercross design, $n - 2$, for backcross and RIL design. In the ordinary regression analysis where the normality of the error terms is assumed, the F-statistic above follows an exact F-distribution with degrees of freedom $df_0 - df$ and df. However, since the error terms in (4.11) and (4.12) do not follow normal distributions, the F-statistic above multiplied by $df_0 - df$ has an asymptotic χ^2-distribution with degrees of freedom $df_0 - df$, i.e., $(df_0 - df)f \sim \chi^2_{df_0 - df}$.

The mapping strategy of the Haley-Knott regression approach is the same as the normal mixture model approach. It simply replaces the LOD score with the F-statistic. The peaks of the F-statistic curve that exceed a predetermined threshold value are declared as QTL.

Another difference of (4.11) and (4.12) from ordinary regression models is that the error terms e_i do not have a constant variance. For example, in (4.11),

$$
\text{Var}(e_i) = \sigma^2 + \sum_{k=0}^{2} p_k(x_i, r, \gamma)\mu_k^2 - \left[\sum_{k=0}^{2} p_k(x_i, r, \gamma)\mu_k\right]^2,
$$

which depends on x_i. An iterated weighted regression analysis can improve the Haley-Knott regression approach, see [204]. In the following, we give a simple procedure for the implementation of the iterated weighted regression analysis. For i with $x_i = j$, take the initial weight as

$$w_i^{(0)} = \left[\frac{1}{n_j} \sum_{x_i = k_1 k_2} (y_i - z_i^\tau \hat{\beta}^{(0)})^2 \right]^{-1},$$

where n_j is the number of i's such that $x_i = j$, and $\hat{\beta}^{(0)}$ is an initial estimate of β. With the initial estimate $\hat{\beta}^{(0)}$, update the estimate of β by

$$\hat{\beta}_W = (Z^\tau W^{(0)} Z)^{-1} Z^\tau W^{(0)} y,$$

where $W^{(0)}$ is the diagonal matrix with its ith diagonal element equal to $w_i^{(0)}$. Iterating the above calculation we set $\hat{\beta}^{(0)} = \hat{\beta}_W$ until the estimate converges, denote the final weight matrix by W, then in the F-statistic, replace RSS by

$$\mathrm{RSS}_W = (y - Z\hat{\beta}_W)^\tau W (y - Z\hat{\beta}_W).$$

The F-statistic still have the same asymptotic distribution. The other aspects of the analysis also remain unchanged. The Haley-Knott regression approach can then be carried out with the F-statistic based on the weighted estimates.

4.2.3 A Remark

In the interval mapping approach, it is implicitly assumed that there is at most one QTL over the whole genome for the quantitative trait of concern. The interval mapping approach is the most useful when this assumption is true. However, usually, a quantitative trait is genetically affected by many loci; that is, there are multiple QTL. When investigating a particular putative QTL, the interval mapping approach ignores the effects of other QTL. This has at least two adverse effects. First, the mapping of QTL can be seriously biased. As noticed in [131] and [106], a non-existent "ghost" QTL might be identified as a real QTL. This phenomenon usually occurs in the case that a non-QTL interval is bracketed by adjacent intervals containing QTL. A "ghost" QTL, which appears to have an even larger effect than a real QTL, might be detected on the non-QTL interval. Statistically, this is the phenomenon of confounding, which occurs when the genotype of the "ghost" QTL is highly correlated with the two real QTL; that is, the groups classified by the genotype of the "ghost" QTL are overlapping with the groups classified by the genotypes of the real QTL. If it happens that the effects of the two QTL are in the same direction, the reinforced effect will be reflected by the "ghost" QTL and the "ghost" will appear. Second, even if other QTL are completely independent with the putative QTL (i.e., they are in different chromosomes and unlinked), the variation in the quantitative trait caused by the other QTL is counted in the variation caused by random error in the interval mapping model; that is, the variance of the random error is artificially inflated. As a consequence, the power for detecting a real QTL will be compromised. A remedy for overcoming the drawback of interval mapping is considered in the next section.

4.3   Composite Interval Mapping

A natural remedy for interval mapping is to adjust for the effects of other QTL while investigating a particular putative QTL. Adjusting for the effects of other factors is a widely used technique in the analysis of clinical trials. In QTL mapping, the only difficulty is that the other QTL are unknown. Surrogates of the QTL, which can account for the effects of the QTL, must be found. In the old tradition, multiple marker regression is used for QTL mapping. The markers are taken as the surrogates of the QTL. The drawback of the multiple marker regression is that it cannot infer the numbers and locations of the QTL. But the markers can account for the effects of the QTL, which is what the adjustment needs. It is then feasible to combine multiple marker regression and interval mapping to overcome the drawback of the original interval mapping. This idea was first explored in [95] and [218]. The combination of the marker regression and interval mapping is called composite interval mapping. In this section, we describe both the normal mixture model version and the Haley-Knott regression version of the composite interval mapping.

4.3.1   Composite Interval Mapping with Normal Mixture Models

The first step of composite interval mapping is to determine the adjusting markers. The adjusting markers can be taken the same for all intervals with some adaptation for each particular interval. We delay the discussion on how to determine the adjusting markers to § 4.3.3. In this and the next sub section, we assume that the adjusting markers have already been determined.

Let $U_k, k = 1, \ldots, K$, be the adjusting markers. Let U denote the matrix of marker genotypes of the n subjects. In the case of intercross design, U is of dimension $n \times 2K$, and in the case of backcross and RIL design, U is of dimension $n \times K$. Let D denote the matrix of the genotypes of the putative QTL. For intercross design, $D = (\delta_1, \delta_2)$. For backcross and RIL design, $D = \delta$. Let $X = (\mathbf{1}, D, U)$. The model for composite interval mapping with normal mixture distribution can be expressed as follows. Given D,

$$y = X\xi + \epsilon, \tag{4.13}$$

where $\xi = (\beta_0, \beta^\tau, b^\tau)^\tau$ and $\epsilon \sim N(0, \sigma^2 I)$. Here β is 2-vector for intercross design and a scaler for backcross and RIL design.

Instead of the LOD score, we consider the likelihood ratio test statistic for testing the hypothesis $H_0 : \boldsymbol{\beta} = 0$ vs. $H_1 : \boldsymbol{\beta} \neq 0$ in model (4.13).

Under H_0, (4.13) is an ordinary normal linear model. Let $X_0 = (\mathbf{1}, U)$. The MLE of the reduced parameter vector $\boldsymbol{\xi}_0 = (\beta_0, \boldsymbol{b}^\tau)^\tau$ under H_0 is given by

$$\hat{\boldsymbol{\xi}}_0 = (X_0^\tau X_0)^{-1} X_0^\tau \boldsymbol{y},$$

and

$$\hat{\sigma}_0^2 = \frac{1}{n} \|\boldsymbol{y} - X_0 \hat{\boldsymbol{\xi}}_0\|_2^2.$$

Therefore, the maximum log likelihood under H_0 is given by

$$\ell(\hat{\boldsymbol{\xi}}_0, \hat{\sigma}_0^2) = -\frac{n}{2}(1 + \ln 2\pi) - \frac{n}{2} \ln \hat{\sigma}_0^2.$$

Under H_1, (4.13) is not an ordinary normal linear model, since the component D of X is unobserved. To obtain the MLE of $\boldsymbol{\xi}$, an EM algorithm is needed for the computation. The same argument as in § 4.2.1 leads to the following E-step and M-step.

E-step: Denote by $\boldsymbol{u}_i = (u_{1i}, \ldots, u_{Ki})^\tau$ the vector of the marker genotypes of subject i. Compute

$$\tilde{\delta}_{ki} = \frac{p_k(x_i, r, \gamma)\phi(y_i; \mu_k^{(0)} + \boldsymbol{u}_i^\tau \boldsymbol{b}^{(0)}, \sigma^{2(0)})}{\sum_k p_0(x_i, r, \gamma)\phi(y_i; \mu_k^{(0)} + \boldsymbol{u}_i^\tau \boldsymbol{b}^{(0)}, \sigma^{2(0)})}.$$

M-step: Let $\tilde{X} = (\mathbf{1}, \tilde{D}, U)$ where $\tilde{D}$ is the matrix D with its elements δ_{ki} replaced by $\tilde{\delta}_{ki}$. Let $\tilde{\Omega}$ be the matrix obtained from $\tilde{X}^\tau \tilde{X}$ by replacing its submatrix $\tilde{D}^\tau \tilde{D}$ with the diagonal matrix $\text{Diag}(\mathbf{1}^\tau \tilde{D})$; that is,

$$\tilde{\Omega} = \begin{pmatrix} n & \mathbf{1}^\tau \tilde{D} & \mathbf{1}^\tau U \\ \tilde{D}^\tau \mathbf{1} & \text{Diag}(\mathbf{1}^\tau \tilde{D}) & \tilde{D}^\tau U \\ U^\tau \mathbf{1} & U^\tau \tilde{D} & U^\tau U \end{pmatrix}.$$

Then update the estimates of the parameters as

$$\begin{aligned} \tilde{\boldsymbol{\xi}} &= \tilde{\Omega}^{-1} \tilde{X}^\tau \boldsymbol{y}, \\ \tilde{\sigma}^2 &= \|\boldsymbol{y} - \tilde{X}\tilde{\boldsymbol{\xi}}\|_2^2. \end{aligned}$$

Note that the maximum log likelihood under H_1 does not have the same form as $\ell(\hat{\boldsymbol{\xi}}_0, \hat{\sigma}_0^2)$. The maximum log likelihood $\ell(\hat{\boldsymbol{\xi}}, \hat{\sigma}^2)$, which must be computed using the marginal distribution of $\boldsymbol{y}$, is given by

$$\ell(\hat{\boldsymbol{\xi}}, \hat{\sigma}^2) = \sum_{i=1}^{n} \ln \left(\sum_k p_k(y_i, x_i, r, \gamma)\phi(y_i; \hat{\mu}_k + \boldsymbol{u}_i^\tau \hat{\boldsymbol{b}}, \hat{\sigma}^2) \right).$$

The LRT statistic is given by

$$\text{LRT} = 2[\ell(\hat{\boldsymbol{\xi}}, \hat{\sigma}^2) - \ell(\hat{\boldsymbol{\xi}}_0, \hat{\sigma}_0^2)].$$

Under the null hypothesis, LRT has an asymptotic χ^2-distribution with the dimension of $\boldsymbol{\beta}$ as its degrees of freedom. The mapping strategy of the composite interval mapping is the same as the original interval mapping with the LOD score replaced by LRT.

4.3.2   Composite Interval Mapping with Haley-Knott Regression

In the method of composite interval mapping with Haley-Knott regression, the models (4.11) and (4.12) are augmented by the adjusting markers as

$$\boldsymbol{y} = \mathbf{1}\beta_0 + Z\boldsymbol{\beta} + U\boldsymbol{b} + \boldsymbol{e}, \qquad (4.14)$$

where Z is the matrix computed from the conditional probabilities of the putative QTL genotypes as given in § 4.2.2, and U is the matrix of the genotypes of the adjusting markers. At each position on a interval, for testing $H_0 : \boldsymbol{\beta} = 0$ vs. $H_1 : \boldsymbol{\beta} \neq 0$, still, the following usual F-statistic is used:

$$F = \frac{(\text{RSS}_0 - \text{RSS})/(df_0 - df)}{\text{RSS}/df}.$$

In the context of (4.14),

$$\text{RSS} = \|\boldsymbol{y} - X\hat{\boldsymbol{\xi}}\|_2^2, \quad \text{RSS}_0 = \|\boldsymbol{y} - X_0\hat{\boldsymbol{\xi}}_0\|_2^2,$$

where $X = (\mathbf{1}, Z, U)$, $X_0 = (\mathbf{1}, U)$, $\hat{\boldsymbol{\xi}} = (X^\tau X)^{-1}X^\tau \boldsymbol{y}$, $\hat{\boldsymbol{\xi}}_0 = (X_0^\tau X_0)^{-1}X_0^\tau \boldsymbol{y}$, and $df = n - p - q - 1$, $df_0 = n - p - 1$ where p is the number of columns of U and q is the number of columns of Z. The F-statistic multiplied by q has an asymptotic χ^2-distribution with degrees of freedom q.

The other aspects of the composite interval mapping with Haley-Knott regression are the same as the ordinary interval mapping with Haley-Knott regression.

4.3.3   Choice of Adjusting Markers

We now consider the choice of adjusting markers. There are a few suggestions in the literature, see [116] and [218]. These suggestions include: (i) use all the markers, (ii) for a given interval, use only unlinked markers (i.e., markers on different chromosomes) and (iii) use selected markers obtained by stepwise regression of the quantitative trait on marker genotypes. The drawbacks of the first two suggestions are obvious. If all the markers are used as adjusting markers, the effects of the QTL will be over-adjusted. When the total number of markers is not small, which is usually the case, there might be spurious correlations between certain markers and the flanking markers of an interval. While those markers are included in the adjusting markers, the effect of the QTL (if exists) on the interval will be accounted for by those markers, which makes the detection of the QTL less powerful. Even if spurious correlation does not exist, fitting too many parameters reduces the accuracy of the estimation on the QTL effect. When only unlinked markers are used as adjusting markers, if there are multiple QTL on the same chromosome, the effects of these QTL will be under-adjusted. In order to make the adjustment appropriate, adjusting markers should be properly selected. The selection of adjusting markers by stepwise regression is a feasible approach. We present another approach as follows.

Using the ordinary interval mapping for a preliminary analysis, first identify the peaks of the LOD score curve (or F-statistic curve) that exceed the 0.05 level (individual level) critical value of the test statistic. Then take the markers flanking the peaks as the total set of adjusting markers. If it happens that the two flanking markers of a particular interval are in the set, exclude them from the set, otherwise use all the markers in the set as adjusting markers for this interval.

In a practical mapping problem, both the stepwise regression and the preliminary interval mapping can be used for the selection of the adjusting markers. The results by using the adjusting markers chosen by these two different methods can then be compared to see whether they are consistent with each other. When inconsistence occurs, it will probably shed some new light for further study.

4.4 Determination of Threshold Values

The determination of threshold values is crucial for interval mapping (including composite interval mapping). In interval mapping, the whole genome is investigated at equally spaced points, and, at each point, a hypothesis testing is conducted. The determination of threshold values for interval mapping is similar to but much more complicated than that for the usual multiple tests. The tests in interval mapping, especially those at the points on the same interval, are closely related. The commonly used Bonferroni adjustment for multiple tests is no longer appropriate since it becomes too conservative in this situation. The overall threshold value pertains to the distribution of the maximum of the test statistic over the genome. This distribution depends on many factors, i.e., the design of the crosses, the number and lengths of chromosomes, the density of markers, the sample sizes, etc.. In the literature, different approaches have been proposed for dealing with this distribution and hence for determining the overall threshold values, including theoretical, simulation and empirical permutation approaches. The first result on theoretical approximation is given in [112]. The result is obtained by treating the test statistic indexed by the position on the genome as a stochastic process. The research along this line has since been carried out and the results are reported in [46], [57], [147] and [148]. A simulation approach based on asymptotic distribution is given in [30]. An empirical permutation approach is proposed in [40]. We summarize these results and approaches in this section.

4.4.1   Theoretical Approximation

Let LRT(d) denote the likelihood ratio test statistic at position d. The overall threshold value that controls the family-wise type I error rate at α is given by c_α such that

$$P(\sup_d \mathrm{LRT}(d) > c_\alpha) \leq \alpha,$$

where the supremum is over all position d in the genome.

For backcross designs, two approximations are proposed in [147] and [57]. We refer to the approximation proposed in [147] as RGM approximation and that in [57] as FBS approximation.

The RGM approximation of c_α is given by c^2 such that c solves the following equation:

$$\frac{\alpha}{2} = 1 - \Phi(c) + \frac{1}{\pi} \exp\left(-\frac{c^2}{2}\right) \sum_{j=1}^{M} \mathrm{Arctan}\left(\sqrt{\frac{\gamma_j}{1-\gamma_j}}\right), \qquad (4.15)$$

where $\Phi(\cdot)$ is the cumulative distribution function of the standard normal distribution, M is the number of intervals and γ_j is the recombination fraction between the two flanking markers of the jth interval.

The FBS approximation of c_α is given by c^2 such that c solves the following equation:

$$\alpha = 1 - \Phi(c) + 0.02Lc\phi(c)\exp(-1.166c\sqrt{\Delta}), \qquad (4.16)$$

where $\phi(\cdot)$ is the density function of the standard normal distribution, L is the total genetic length of the genome and Δ is the average genetic length of the intervals, both L and Δ are in units of centi-Morgan (cM). The above equation, which is given in [147], is adapted from the approximation in [57].

There are some implicit assumptions in deriving (4.15) and (4.16). Equation (4.15) is derived by assuming that the markers are sparse while (4.16) is obtained by assuming that the markers are dense. The simulation studies reported in [147] show that the FBS approximation is more accurate than the RGM approximation when the average interval length is less than or equal to 1 cM and is less accurate than the RGM approximation when the average interval length is larger than 1 cM.

Approximations for intercross designs are considered in [46] and [147]. The approximation given in [147] is too complicated, which renders its application difficult. In the following, we give the approximation considered in [46]. The approximation is provided by

$$P(\sup_d \mathrm{LRT}(d) > a) \approx 1 - \exp\left[-\left(N_c + 0.03Lc^2e^{-0.1428c\sqrt{\Delta}}\right)e^{-c^2/2}\right], \quad (4.17)$$

where $a = c^2$, N_c is the number of chromosomes, L and Δ are the same as those in (4.16). Thus the threshold value c_α is approximated by c^2 such that c solves

$$\alpha = 1 - \exp\left[-\left(N_c + 0.03Lc^2e^{-0.1428c\sqrt{\Delta}}\right)e^{-c^2/2}\right].$$

Like (4.16), dense markers are assumed for (4.17).

4.4.2   A Simulation Approach

It is shown in [30] that the asymptotic distribution of $\sup_d \mathrm{LRT}(d)$ in the case of backcross designs is the same as the following random variable:

$$T = \max_{1 \leq k \leq M} \sup_{0 \leq r_k \leq \gamma_k} \left[\sum_{j=1}^{4} \sqrt{q(j, \gamma_k)}[p(j, r_k, \gamma_k) - 1/2]Z_{kj}/\tau(r_k, \gamma_k)\right]^2,$$

$$(4.18)$$

where M is the number of intervals, γ_k is the recombination fraction between the flanking markers of the kth interval, $q(\cdot, \cdot)$ is defined by

$$q(1, \gamma) = q(4, \gamma) = \frac{1 - \gamma}{2}, \quad q(2, \gamma) = q(3, \gamma) = \frac{\gamma}{2},$$

$\tau^2(\cdot, \cdot)$ is defined by

$$\tau^2(r, \gamma) = \sum_{j=1}^{4} q(j, \gamma)\{p(j, r, \gamma) - 1/2\}^2,$$

$p(j, r_k, \gamma_k)$ is the probability of the putative QTL genotype QQ as defined in § 4.2.1 and Z_{kj}'s are normal random variables with mean 0, variance 1 and covariances

$$\mathrm{Cov}(Z_{ki}, Z_{lj}) = \begin{cases} 0, & \text{if } k = l, \\ P(X_k = i, X_l = j), & \text{if } k \neq l. \end{cases}$$

Here, X_k is the random code of the genotype combination of the flanking markers of the kth interval. If intervals k and l are not in the same chromosome,

$$P(X_k = i, X_l = j) = P(X_k = i)P(X_l = j) = q(i, \gamma_k)q(j, \gamma_l).$$

In the case that k and l are in the same chromosome, suppose that interval k is to the left of interval l and the recombination fraction between the right marker of interval k and the left marker of interval l is γ_{kl}. Then $P(X_k = i, X_l = j)$ can be determined from the within and between interval recombination fractions. For example,

$$P(X_k = 1, X_l = 1) = q(1, \gamma_k)(1 - \gamma_{kl})(1 - \gamma_l),$$
$$P(X_k = 1, X_l = 2) = q(1, \gamma_k)(1 - \gamma_{kl})\gamma_l.$$

The above asymptotic distribution is valid no matter whether the markers are sparse or dense. The problem is that the distribution of the random variable T in (4.18) does not have a closed form. However, since the random variable is completely specified, its distribution can be simulated. The simulation can be done quite straightforwardly. First, the covariance matrix of the normal random variables Z_{kj}'s is computed from the marker map. Then, Z_{kj}'s are generated repeatedly from the multivariate normal distribution with mean vector zero and the covariance matrix. for each generated set of Z_{kj}'s, T is computed through the maximization of each of its components since, for each k, r_k only appears in the kth component. Let T be simulated for a large number of times. The $1 - \alpha$ quantile of the empirical distribution of the simulated T values is used as the approximation to c_α. As demonstrated in the simulation study of [30], the simulated c_α provides a better approximation than both RGM and FBS approximations.

4.4.3   Permutation Test

The theoretical approximations in § 4.4.1 provide neat formulae for computing the threshold values. But they are obtained by assuming certain conditions, i.e., the density of the markers, the type of the crosses, etc.. The simulation method in § 4.4.2 is only for backcross designs. Though the same simulation method can be developed for intercross and RIL designs, such methods are not available yet. Furthermore, these methods are confined to interval mapping. Adaptation of the methods to composite interval mapping is possible but has not been done yet. A permutation test provides a general method for the determination of threshold values. It makes no assumption at all regarding the distribution of the quantitative trait, the designs, markers, the mapping procedure and so on. A permutation test is an old idea. It is first proposed by Fisher in [61]. The theory of a permutation test can be found in many textbooks, e.g. [115]. The use of a permutation test for the determination of threshold values in QTL mapping is first considered in [40]. We describe the method of a permutation test in this sub section.

The idea of a permutation test is to shuffle the trait values across the experiment subjects while the genetic maps of the subjects are retained (that is, their typed markers and marker genotypes are unchanged), and then apply whatever methods, i.e., single marker analysis, interval mapping or composite interval mapping, to the shuffled data.

Let the experiment subjects be indexed from 1 to n. A permutation of $\boldsymbol{n} =$

$(1, \ldots, n)$ is a random shuffle of the n indices. Denote by $s(\boldsymbol{n}) = (s_1, \ldots, s_n)$ the randomly permuted vector. For example, the following are two realizations of $s((1, 2, 3, 4, 5, 6, 7, 8, 9, 10))$:

$$(7, 4, 1, 3, 6, 2, 9, 10, 8, 5), \quad (1, 4, 2, 6, 5, 7, 9, 10, 3, 8).$$

In the first realization, $s_1 = 7, s_2 = 4$, and so on. Denote by $\{(y_i, x_{i1}, \ldots, x_{iM}) : i = 1, \ldots, n\}$ the original data, where y_i is the trait value and $(x_{i1}, \ldots, x_{iM})$ are the genotype data of the ith subject. Let $T(\boldsymbol{y}, d)$ be the test statistics based on the original data, e.g., $T(\boldsymbol{y}, d) = \text{LRT}(\boldsymbol{y}, d)$, the likelihood ratio statistic. Let $T_{\text{MAX}}(\boldsymbol{y}) = \sup_d T(\boldsymbol{y}, d)$. Denote by $\{(y_{s_i}, x_{i1}, \ldots, x_{iM}) : i = 1, \ldots, n\}$ the permuted data, and $T_{\text{MAX}}(\boldsymbol{y}_s)$ the maximum of the test statistic computed from the permuted data.

The procedure of the permutation test is simply as follows. First, permute the trait values and compute the test statistics based on the permuted data for a large number, say K, of times. Denote the maximum of the statistic obtained in the kth permutation by $T_{\text{MAX}}(\boldsymbol{y}_s^{(k)})$. Then take the empirical distribution of $\{T_{\text{MAX}}(\boldsymbol{y}_s^{(k)}) : k = 1, \ldots, K\}$ as the approximation to the distribution of $T_{\text{MAX}}(\boldsymbol{y})$. The overall threshold value at level α is approximated by the $(1 - \alpha)$th quantile of the empirical distribution of $\{T_{\text{MAX}}(\boldsymbol{y}_s^{(k)}) : k = 1, \ldots, K\}$.

The permutation test described above is valid for all the mapping procedures discussed in § 4.1 through 4.3. The following is a heuristic argument. The shuffling of the trait values across all the subjects effectively de-links the connection of the trait value with any specific locus whether it is a QTL or unlinked marker. If there are some QTL on the genome, the shuffled trait values follow a mixture distribution unrelated to any loci. If there is no QTL at all, the shuffled trait values follow a a single (non-mixture) distribution. In QTL mapping, we can consider two kinds of null hypotheses at a specific locus. The first one is that there is no QTL at all on the genome. If this is indeed the case, then the shuffling of the trait values does not change their distribution. The second one is that there are some QTL on the genome but the concerned locus is not a QTL or linked to a QTL. In fact, the second null hypothesis is more reasonable, since, in any QTL mapping experiment, the existence of QTL is already confirmed by some means. Thus, in the case of the existence of QTL, the distribution of shuffled trait values is exactly the distribution under the null hypothesis.

The last issue on the permutation test is how to determine K, the number of permutations. It is recommended in [40] that K should be at least 1,000 for $\alpha = 0.05$, and $K \geq 10,000$ might be needed for $\alpha = 0.01$. We provide a more rigorous method for the determination of K in the following. Note that $\{T_{\text{MAX}}(\boldsymbol{y}_s^{(k)}) : k = 1, \ldots, K\}$ can be considered as a random sample from the distribution of $T_{\text{MAX}}(\boldsymbol{y})$. As an approximation to the threshold value c_α, the $(1 - \alpha)$th quantile of the sample has an asymptotic variance $\nu^2 = \frac{\alpha(1-\alpha)}{K f^2(c_\alpha)}$ where $f(\cdot)$ is the density function of $T_{\text{MAX}}(\boldsymbol{y})$. Thus, to have an approximation with an asymptotic variance ν^2, K must be such that $K \approx \frac{\alpha(1-\alpha)}{\nu^2 f^2(c_\alpha)}$. This formula cannot be directly used for the computation of K since $f^2(c_\alpha)$ is unknown. An estimate of $f^2(c_\alpha)$ is needed. To get the estimate of $f^2(c_\alpha)$, we recommend the method below. First, use an initial K, i.e., the one recommended in [40], as given above. Then estimate f based on the simulated sample $\{T_{\text{MAX}}(\boldsymbol{y}_s^{(k)}) :$ $k = 1, \ldots, K\}$. Such an estimate, say $\hat{f}$, can be obtained by the method of kernel density estimation using the R function `density`, see [142] and [165]. Let $\hat{c}_\alpha$ be the estimate of c_α from this sample. Then estimate $f^2(c_\alpha)$ by $\hat{f}^2(\hat{c}_\alpha)$. Substituting this estimate into the formula for K. If the computed value is larger than the initial K then do more permutations to meet the requirement for K. Otherwise take the initial K as it is.

The methods presented in § 4.4.1 through 4.4.3 all have their pros and cons. The permutation test is a universal method. But it needs more computation than all the other methods including the simulation method in § 4.4.2. The simulation method provides more accurate approximation than the theoretical approximations in § 4.4.1. But the inroad is made at the cost of more computation. The theoretical approximations can be computed by a simple algorithm, which needs only a minimal amount of computation. More importantly, since the theoretical approximation does not depend on the sample size, it can be used in the determination of sample sizes which we discuss in the next section.

4.5 Determination of Sample Sizes

The threshold determined for a QTL mapping study controls the overall false discovery rate of the study. But controlling the false discovery rate is only one side of the coin in a QTL mapping study. To guarantee a certain power for detecting true QTL with a specified minimum effect is the other side. In order to achieve a certain power, a sufficient sample size is required. The required sample size depends on a number of factors, including the threshold value, the minimum effect one wishes to detect, the specified power, the non-genetic variance, the type of crosses, the marker map, etc. The determination of the required sample size is a complicated problem. We start our discussion with the ideal but impractical case that the QTL genotypes are known. Then we consider how the result for the ideal case can be adapted to the realistic QTL mapping studies.

4.5.1 The Case with Known QTL Genotypes

In the case of backcross design, when the genotypes of the QTL are known, both the likelihood approach and regression approach are equivalent to a two-sided z-test. Let n_{QQ} and n_{Qq} denote the numbers of individuals with genotype

QQ and Qq in a sample of size n respectively. Their expectation are the same and equal to $n/2$. Without loss of generality, let $n_{QQ} = n_{Qq} = n/2$. The test statistic of the z-test is given by

$$Z_0 = \frac{\bar{Y}_{QQ} - \bar{Y}_{Qq}}{s\sqrt{4/n}},\tag{4.19}$$

where $\bar{Y}_{QQ}$ and $\bar{Y}_{Qq}$ are, respectively, the average trait values of the individuals with genotype QQ and Qq, s^2 is the pooled estimate of the non-genetic variance σ^2 from the two subsamples. When $\Delta = \mu_{QQ} - \mu_{Qq} = 0$, Z_0 has an asymptotic standard normal distribution. Let T be the threshold value of the test, and Δ the minimum effect we wish to detect with a power β. The required n must satisfy

$$P(|Z_0| > T|\Delta) = \beta.$$

Replacing s by σ in (4.19) does not change the asymptotic distribution of Z_0. Hence, asymptotically,

$$
\begin{aligned}
&P(|Z_0| > T|\Delta)\\
=\ & P\left(\left|\frac{\bar{Y}_{QQ} - \bar{Y}_{Qq}}{\sigma\sqrt{4/n}}\right| > T\right)\\
=\ & P\left(\frac{\bar{Y}_{QQ} - \bar{Y}_{Qq}}{\sigma\sqrt{4/n}} > T\right) + P\left(\frac{\bar{Y}_{QQ} - \bar{Y}_{Qq}}{\sigma\sqrt{4/n}} < -T\right)\\
=\ & P\left(\frac{\bar{Y}_{QQ} - \bar{Y}_{Qq} - \Delta}{\sigma\sqrt{4/n}} > T - \frac{\Delta}{\sigma\sqrt{4/n}}\right)\\
& + P\left(\frac{\bar{Y}_{QQ} - \bar{Y}_{Qq} - \Delta}{\sigma\sqrt{4/n}} < -T - \frac{\Delta}{\sigma\sqrt{4/n}}\right)\\
=\ & P\left(Z > T - \frac{\Delta}{\sigma\sqrt{4/n}}\right) + P\left(Z < -T - \frac{\Delta}{\sigma\sqrt{4/n}}\right)\\
\approx\ & P\left(Z > T - \frac{\Delta}{\sigma\sqrt{4/n}}\right),
\end{aligned}
$$

where Z denotes the standard normal variable. Let z_β denote the $(1 - \beta)$-quantile of the standard normal distribution. Then the required n must satisfy

$$T - \frac{\Delta}{\sigma\sqrt{4/n}} = z_\beta,$$

i.e.,

$$n = (T - z_\beta)^2 \sigma^2 / (\Delta^2/4).$$

To determine n, one must have an estimate of σ^2 and specify both Δ and β.

The determination of sample size in the case of RIL design is the same as in the case of backcross design. One only needs to replace $\Delta = \mu_{QQ} - \mu_{Qq}$ by $\Delta = \mu_{QQ} - \mu_{qq}$.

Now consider the case of intercross design. In a sample of size n, the numbers of individuals with genotype QQ, Qq and qq, n_{QQ}, n_{Qq} and n_{qq}, have expectations $n/4, n/2$ and $n/4$ respectively. Without loss of generality, let these numbers equal their expectations. To test the null hypothesis $\mu_{QQ} = \mu_{Qq} = \mu_{qq}$, the Wald statistic, which is equivalent to the test in the QTL mapping analysis, is formed as follows. Let $Y = (\bar{Y}_{QQ}, \bar{Y}_{Qq}, \bar{Y}_{qq})^{\tau}$. The variance-covariance matrix is given by

$$\frac{\sigma^2}{n} \begin{pmatrix} 4 & 0 & 0 \\ 0 & 2 & 0 \\ 0 & 0 & 4 \end{pmatrix} = \frac{\sigma^2}{n} A, \text{ say.}$$

Let

$$C = \begin{pmatrix} 1 & -1 & 0 \\ 0 & 1 & -1 \end{pmatrix}.$$

The Wald statistic is given by

$$\chi^2 = n Y^{\tau} C^{\tau} (CAC^{\tau})^{-1} CY / s^2,$$

where s^2 is an estimate of σ^2. Let $\delta = \mu^{\tau} C^{\tau} (CAC^{\tau})^{-1} C\mu$ where $\mu = (\mu_{QQ}, \mu_{Qq}, \mu_{qq})^{\tau}$. Asymptotically, the Wald statistic has a non-central χ^2-distribution with degrees of freedom 2 and non-centrality parameter $\Delta = n\delta/\sigma^2$. When the null hypothesis is true, i.e., $\delta = 0$, the asymptotic distribution is a central χ^2-distribution. Let T be the threshold value, δ the minimum relative effect which one wishes to detect, and β the specified power. The required sample size n is determined by

$$P(n Y^{\tau} C^{\tau} (CAC^{\tau})^{-1} CY / s^2 \geq T | \delta) = \beta. \tag{4.20}$$

Unlike the case of z-test, there is no explicit form for the required n. The left hand side of (4.20) for a given n, δ and σ^2, can be computed using the R-function pchisq. By using this function, the required n can be computed by the R function defined as follows:

```
chisq.sample.size = function(T, sigma, delta, beta, df, n.start) {
  # Input
  # T:  threshold value;
  # sigma: error standard deviation;
  # delta: as defined in the text;
  # beta: power of the test at assumed Delta;
  # df: degrees of freedom of the chi-square distribution;
  # n.start: a starting value for the sample size.
  #
  # Output
```

```
# n:   required sample size.

#determination of the bounds of n
  flag1 = 0
  flag2 = 0
  while (flag1 == 0 | flag2 == 0) {
     n = n.start
     D = (n * delta)/sigma^2
     power = 1 - pchisq(T, df, D)
     if (power >= beta) {
        n.u = n
        flag1 = 1
        n.start = floor(n*0.618)
     } else {
        n.l = n
        flag2 = 1
        n.start = floor(n/0.618)
     }
  }
}
# search for the required n
  flag3 = 0
  while (flag3 == 0) {
     n.tmp = floor( n.l + (n.u-n.l)*0.618 )
     n = n.tmp
     D = (n * delta)/sigma^2
     F1 = qf(1-alpha, df1, df2)
     power = 1 - pchisq(T, df, D)
     if (power >= beta) {
        n.u = n.tmp
     } else {
        n.l = n.tmp
     }
     if (n.u-n.l <= 1 ) flag3 = 1
  }
  n = n.tmp
  n
}
```

4.5.2   The Case of Interval Mapping

In this sub section, we focus on backcross design by using an approximate approach. The RIL design can be treated similarly. For intercross design, in principle, a method similar to the Wald test can be developed. But the exact development is too complicated to be necessary. We will recommend an ad hoc rule for intercross design after we derive the formula for backcross and RIL designs.

Before we tackle the problem with interval mapping, let us briefly consider the case of mapping by single markers. In this case, the individuals in a sample are divided into two groups by the marker genotypes rather than the QTL genotypes. If we take the two groups as if they have the same variance (the variances of the two groups are in fact different), then the case reduces to that of known QTL genotypes, except that the difference between the two expected group means is not the same. As argued in § 4.1, the difference between the two groups is $\mu_{MM} - \mu_{Mm} = (1 - 2\theta)(\mu_{QQ} - \mu_{Qq})$, where θ is the recombination fraction between the QTL and the marker. By the same argument as in § 4.5.1, the sample size required to detect the QTL with minimum effect $\Delta = \mu_{QQ} - \mu_{Qq}$ is given by

$$n = (T - z_\beta)^2 \sigma^2 / [(1 - 2\theta)^2 \Delta^2 / 4] = n_0 / (1 - 2\theta)^2,$$

where n_0 is the required sample size in the case of known QTL genotypes. That is, the required sample size for single marker analysis is increased by a factor $1/(1 - 2\theta)^2$. In the sample size determination, the θ can be taken as the average recombination fraction between the middle point of an interval to either of its flanking markers in the marker map.

Let us turn to the interval mapping with backcross design now. Let γ be the average recombination fraction between the two flanking markers of an interval in the marker map. In the case of backcross design, a sample of size n is divided into four groups by the genotypes of the two flanking markers with the following expected proportions:

Genotype	LL-UU	LL-Uu	Ll-UU	Ll-Uu
Proportion	$\frac{1-\gamma}{2}$	$\frac{\gamma}{2}$	$\frac{\gamma}{2}$	$\frac{1-\gamma}{2}$

Since γ is small (for instance, $\gamma = 0.09$ when the interval length is 10 cM), the individuals with recombined markers only count for a small portion. We can approximate the interval mapping by ignoring the individuals with recombined markers. The problem is then reduced to the comparison of the two groups without marker recombination, i.e., the groups with marker genotypes LL-UU and Ll-Uu. Assume that the putative QTL is at the middle point of the interval. Let θ denote the recombination fraction between the middle point and either one of the flanking markers. By using Table 3.4, we have that the expected difference of means between LL-UU and Ll-Uu is given by

$$\left[\frac{(1 - \theta)^2}{1 - \gamma} - \frac{\theta^2}{1 - \gamma} \right] (\mu_{QQ} - \mu_{Qq})$$
$$= \frac{1 - 2\theta}{1 - \gamma} \Delta.$$

Note that the effective sample size for the two groups is $(1 - \gamma)n$. The same argument as that for the single marker case yields that the required sample size is determined by

$$(1 - \gamma)n = (T - z_\beta)^2 \sigma^2 / \left[\left(\frac{1 - 2\theta}{1 - \gamma} \right)^2 \Delta^2/4 \right],$$

i.e.,

$$n = \frac{1 - \gamma}{(1 - 2\theta)^2} n_0. \tag{4.21}$$

This formula shows that, to achieve the same power, the required sample size for interval mapping is less than that required for single marker analysis by a factor $1 - \gamma$. Hence, interval mapping is more efficient than single marker analysis.

For RIL design, a similar argument using Table 3.6 yields that the required sample size is given by

$$n = \frac{1 - \Gamma}{(1 - 2\Theta)^2} n_0,$$

where Γ is the average zegote recombination fraction between blanking markers of the intervals and Θ is the zygote recombination fraction between the middle point of the interval and either of its flanking markers.

For intercross design, we recommmend the following ad hoc rule for the determination of sample size. First, using the R function given in § 4.5.1 to derive the required sample size n_0 for the case of known QTL genotypes. Then, multiply n_0 by the factor $\frac{1-\gamma}{(1-2\theta)^2}$ to get the required sample size.

The same formulae can be used for composite interval mapping. The only difference which might arise is in the estimation of σ^2. In the case of interval mapping, the QTL effect under consideration is not adjusted for the effect of other possible QTL. The σ^2 consists of not only the variation caused by non-genetic random factors but also those caused by other possible QTL. In the case of composite interval mapping, since the effect of other possible QTL has been adjusted for, the σ^2 is essentially the variance of non-genetic random factors. Hence the estimate of σ^2 in the case of composite interval mapping should be smaller than that in the case of interval mapping.

4.6 Selective Genotyping

In this section, we consider the strategy of *selective genotyping* to increase the power of QTL mapping. The strategy is applicable when the cost of growing experimental progeny is less than the cost of genotyping. It is observed in [112] that more linkage information is provided by the progeny whose trait values deviate most from the population mean. Based on this fact, several selective genotyping methods have been proposed in the literature.

A truncation selection (TS) approach is proposed in [169]. The TS approach draws a sample from individuals whose trait values exceed a predetermined threshold and draws another sample at random from the population. Only the individuals in the samples are genotyped. The two samples are then compared in the QTL mapping analysis. The TS approach considers individuals at only one extreme.

The TS approach is extended in [34] to an approach dubbed as TS-II that considers individuals at both extremes. In addition to drawing a sample from individuals whose trait values exceed a pre-determined threshold, the TS-II approach draws another sample from individuals whose trait values fall below another pre-determined threshold, and compares these two samples. The ideal threshold values are the upper and lower quantiles of the trait distribution determined by a pre-determined intensity of selection. For example, if the intensity of selection is 30%, then the upper and lower quantiles are respectively the 0.85 and 0.15 quantiles of the distribution. With the TS approaches, a prescreening process is necessary for the estimation of the cutoff quantiles if they are not known a priori, which is usually the case in practice. The selection procedure can only be performed after the estimated cutoff quantiles are obtained. In certain cases, this causes difficulty in the implementation of the TS approaches.

To ease the difficulty of TS approaches, an extreme rank selection (ERS) approach is proposed in [34]. The ERS approach is as follows. Let k be a specified integer. For each selection, k individuals are chosen at random from the population. The trait values of these k individuals are measured and ordered from the smallest to the largest, the individual with rank 1 is selected as a member of the lower sample and the individual with rank k is selected as a member of the upper sample. Then the lower sample and upper sample are compared. The integer k is determined such that $2/k$ equals to the intensity of selection.

The selected samples can be analyzed in different ways. The first way is to treat one sample as a sample of cases and the other as a sample of controls, then ignore the trait values and analyze the allele frequencies of the putative QTL in the two samples the same as in case-control studies, see, e.g., [169], [34]. Another way is to combine the two samples together and treat the combined sample as if it is a random sample, then apply likelihood analysis or regression analysis, see, e.g., [203], [192]. In other words, the methods discussed in this chapter can be used to analyze the combined sample. Since the second way makes use of all the information in the samples, while the first way ignores the information contained in the trait values, the second way of analysis is in general more efficient than the first way.

It has been demonstrated by simulation studies in [34] and [192] that, among the three selective genotyping approaches, TS-II and ERS are more powerful than the one-directional TS, TS-II is slightly better than ERS, but ERS is much easier to implement and is still comparable with TS-II in terms of power.

Now, we take backcross design as an example to show how selective genotyping improves efficiency. We only consider ERS. The argument for other selective genotyping approaches is similar. Suppose that there is only one QTL with genotypes QQ and Qq. Without loss of generality, assume that the quantitative trait Y follows the model:

$$Y = -c\delta_1 + c\delta_2 + \epsilon,$$

where

$$\delta_1 = \begin{cases} 1, & \text{if the genotype is Qq,} \\ 0, & \text{otherwise,} \end{cases}$$

$$\delta_2 = \begin{cases} 1, & \text{if the genotype is QQ,} \\ 0, & \text{otherwise,} \end{cases}$$

and $\epsilon \sim N(0, \sigma^2)$. This is equivalent to the assumption that the genotypic values are $-c$ and c when the genotypes of the QTL are Qq and QQ respectively. Denote by Φ the cumulative distribution function (CDF) of ϵ. Let F_1 and F_2 be the CDF of Y given genotype Qq and QQ respectively, and f_1, f_2 their corresponding PDF. Then $F_1(y) = \Phi(y + c)$ and $F_2(y) = \Phi(x - c)$. Consider the random vector (Y, δ_1, δ_2) (note that the δ's are unobserved). The joint PDF of (Y, δ_1, δ_2) is given by

$$g(y, \delta_1, \delta_2) = q\delta_1 f_1(y) + p\delta_2 f_2(y),$$

where p is the probability that a random individual from the backcross population has genotype QQ and $q = 1 - p$. In fact, $p = q = 1/2$ in the case of backcross design. The marginal PDF of Y is given by

$$f(y) = qf_1(y) + pf_2(y).$$

The corresponding CDF is given by $F(y) = qF_1(y) + pF_2(y)$. The conditional distribution of $\delta_l, l = 1, 2$, given Y is as follows:

$$P(\delta_1 = 1|Y = y) = \frac{qf_1(y)}{f(y)}, \quad P(\delta_2 = 1|Y = y) = \frac{pf_2(y)}{f(y)}.$$

In the ERS scheme, random sets of k individuals are taken from the experiment population. The individuals with the smallest and the largest trait values are selected to form the lower and upper samples. Let $Y_{(1)}$ and $Y_{(k)}$ denote the smallest and the largest trait values from the same random set. Indeed, they are, respectively, the smallest and the largest order statistics of a random sample of size k. Let $p_{(1)}$ and $p_{(k)}$ denote, respectively, the probability that an individual has genotype QQ in the lower sample and the upper sample. Denote $q_{(r)} = 1 - p_{(r)}, r = 1, k$. We have

$$
\begin{aligned}
p_{(k)} &= P(\delta_{2(k)} = 1) = E[E(P(\delta_{2(k)} = 1|Y_{(k)}))] \\
&= E\frac{pf_2(Y_{(k)})}{f(Y_{(k)})} \\
&= \int \frac{pf_2(y)}{f(y)} kF^{k-1}(y)f(y)dy \\
&= p\int kF^{k-1}(y)[f(y) + q(f_2(y) - f_1(y))]dy \\
&= p + pq\int kF^{k-1}(y)[f_2(y) - f_1(x)]dy, \\
&= p + \Delta, \text{ say.}
\end{aligned}
$$

Since $F_1(y) > F_2(y)$ and $F^{k-1}(y)$ is increasing, we have

$$
\begin{aligned}
&\int F^{k-1}(y)[f_2(y) - f_1(y)]dy \\
&= \int F^{k-1}(y)dF_2(y) - \int F^{k-1}(y)dF_1(y) \\
&= \int_0^1 [F^{k-1}(F_2^{-1}(x)) - F^{k-1}(F_1^{-1}(x))]dx \\
&> \quad 0, \text{ since } F_1(y) > F_2(y) \text{ and hence } F_1^{-1}(x) \leq F_2^{-1}(x).
\end{aligned}
$$

That is, $\Delta > 0$. By the relationship of $p_{(k)}$ with $q_{(k)}$ and an argument of symmetry, we obtain

$$
q_{(k)} = q - \Delta, \quad p_{(1)} = p - \Delta, \quad q_{(1)} = q + \Delta.
$$

Denote by μ_1 and μ_2 the means of Y given genotype Qq and QQ respectively in the backcross population. Let $\mu_{(k)1}, \mu_{(k)2}$ and $\mu_{(1)1}, \mu_{(1)2}$ denote the corresponding means of $Y_{(k)}$ and $Y_{(1)}$ respectively. Mix $Y_{(1)}$ and $Y_{(k)}$ together. The mean of the trait value given genotype Qq in the mixed population is given by

$$
\tilde{\mu}_1 = \frac{q_{(1)}}{q_{(1)} + q_{(k)}}\mu_{(1)1} + \frac{q_{(k)}}{q_{(1)} + q_{(k)}}\mu_{(k)1},
$$

and that given genotype QQ is given by

$$
\tilde{\mu}_2 = \frac{p_{(1)}}{p_{(1)} + p_{(k)}}\mu_{(1)2} + \frac{p_{(k)}}{p_{(1)} + p_{(k)}}\mu_{(k)2}.
$$

It follows from symmetry that

$$
\frac{1}{2}[\mu_{(k)1} + \mu_{(1)1}] = \mu_1, \quad \frac{1}{2}[\mu_{(k)2} + \mu_{(1)2}] = \mu_2.
$$

Since $p_{(1)} < p_{(k)}$ and $q_{(1)} > q_{(k)}$, it is easy to see that $\tilde{\mu}_1 < \mu_1$ and $\tilde{\mu}_2 > \mu_2$. Hence

$$
\tilde{\mu}_2 - \tilde{\mu}_1 > \mu_2 - \mu_1.
$$

This indicates that the signal in the mixed selected population is larger than the signal in the original population. Furthermore, we can derive that the variance in the mixed selected population is smaller than that in the original population, which we omit to avoid unnecessary technical details. In summary, the selective genotyping using ERS results in a larger signal-to-noise ratio, $|\Delta|/\sigma$, than that without selective genotyping. We have seen in § 4.5 that the larger the signal-to-noise ratio the less sample size is needed to achieve the same power. Hence, selective genotyping is more efficient than non-selective genotyping if the cost of growing the experiment subjects can be ignored compared with the cost of genotyping.

Chapter 5

Multiple Interval Mapping

The one-dimensional QTL mapping approaches discussed in Chapter 4 are simple and easy to implement. But those approaches are based on single-QTL models; that is, the approaches assume that there is at most one QTL in the genome, which is unrealistic. Usually, a quantitative trait is affected by multiple QTL. By ignoring the presence of multiple QTL, the power of one-dimensional QTL mapping approaches for detecting true QTL is compromised. Furthermore, the effects of the multiple QTL are not necessarily additive. Interaction effects among multiple QTL might be present. One-dimensional mapping approaches are unable to accommodate interaction effects. More reasonable alternatives are based on multiple-QTL models. In this chapter, we consider one such approach — multiple interval mapping.

The multiple interval mapping procedure mimics the traditional stepwise regression. However, the models involved in multiple interval mapping are mixtures of linear or generalized linear models. We first consider the case of Gaussian traits, i.e., traits with normal distributions. Then we extend the consideration to the general case where traits follow exponential family distributions. In § 5.1, we describe the Gaussian mixture models for mapping Gaussian traits. An EM algorithm for the computation of Gaussian mixture models is described in § 5.2. In § 5.3, the mixture generalized linear models for traits with exponential family distributions are formulated. A general EM algorithm for the computation of the mixture generalized linear models is presented in § 5.4. A general multiple interval mapping procedure is given in § 5.5. The application of the multiple interval mapping procedure to a real data example is provided in § 5.6. Two special types of traits: categorical traits and traits with a spike distribution, are treated in § 5.7 and § 5.8.

The multiple interval mapping approach for Gaussian traits is first proposed in [100] and [99]. It was extended in [33] to generalized linear models for traits with exponential family distributions. EM algorithms for the computation of mixture models in multiple interval mapping are developed in [98], [29] and [33]. Multiple interval mapping for categorical traits is treated in [117] and [32]. Multiple interval mapping for traits with a spike distribution is studied in [118].

5.1   Gaussian Mixture Models for QTL Mapping

The data for multiple interval mapping is the same as that for single interval mapping. It contains, for each individual, the trait value and the genotypes at markers spread over the genome. Suppose that the markers divide the genome into M non-overlapping intervals. For a generic individual, Let Y be the quantitative trait value, x_j be the code of the marker genotype combination of interval j, δ_j and $(\delta_{j1}, \delta_{j2})$ be the dummy variables for the putative QTL gentotypes on interval j, respectively, for backcross (or RIL) design and intercross design. For their definition, see § 4.2.1. Later, for a particular individual i, these notations are augmented by the index i. Consider the model with m $(1 \leq m \leq M)$ intervals. We first describe the model for backcross design and then for intercross design. The model for RIL design is the same as that for backcross design except that the allele recombination fraction in backcross design is replaced by zygote recombination fraction in RIL design.

Gaussian mixture models with backcross design. It is conventional to assume that there is at most one QTL on each interval. If QTLs on the m intervals act independently, we have an additive genetic model:

$$
\begin{aligned}
Y &= \mu(\delta_1, \ldots, \delta_m) + \epsilon \\
&= \beta_0 + \sum_{j=1}^{m} \beta_j \delta_j + \epsilon,
\end{aligned}
\tag{5.1}
$$

where $\epsilon \sim N(0, \sigma^2)$ and β_j is the effect of the QTL on inteval j. If interactions among the QTLs exist, they can be incorporated into the above model by including product terms of the δ_j's. The model with main effects and two-factor interactions is given below.

$$
\begin{aligned}
Y &= \mu(\delta_1, \ldots, \delta_m) + \epsilon \\
&= \beta_0 + \sum_{j=1}^{m} \beta_j \delta_j + \sum_{j,k} \xi_{jk} \delta_j \delta_k + \epsilon.
\end{aligned}
\tag{5.2}
$$

The δ_j's are unobserved random variables. As argued in § 4.2.1, each δ_j follows a Bernoulli distribution. Furthermore, given $(x_1, \ldots, x_m)$, the marker genotype combinations of the m intervals, $\delta_1, \ldots, \delta_m$ are independent. Let γ_j and r_j denote, respectively, the recombination fractions between the two flanking markers and between the left marker and the putative QTL for interval j. The joint probability density of the δ_j's is

$$
\prod_{j=1}^{m} p(x_j, r_j, \gamma_j)^{\delta_j} [1 - p(x_j, r_j, \gamma_j)]^{1-\delta_j},
$$

where $p(x_j, r_j, \gamma_j)$ is the probability of genotype $Q_j Q_j$ given x_j; see Table 3.4. The joint density of $(Y, \delta_1, \ldots, \delta_m)$ given $(x_1, \ldots, x_m)$ is

$$f(y, \delta_1, \ldots, \delta_m | \boldsymbol{\theta}) \tag{5.3}$$
$$= \prod_{j=1}^{m} p(x_j, r_j, \gamma_j)^{\delta_j} [1 - p(x_j, r_j, \gamma_j)]^{1-\delta_j} \phi(y, \mu(\delta_1, \ldots, \delta_m), \sigma^2),$$

where $\boldsymbol{\theta}$ is the totality of unknown parameters containing the QTL location parameters r_j's and effect parameters β_j's (as well as ξ_{jk}'s when interactions are considered). It suffices to consider this conditional joint density, since the distribution of $(x_1, \ldots, x_m)$ does not involve any unknown parameters.

Finally, we arrive at the (marginal) probability density of Y:

$$f(y | \boldsymbol{\theta}) \tag{5.4}$$
$$= \sum_{(k_1, \ldots, k_m)} \prod_{j=1}^{m} p(x_j, r_j, \gamma_j)^{k_j} [1 - p(x_j, r_j, \gamma_j)]^{1-k_j} \phi(y, \mu(k_1, \ldots, k_m), \sigma^2),$$

where the sum is over the set $\{(k_1, \cdots, k_m) : k_j = 0 \text{ or } 1, j = 1, \ldots, m\}$. The density $f(y | \boldsymbol{\theta})$ is a mixture of 2^m components of normal densities.

Thus, with observations $\{(y_i, x_{i1}, \ldots, x_{im}) : i = 1, \ldots, n\}$, the likelihood function for the model with the m intervals is given by

$$L(\boldsymbol{\theta} | \boldsymbol{y}) = \prod_{i=1}^{n} f(y_i | \boldsymbol{\theta})$$
$$= \prod_{i=1}^{n} \{ \sum_{(k_1, \ldots, k_m)} \prod_{j=1}^{m} p(x_{ij}, r_j, \gamma_j)^{k_j} [1 - p(x_{ij}, r_j, \gamma_j)]^{1-k_j}$$
$$\times \phi(y_i, \mu(k_1, \ldots, k_m), \sigma^2) \}. \tag{5.5}$$

This likelihood function is the core in multiple interval mapping.

Gaussian mixture models with intercross design. We now turn to the case of intercross design. The additive genetic model with m intervals now has the form:

$$Y = \mu(\boldsymbol{\delta}_1, \ldots, \boldsymbol{\delta}_m) + \epsilon$$
$$= \beta_0 + \sum_{j=1}^{m} (\beta_{1j} \delta_{1j} + \beta_{2j} \delta_{2j}) + \epsilon, \tag{5.6}$$

where $\boldsymbol{\delta}_j = (\delta_{1j}, \delta_{2j})^\tau$. The model with two-factor interactions is given by

$$Y = \mu(\boldsymbol{\delta}_1, \ldots, \boldsymbol{\delta}_m) + \epsilon \tag{5.7}$$
$$= \beta_0 + \sum_{j=1}^{m} (\beta_{1j} \delta_{1j} + \beta_{2j} \delta_{2j})$$
$$+ \sum_{j,k} (\xi_{11jk} \delta_{1j} \delta_{1k} + \xi_{12jk} \delta_{1j} \delta_{2k} + \xi_{21jk} \delta_{2j} \delta_{1k} + \xi_{22jk} \delta_{2j} \delta_{2k}) + \epsilon.$$

Given $(x_1, \ldots, x_m)$, the $\boldsymbol{\delta}_j$'s are independent. Each of them has a trinomial distribution. Their joint probability density is

$$\prod_{j=1}^{m} p_0(x_j, r_j, \gamma_j)^{1-\delta_{1j}-\delta_{2j}} p_1(x_j, r_j, \gamma_j)^{\delta_{1j}} p_2(x_j, r_j, \gamma_j)^{\delta_{2j}}$$

where $p_0(x_j, r_j, \gamma_j) = 1 - p_1(x_j, r_j, \gamma_j) - p_2(x_j, r_j, \gamma_j)$. The joint probability density of $(Y, \boldsymbol{\delta}_1, \ldots, \boldsymbol{\delta}_m)$ given $(x_1, \ldots, x_m)$ is

$$f(y, \boldsymbol{\delta}_1, \ldots, \boldsymbol{\delta}_m | \boldsymbol{\theta}) \tag{5.8}$$

$$= \prod_{j=1}^{m} p_0(x_j, r_j, \gamma_j)^{1-\delta_{1j}-\delta_{2j}} p_1(x_j, r_j, \gamma_j)^{\delta_{1j}} p_2(x_j, r_j, \gamma_j)^{\delta_{2j}}$$

$$\times \phi(y, \mu(\boldsymbol{\delta}_1, \ldots, \boldsymbol{\delta}_m), \sigma^2).$$

The marginal probability density of Y is then given by

$$f(y | \boldsymbol{\theta}) \tag{5.9}$$

$$= \sum_{(\boldsymbol{k}_1, \ldots, \boldsymbol{k}_m)} \{ \prod_{j=1}^{m} p_0(x_j, r_j, \gamma_j)^{1-k_{1j}-k_{2j}} p_1(x_j, r_j, \gamma_j)^{k_{1j}} p_2(x_j, r_j, \gamma_j)^{k_{2j}}$$

$$\times \phi(y, \mu(\boldsymbol{k}_1, \ldots, \boldsymbol{k}_m), \sigma^2) \},$$

where the sum is over the set $\{(\boldsymbol{k}_1, \cdots, \boldsymbol{k}_m) : \boldsymbol{k}_j = (1,0)^\tau \text{ or } (0,1)^\tau, j = 1, \ldots, m\}$. The density $f(y | \boldsymbol{\theta})$ is a mixture of 3^m components of normal densities. The likelihood function for the model with m intervals and observations $\{(y_i, x_{i1}, \ldots, x_{im}) : i = 1, \ldots, n\}$ is given by

$$L(\boldsymbol{\theta} | \boldsymbol{y}) \tag{5.10}$$

$$= \prod_{i=1}^{n} \{ \sum_{(\boldsymbol{k}_1, \ldots, \boldsymbol{k}_m)} \prod_{j=1}^{m} p_0(x_{ij}, r_j, \gamma_j)^{1-k_{1j}-k_{2j}} p_1(x_{ij}, r_j, \gamma_j)^{k_{1j}} p_2(x_{ij}, r_j, \gamma_j)^{k_{2j}}$$

$$\times \phi(y_i, \mu(\boldsymbol{k}_1, \ldots, \boldsymbol{k}_m), \sigma^2) \}.$$

In multiple interval mapping, for each model, we need to obtain the MLE of the parameter $\boldsymbol{\theta}$ (including components of QTL effects and QTL positions) based on (5.5) or (5.10). The computation for obtaining the MLE is handled by the EM algorithms considered in the next and later sections.

5.2 An EM algorithm for Gaussian Mixture Models

The first EM algorithm for the Gaussian mixture models in multiple interval mapping is developed in [98]. However this EM algorithm is not a full algorithm for the computation of the MLEs of mixture Gaussian models. It only computes the MLEs of the QTL effects, i.e., the $\boldsymbol{\beta}$ parameter, while the QTL positions, i.e., the $\boldsymbol{r}$ parameter, are fixed. This algorithm entails a multi-dimensional search over the QTL positions. A search strategy is prescribed in [98], but the strategy is not an accurate multi-dimensional search. An accurate multi-dimensional search is computationally intractable. This motivated the development of a full EM algorithm for the simultaneous computation of both the QTL effects and QTL positions in [29]. In this section, we describe the EM algorithm developed in [29] but focus on the case of backcross design. The general case will be dealt with when we consider the EM algorithm for mixture generalized linear models in § 5.4.

To maximize (5.5), the EM algorithm treats the unobserved QTL geno-types as missing data and makes use of the likelihood function based on $\mathcal{T} = \{(y_i; \delta_{i1}, \ldots, \delta_{im}) : i = 1, \ldots, n\}$ as if we have observed $\{(\delta_{i1}, \ldots, \delta_{im}) : i = 1, \ldots, n\}$. This likelihood function is given as follows.

$$
\begin{aligned}
&L(\boldsymbol{\theta}|\mathcal{T}) \hspace{6cm} (5.11)\\
&= \prod_{i=1}^{n} \prod_{j=1}^{m} p(x_{ij}, r_j, \gamma_j)^{\delta_{ij}} [1 - p(x_{ij}, r_j, \gamma_j)]^{1-\delta_{ij}} \phi(y_i, \mu(\delta_{i1}, \ldots, \delta_{im}), \sigma^2).
\end{aligned}
$$

The EM algorithm iterates between an E-step and an M-step. The E-step computes the conditional expectation of $\ell(\boldsymbol{\theta}|\mathcal{T}) = \ln L(\boldsymbol{\theta}|\mathcal{T})$ conditioning on $\{y_i; i = 1, \ldots, n\}$. The M-step maximizes the conditional expectation to update the estimated $\boldsymbol{\theta}$.

Let $\boldsymbol{\delta}_i = (\delta_{i1}, \ldots, \delta_{im})^\tau$. Then

$$
\mu(\delta_{i1}, \ldots, \delta_{im}) \equiv \mu(\boldsymbol{\delta}_i) = \beta_0 + \boldsymbol{\delta}_i^\tau \boldsymbol{\beta},
$$

where $\boldsymbol{\beta} = (\beta_1, \ldots, \beta_m)^\tau$ is the vector of QTL effects. Let $\boldsymbol{y} = (y_1, \ldots, y_n)^\tau$, $Z^\tau = (\boldsymbol{\delta}_1, \ldots, \boldsymbol{\delta}_n)$ and $\mathbf{1}$ be a vector of elements 1. We can write

$$
\ell(\boldsymbol{\theta}|\mathcal{T}) = \ell_1(\boldsymbol{\theta}|\mathcal{T}) + \ell_2(\boldsymbol{\theta}|\mathcal{T}),
$$

where

$$
\ell_1(\boldsymbol{\theta}|\mathcal{T}) = \sum_{i=1}^{n} \sum_{j=1}^{m} \left[\delta_{ij} \ln\left(\frac{p_j(x_{ij}, r_j, \gamma_j)}{1 - p_j(x_{ij}, r_j, \gamma_j)}\right) + \ln(1 - p_j(x_{ij}, r_j, \gamma_j)) \right],
$$

and

$$
\begin{aligned}
&\ell_2(\boldsymbol{\theta}|\mathcal{T}) \\
&= \sum_{i=1}^{n} \ln \phi(y_i, \beta_0 + \boldsymbol{\delta}_i^\tau \boldsymbol{\beta}, \sigma^2) \\
&= -\frac{1}{2\sigma^2}[(\boldsymbol{y} - \mathbf{1}\beta_0)^\tau (\boldsymbol{y} - \mathbf{1}\beta_0) - 2(\boldsymbol{y} - \mathbf{1}\beta_0)^\tau Z\boldsymbol{\beta} + \boldsymbol{\beta}^\tau Z^\tau Z\boldsymbol{\beta}] - \frac{n}{2} \ln(2\pi\sigma^2).
\end{aligned}
$$

The above expression has two implications: (1) in the E-step, the computation of the conditional expectation of $\ell(\boldsymbol{\theta}|\mathcal{T})$ is reduced to the computation of the conditional expectations of δ_{ij}'s and their products which appear in $Z^\tau Z$; (2) in the M-step, the maximization of $\ell(\boldsymbol{\theta}|\mathcal{T})$ can be broken into two separate sub problems — the maximization of $\ell_1(\boldsymbol{\theta}|\mathcal{T})$ with respect to the r_j's (the QTL location parameters) and the maximization of $\ell_2(\boldsymbol{\theta}|\mathcal{T})$ with respect to the β_j's (the QTL effect parameters) and σ^2. The following are the details of the E-step and M-step.

E-step: Let $\boldsymbol{\theta}^{(0)}$ be the most updated value of $\boldsymbol{\theta}$ up to the current step. By using Bayes' formula, the conditional expectation of δ_{ij} at the current step is computed as

$$E(\delta_{ij}|y_i, \boldsymbol{\theta}^{(0)}) = \frac{p(x_{ij}, r_j^{(0)}, \gamma_j)E_{-j}}{f(y_i|\boldsymbol{\theta}^{(0)})}, \tag{5.12}$$

where the form of $f(y_i|\boldsymbol{\theta}^{(0)})$ is given by (5.4) and

$$E_{-j} = \sum_{\{k_1\cdots k_m\}/\{k_j\}} \{\prod_{h\neq j}[p(x_{ih}, r_h^{(0)}, \gamma_h)]^{k_h}[1 - p(x_{ih}, r_h^{(0)}, \gamma_h)]^{1-k_h}$$
$$\times \phi(y_i, \beta_0^{(0)} + \beta_j^{(0)} + \boldsymbol{k}_{-j}^\tau \boldsymbol{\beta}_{-j}^{(0)} \sigma^{2(0)})\}.$$

Here $\boldsymbol{k}$ denotes the vector $(k_1, \ldots, k_m)^\tau$. For any vector $\boldsymbol{v}$, the notation $\boldsymbol{v}_{-j}$ denotes the vector derived from $\boldsymbol{v}$ by deleting its jth component.

Similarly, the conditional expectation of $\delta_{ij}\delta_{il}$ is computed as

$$E(\delta_{ij}\delta_{il}|y_i, \boldsymbol{\theta}^{(0)}) = \frac{p(x_{ij}, r_j^{(0)}, \gamma_j)p(x_{il}, r_l^{(0)}, \gamma_l)E_{-(j,l)}}{f(y_i|\boldsymbol{\theta}^{(0)})} \tag{5.13}$$

with

$$E_{-(j,l)} = \sum_{\{k_1\cdots k_m\}/\{k_j, k_l\}} \{\prod_{h\neq j,l}[p(x_{ih}, r_h^{(0)}, \gamma_h)]^{k_h}[1 - p(x_{ih}, r_h^{(0)}, \gamma_h)]^{1-k_h}$$
$$\times \phi(y_i, \beta_0^{(0)} + \beta_j^{(0)} + \beta_l^{(0)} + \boldsymbol{k}_{-(j,l)}^\tau \boldsymbol{\beta}_{-(j,l)}^{(0)} \sigma^{2(0)})\}.$$

The matrix $Z^\tau Z$ can be written as

$$Z^\tau Z = \sum_{i=1}^n \boldsymbol{\delta}_i \boldsymbol{\delta}_i^\tau \equiv \sum_{i=1}^n \Delta_i, \text{ say.}$$

The jth diagonal element of Δ_i is $\delta_{ij}^2 = \delta_{ij}$ and the (j, l)th element is $\delta_{ij}\delta_{il}$. By using formulas (5.12) and (5.13), we obtain the conditional expectations of δ_{ij}, Z and $Z'Z$ and denote them, respectively, by $\tilde{\delta}_{ij}$, $\tilde{Z}$ and $\tilde{\Delta}$. Note that $\tilde{\Delta} \neq \tilde{Z}^\tau \tilde{Z}$. By replacing δ_{ij}, Z and $Z'Z$, respectively, with $\tilde{\delta}_{ij}$, $\tilde{Z}$ and $\tilde{\Delta}$ in $\ell(\boldsymbol{\theta}|\mathcal{T})$, we obtain the conditional expectation of the log likelihood based on $\mathcal{T}$, which is to be maximized in the M-step.

M-step: Denote the conditional expectation of $\ell(\boldsymbol{\theta}|\mathcal{T})$ by $\tilde{\ell}(\boldsymbol{\theta}|\mathcal{T})$. As commented earlier, the maximization of $\tilde{\ell}(\boldsymbol{\theta}|\mathcal{T})$ can be broken into the maximizations of $\tilde{\ell}_1(\boldsymbol{\theta}|\mathcal{T})$ and $\tilde{\ell}_2(\boldsymbol{\theta}|\mathcal{T})$ separately. Maximizing $\tilde{\ell}_1(\boldsymbol{\theta}|\mathcal{T})$ with respect to the r_j's is equivalent to solving the following equations

$$\sum_{l=1}^{4} \frac{p'(l, r_j, \gamma_j)}{p(l, r_j, \gamma_j)[1 - p(l, r_j, \gamma_j)]} \sum_{i:x_{ij}=l} [\tilde{\delta}_{ij} - p(l, r_j, \gamma_j)] = 0, \quad (5.14)$$
$$j = 1, \ldots, m,$$

where $p'(l, r_j, \gamma_j)$ is the first derivative of $p(l, r_j, \gamma_j)$ as a function of r_j. A straightforward calculus yields that the β_j's and σ^2 that maximize $\tilde{\ell}_2(\boldsymbol{\theta}|\mathcal{T})$ are given as follows.

$$\hat{\boldsymbol{\beta}} = (\tilde{\Delta} - \frac{1}{n}\tilde{Z}^\tau \mathbf{1}\mathbf{1}^\tau \tilde{Z})^{-1}\tilde{Z}^\tau (I - \frac{1}{n}\mathbf{1}\mathbf{1}^\tau)\boldsymbol{y}, \quad (5.15)$$

$$\hat{\beta}_0 = \bar{Y} - \frac{1}{n}\mathbf{1}^\tau \tilde{Z}\hat{\boldsymbol{\beta}}, \quad (5.16)$$

$$\hat{\sigma}^2 = \frac{1}{n}[(\boldsymbol{y} - \mathbf{1}\hat{\beta}_0)^\tau (\boldsymbol{y} - \mathbf{1}\hat{\beta}_0) - \hat{\boldsymbol{\beta}}^\tau \tilde{\Delta}\hat{\boldsymbol{\beta}}]. \quad (5.17)$$

The EM algorithm is summarized as follows.

Initialization
   Specify precision ξ and initial values:

$$\boldsymbol{\theta}^{(0)} = (r_1^{(0)}, \ldots, r_m^{(0)}, \beta_0^{(0)}, \boldsymbol{\beta}^{(0)}, \sigma^{2(0)})^\tau.$$

Iteration

   E step: Compute $E(\delta_{ij}|y_i, \boldsymbol{\theta}^{(0)})$ and $E(\delta_{ij}\delta_{il}|y_i, \boldsymbol{\theta}^{(0)})$ according to formulas (5.12) and (5.13) respectively. Form matrices $\tilde{Z}$ and $\tilde{\Delta}$.

   M step: Compute $\hat{\boldsymbol{\theta}} = (\hat{r}_1, \ldots, \hat{r}_m, \hat{\beta}_0, \hat{\boldsymbol{\beta}}, \hat{\sigma}^2)^\tau$ according to formulas (5.14) through (5.17).

   Check convergence: If $\|\hat{\boldsymbol{\theta}} - \boldsymbol{\theta}^{(0)}\|_2 < \xi$, stop; otherwise let $\boldsymbol{\theta}^{(0)} = \hat{\boldsymbol{\theta}}$ and go to E step.

The initial value $r_j^{(0)}$ can be taken as the recombination fraction between the left flanking marker and the middle point of interval j, $\beta_0^{(0)}$ and $\sigma^{2(0)}$ can be taken as the sample mean and sample variance of the trait values respectively, and $\boldsymbol{\beta}^{(0)}$ can be taken as an zero vector.

5.3   Mixture Generalized Linear Models

   Many traits do not follow normal distributions. There are continuous traits which follow non-normal distributions. There are also non-continuous traits such as binary, ordinal or count traits. The mapping of these types of traits has been dealt with using other methods in the literature. The binary traits have been studied in, e.g., [205], [191], [206], [211]. The count traits have been considered in [108], [127], [160] and [109]. The ordinal traits have been considered in [78] and [117]. In this section, we unify the treatment of these types of traits under the roof of mixture generalized linear models and consider the multiple interval mapping approach developed in [33].

We introduced exponential families in § 2.1.3. In this section, we consider the exponential family with the following probability density function:

$$f(y) = \exp\{\frac{y\theta - b(\theta)}{\tau} + c(y, \tau)\}I_A(y), \tag{5.18}$$

where the canonical parameter θ is a scaler parameter. This family includes the Bernoulli distribution for binary response, Poisson distribution for count response as well as some continuous non-normal distributions such as gamma distributions. The case where θ is a vector parameter, which includes polytomous distributions for categorical response, will be treated in § 2.5.

First, we briefly review the generalized linear models (GLIM) introduced in § 2.5.2. Consider a response variable Y and a covariate vector z. A GLIM assumes: (i) The response variable Y given the covariate vector z follows an exponential family distribution with probability density function of form (5.18) where $\theta = \theta(z)$ is determined by z but the dispersion parameter τ is independent of z. Under the GLIM, $\mu(z) = E(Y|z) = b'(\theta(z))$ and $\text{Var}(Y|z) = \tau b''(\theta(z))$, where b' and b'' are the first and second derivatives of b with respect to the canonical parameter. (ii) The covariate vector z affects Y through a linear predictor $\eta(z) = \beta_0 + \boldsymbol{\beta}^T z$. (iii) The linear predictor is related to the mean of Y through a link function g such that $\eta(z) = g(\mu(z))$.

We now turn to the mixture GLIM. Except for the distribution of the quantitative trait, all the assumptions for mixture GLIM are the same as those for Gaussian mixture models in § 5.1. Let $\eta(\delta_1, \ldots, \delta_m)$ denote a linear form of the unobserved QTL genotype variables $\delta_1, \ldots, \delta_m$. For convenience, δ_j is used to denote both the scaler genotype variable in the case of backcross design and the vector genotype variable in the case of intercross design, and the vector $(\delta_1, \ldots, \delta_m)$ is denoted by $\boldsymbol{\delta}$. The linear form $\eta(\boldsymbol{\delta})$ takes the form of $\mu(\boldsymbol{\delta})$ in (5.1), (5.2), (5.6) and (5.7) according to whether the design is backcross or intercross and whether the genetic model assumes no interaction or interaction among the QTLs. Under the GLIM assumption, the QTLs affect the quantitative trait Y through the linear form $\eta(\boldsymbol{\delta})$ and the probability density of Y given the QTL genotypes is as follows.

$$f(y|\boldsymbol{\delta}) = \exp\left\{[y\theta(\boldsymbol{\delta}) - b(\theta(\boldsymbol{\delta}))]/\tau + c(\tau, y)\right\},$$

where $\theta(\boldsymbol{\delta})$ is related to $\eta(\boldsymbol{\delta})$ by the relationships $\mu(\boldsymbol{\delta}) = b'(\theta(\boldsymbol{\delta}))$ and $g(\mu(\boldsymbol{\delta})) = \eta(\boldsymbol{\delta})$ and g is the link function.

Following the same derivation as in § 5.1, by replacing $\phi(y, \mu(\boldsymbol{\delta}), \sigma^2)$ with

$\exp\{[y\theta(\boldsymbol{\delta}) - b(\theta(\boldsymbol{\delta}))]/\tau + c(\tau, y)\}$, we obtain the following likelihood functions based on $\boldsymbol{y}$ and $\mathcal{T}$. In the case of backcross design,

$$
\begin{aligned}
L(\boldsymbol{\theta}|\boldsymbol{y}) &= \prod_{i=1}^{n} \sum_{(k_1,\ldots,k_m)} \prod_{j=1}^{m} p(x_{ij}, r_j, \gamma_j)^{k_j} [1 - p(x_{ij}, r_j, \gamma_j)]^{1-k_j} \\
&\qquad \times \exp\{[y_i\theta(\boldsymbol{k}) - b(\theta(\boldsymbol{k}))]/\tau + c(y_i, \tau)\}, \quad (5.19) \\
L(\boldsymbol{\theta}|\mathcal{T}) &= \prod_{i=1}^{n}\prod_{j=1}^{m} p(x_{ij}, r_j, \gamma_j)^{\delta_{ij}} [1 - p(x_{ij}, r_j, \gamma_j)]^{1-\delta_{ij}} \\
&\qquad \times \exp\{[y_i\theta(\boldsymbol{\delta}_i) - b(\theta(\boldsymbol{\delta}_i))]/\tau + c(y_i, \tau)\}\}. \quad (5.20)
\end{aligned}
$$

In the case of intercross design,

$$
\begin{aligned}
L(\boldsymbol{\theta}|\boldsymbol{y}) &= \prod_{i=1}^{n} \sum_{(k_1,\ldots,k_m)} \prod_{j=1}^{m} p_0(x_{ij}, r_j, \gamma_j)^{1-k_{1j}-k_{2j}} p_1(x_{ij}, r_j, \gamma_j)^{k_{1j}} \\
&\qquad \times p_2(x_{ij}, r_j, \gamma_j)^{k_{2j}} \exp\{[y_i\theta(\boldsymbol{k}) - b(\theta(\boldsymbol{k}))]/\tau + c(y_i, \tau)\}, \quad (5.21) \\
L(\boldsymbol{\theta}|\mathcal{T}) &= \prod_{i=1}^{n}\prod_{j=1}^{m} \{p_0(x_{ij}, r_j, \gamma_j)^{1-\delta_{1j}-\delta_{2j}} p_1(x_{ij}, r_j, \gamma_j)^{\delta_{1j}} p_2(x_{ij}, r_j, \gamma_j)^{\delta_{2j}} \\
&\qquad \times \exp\{[y_i\theta(\boldsymbol{\delta}_i) - b(\theta(\boldsymbol{\delta}_i))]/\tau + c(y_i, \tau)\}\}. \quad (5.22)
\end{aligned}
$$

For the sake of convenience, we have used k_j to denote both the scaler k_j in the backcross design and the vector $k_j = (k_{j1}, k_{j2})^\tau$ in the intercross design. The likelihood function $L(\boldsymbol{\theta}|\boldsymbol{y})$ is to be maximized to obtain the MLE of $\boldsymbol{\theta}$. Since the direct maximization of $L(\boldsymbol{\theta}|\boldsymbol{y})$ by traditional numerical methods such as the Newton method is difficult, EM algorithms are needed for the maximization. The likelihood function $L(\boldsymbol{\theta}|\mathcal{T})$ is needed in the EM algorithms. The EM algorithm for computing the MLE of $\boldsymbol{\theta}$ in the mixture generalized linear models is developed in the next section.

5.4 General EM Algorithms for Mixture Generalized Linear Models

The general EM algorithm for mixture GLIM developed in [33] is described in this section. Let $p_{ik_1\cdots k_m}$ denote $\prod_{j=1}^{m} p(x_{ij}, r_j, \gamma_j)^{k_j} [1 - p(x_{ij}, r_j, \gamma_j)]^{1-k_j}$ in the backcross design or $\prod_{j=1}^{m} p_0(x_{ij}, r_j, \gamma_j)^{1-k_{1j}-k_{2j}} p_1(x_{ij}, r_j, \gamma_j)^{k_{1j}}$ $p_2(x_{ij}, r_j, \gamma_j)^{k_{2j}}$ in the intercross design. From (5.19) and (5.21), the log likelihood function of the mixture GLIM with m intervals is given by

$$\ell(\boldsymbol{\theta}|\boldsymbol{y}) = \sum_{i=1}^{n} \log \sum_{(k_1,\ldots,k_m)} p_{ik_1\cdots k_m} \exp\left\{\frac{y_i\theta(\boldsymbol{k}) - b(\theta(\boldsymbol{k}))}{\tau}\right\}$$

$$+ \sum_{i=1}^{n} c(\tau, y_i). \tag{5.23}$$

It is conventional in the ordinary GLIM that the dispersion parameter is estimated by a moment-type estimator and the maximum likelihood estimates (MLEs) of the other parameters are derived by solving the score equations with respect to those parameters. The same strategy is adopted here. Thus, the second sum in the log likelihood above can be omitted.

We first consider the EM algorithm for backcross design, then adapt the algorithm for intercross design. Finally, we discuss a modification of the algorithm which is necessary for the mixture GLIM with Bernoulli distributions.

5.4.1   EM Algorithm for Backcross Design

Instead of working with the log likelihood (5.23), the EM algorithm makes use of the likelihood (5.20) based on $\mathcal{T}$. Ignoring the terms $c(\tau, y_i)$'s, the log of this likelihood is

$$\ell(\boldsymbol{\theta}|\mathcal{T}) = \sum_{j=1}^{m}\sum_{i=1}^{n}[\delta_{ij}\ln(p(x_{ij}, r_j, \gamma_j)) + (1 - \delta_{ij})\ln(1 - p(x_{ij}, r_j, \gamma_j))]$$

$$+ \frac{1}{\tau}\sum_{i=1}^{n}[y_i\theta(\boldsymbol{\delta}_i) - b(\theta(\boldsymbol{\delta}_i))]$$

$$= \ell_1(\boldsymbol{r}|\mathcal{T}) + \frac{1}{\tau}\ell_2(\boldsymbol{\beta}|\mathcal{T}), \text{ say,} \tag{5.24}$$

where $\boldsymbol{r} = (r_1,\ldots,r_m)^{\tau}$ is the vector of QTL position parameters and $\boldsymbol{\beta}$ is the vector of QTL effect parameters including both main effects and interaction effects if any.

Unlike in the Gaussian mixture models, the second component, $\ell_2(\boldsymbol{\beta}|\mathcal{T})$, of the above log likelihood is, in general, not a linear form of the δ_{ij}'s and their products. Thus the E step cannot be simplified to the computation of the conditional expectations of the δ_{ij}'s and their products. To overcome the difficulty caused, we express $\ell_2(\boldsymbol{\beta}|\mathcal{T})$ in the following form:

$$\ell_2(\boldsymbol{\beta}|\mathcal{T}) = \sum_{(k_1\cdots k_m)}\left[\theta_{k_1\cdots k_m}\sum_{i=1}^{n}y_i\Delta_{ik_1\cdots k_m} - b(\theta_{k_1\cdots k_m})\sum_{i=1}^{n}\Delta_{ik_1\cdots k_m}\right],$$

$$\tag{5.25}$$

where the sum is taken over the set $\{(k_1,\ldots,k_m) : k_j = 0 \text{ or } 1, j = 1,\ldots,m\}$ and $\theta_{k_1\cdots k_m} = \theta(\boldsymbol{k})$, $\Delta_{ik_1\cdots k_m} = \prod_{j=1}^{m}I\{\delta_{ij} = k_j\}$.

With expression (5.25), in the E step, the computation of the conditional expectation of $\ell_2(\beta|\mathcal{T})$ is reduced to the computation of the conditional expectations of the $\Delta_{ik_1\cdots k_m}$'s. For convenience, we denote the conditional expectation of a random quantity given $\boldsymbol{y}$ by putting a $\tilde{}$ at the top of the notation for that quantity. By the Bayes formula, the conditional expectation of $\Delta_{ik_1\cdots k_m}$ is given by

$$\tilde{\Delta}_{ik_1\cdots k_m} = \frac{p_{ik_1\cdots k_m}\exp\left\{[y_i\theta^{(0)}_{k_1\cdots k_m} - b(\theta^{(0)}_{k_1\cdots k_m})]/\tau^{(0)}\right\}}{\sum_{(\nu_1,\ldots,\nu_m)} p_{i\nu_1\cdots\nu_m}\exp\left\{[y_i\theta^{(0)}_{\nu_1\cdots\nu_m} - b(\theta^{(0)}_{\nu_1\cdots\nu_m})]/\tau^{(0)}\right\}}, \quad (5.26)$$

where $\sum_{(\nu_1,\ldots,\nu_m)}$ is over the set $\{(\nu_1,\ldots,\nu_m) : \nu_j = 0,1, \ j=1,\ldots,m\}$, and $p_{ik_1\cdots k_m} = \prod_{j=1}^m p^{k_j}(x_{ij},r_j,\gamma_j)[1-p(x_{ij},r_j,\gamma_j)]^{1-k_j}$. The $\tilde{\delta}_{ij}$ is then given by

$$\tilde{\delta}_{ij} = \sum_{k_j=1} \tilde{\Delta}_{ik_1\cdots k_m},$$

where $\sum_{k_j=1}$ is over the subset $\{(k_1,\ldots,k_m) : k_j = 1; \ k_l = 0,1, l \neq j\}$.

In the M step, the maximization of $\ell(\boldsymbol{\theta}|\mathcal{T})$ is again broken into two separate problems: the maximization of $\ell_1(\boldsymbol{r}|\mathcal{T})$ with respect to $\boldsymbol{r}$ and the maximization of $\ell_2(\boldsymbol{\beta}|\mathcal{T})$ with respect to $\boldsymbol{\beta}$. The maximization of $\ell_1(\boldsymbol{r}|\mathcal{T})$ is easy. It can be done separately for each j by maximizing

$$\sum_{i=1}^n [\delta_{ij}\ln(p(x_{ij},r_j,\gamma_j)) + (1-\delta_{ij})\ln(1-p(x_{ij},r_j,\gamma_j))]$$

with respect to r_j, which can be fulfilled by using a grid point search procedure.

The maximization of $\ell_2(\boldsymbol{\beta}|\mathcal{T})$ is more sophisticated. The Newton method with Fisher scoring is used. Let

$$A(\boldsymbol{\beta}) = \frac{\partial^2 \ell_2(\boldsymbol{\beta}|\mathcal{T})}{\partial\boldsymbol{\beta}\partial\boldsymbol{\beta}'}, \qquad \boldsymbol{u}(\boldsymbol{\beta}) = \frac{\partial \ell_2(\boldsymbol{\beta}|\mathcal{T})}{\partial\boldsymbol{\beta}}.$$

The Newton method with Fisher scoring proceeds by iteratively solving

$$-E[A(\boldsymbol{\beta}^{\text{OLD}})](\boldsymbol{\beta}^{\text{NEW}} - \boldsymbol{\beta}^{\text{OLD}}) = \boldsymbol{u}(\boldsymbol{\beta}^{\text{OLD}}).$$

The iteration can be carried out by an iterated weighted least square (IWSL) procedure. We derive this IWSL procedure in the following.

Let $\mu_{k_1\cdots k_m}$ be the mean value corresponding to $\theta_{k_1\cdots k_m}$. Let

$$\Delta^{[a]}_{k_1\cdots k_m} = \sum_{i=1}^n y_i^a \Delta_{ik_1\cdots k_m}, \ a=0,1,2,$$

and

$$w_{k_1\cdots k_m} = \Delta^{[0]}_{k_1\cdots k_m}/[\{g'(\mu_{k_1\cdots k_m})\}^2 b''(\theta_{k_1\cdots k_m})].$$

Let $\boldsymbol{x}_{k_1\cdots k_m}$ be the row vector of the covariate values in $\eta(\boldsymbol{\delta})$ with $\boldsymbol{\delta} = (k_1,\ldots,k_m)$. For example, if there is no interaction, $\boldsymbol{x}_{k_1\cdots k_m} = (1, k_1, \ldots, k_m)^\tau$. If, in addition, there is an interaction between QTL 1 and QTL 2, $\boldsymbol{x}_{k_1\cdots k_m}^\tau = (1, k_1, \ldots, k_m, k_1 k_2)^\tau$. Let X be the matrix obtained by stacking the $\boldsymbol{x}_{k_1\cdots k_m}$'s one above another in lexicographical order, i.e., the indices $(k_1, \ldots, k_m)$'s are in the order

$$(00\cdots00),$$
$$(00\cdots01),$$
$$(00\cdots10),$$
$$(00\cdots11),$$
$$\cdots$$
$$(11\cdots11).$$

Define $W(\boldsymbol{\beta})$ as the diagonal matrix with diagonal elements $w_{k_1\cdots k_m}$ arranged in the same order. Define $\boldsymbol{z}(\boldsymbol{\beta}) = X\boldsymbol{\beta} + \boldsymbol{v}(\boldsymbol{\beta})$, where $\boldsymbol{v}(\boldsymbol{\beta})$ is the vector with $g'(\mu_{k_1\cdots k_m})\left(\Delta^{[1]}_{k_1\cdots k_m}/\Delta^{[0]}_{k_1\cdots k_m} - \mu_{k_1\cdots k_m}\right)$ as its components.

By simple calculus, it can be derived that

$$\boldsymbol{u}(\boldsymbol{\beta}) = \sum_{(k_1,\ldots,k_m)} \boldsymbol{x}_{k_1\ldots k_m} \frac{\tilde{\Delta}^{[1]}_{k_1\ldots k_m} - \mu_{k_1\cdots k_m}\tilde{\Delta}^{[0]}_{k_1\ldots k_m}}{g'(\mu_{k_1\cdots k_m})b''(\theta_{k_1\cdots k_m})}$$

$$= \sum_{(k_1,\ldots,k_m)} \boldsymbol{x}_{k_1\ldots k_m} w_{k_1\ldots k_m} \frac{g'(\mu_{k_1\cdots k_m})}{\tilde{\Delta}^{[0]}_{k_1\ldots k_m}} [\tilde{\Delta}^{[1]}_{k_1\ldots k_m} - \mu_{k_1\cdots k_m}\tilde{\Delta}^{[0]}_{k_1\ldots k_m}],$$

and

$$A(\boldsymbol{\beta}) = \sum_{(k_1,\ldots,k_m)} \boldsymbol{x}_{k_1\ldots k_m} \left\{ \frac{\partial}{\partial\boldsymbol{\beta}} \left(w_{k_1\ldots k_m} \frac{g'(\mu_{k_1\cdots k_m})}{\tilde{\Delta}^{[0]}_{k_1\ldots k_m}} \right) [\tilde{\Delta}^{[1]}_{k_1\ldots k_m} - \mu_{k_1\cdots k_m}\tilde{\Delta}^{[0]}_{k_1\ldots}} \right.$$
$$\left. - \left(w_{k_1\ldots k_m} \frac{g'(\mu_{k_1\cdots k_m})}{\tilde{\Delta}^{[0]}_{k_1\ldots k_m}} \right) b''(\theta_{k_1\cdots k_m})\tilde{\Delta}^{[0]}_{k_1\ldots k_m} \frac{\partial\theta_{k_1\cdots k_m}}{\partial\boldsymbol{\beta}} \right\} \boldsymbol{x}^\tau_{k_1\ldots k_m}.$$

Note that

$$E[\tilde{\Delta}^{[1]}_{k_1\ldots k_m} - \mu_{k_1\cdots k_m}\tilde{\Delta}_{k_1\ldots k_m}] = \sum_{i=1}^{n} E[(Y_i - \mu_{k_1\cdots k_m})E(\Delta_{ik_1\ldots k_m}|\boldsymbol{Y})]$$

$$= \sum_{i=1}^{n} E[(Y_i - \mu_{k_1\cdots k_m})\Delta_{ik_1\ldots k_m}] = \sum_{i=1}^{n} E\{\Delta_{ik_1\ldots k_m}E[(Y_i - \mu_{k_1\cdots k_m})|\Delta_{ik_1\ldots k_m}]$$

The term $E[(Y_i - \mu_{k_1\cdots k_m})|\Delta_{ik_1\ldots k_m}]$ is approximately zero when the EM algorithm converges, and hence the first sum of $A(\boldsymbol{\beta})$ can be ignored while taking its expectation. Noticing that

$$\frac{\partial\theta_{k_1\cdots k_m}}{\partial\boldsymbol{\beta}} = \boldsymbol{x}_{k_1\ldots k_m}[g'(\mu_{k_1\cdots k_m})b''(\theta_{k_1\cdots k_m})]^{-1},$$

thus we obtain

$$-EA(\boldsymbol{\beta}) = \sum_{(k_1,\ldots,k_m)} \boldsymbol{x}_{k_1\ldots k_m} w_{k_1\ldots k_m} \boldsymbol{x}'_{k_1\ldots k_m} = X'W(\boldsymbol{\beta})X.$$

By using the vector $\boldsymbol{v}(\boldsymbol{\beta})$, we also have

$$\boldsymbol{u}(\boldsymbol{\beta}) = X'W(\boldsymbol{\beta})\boldsymbol{v}(\boldsymbol{\beta}).$$

The iterative equation of the Newton method with Fisher scoring becomes

$$X'W(\boldsymbol{\beta}^{\mathrm{OLD}})X(\boldsymbol{\beta}^{\mathrm{NEW}} - \boldsymbol{\beta}^{\mathrm{OLD}}) = X'W(\boldsymbol{\beta}^{\mathrm{OLD}})\boldsymbol{v}(\boldsymbol{\beta}^{\mathrm{OLD}}),$$

which is equivalent to

$$X'W(\boldsymbol{\beta}^{\mathrm{OLD}})X\boldsymbol{\beta}^{\mathrm{NEW}} = X'W(\boldsymbol{\beta}^{\mathrm{OLD}})\boldsymbol{z}(\boldsymbol{\beta}^{\mathrm{OLD}}).$$

Once the IWSL procedure converges, the dispersion parameter τ is updated as

$$\hat{\tau} = \frac{1}{n-q} \sum_{(k_1\cdots k_m)} \sum_{i=1}^{n} \frac{[Y_i - \mu_{k_1\cdots k_m}]^2 \tilde{\Delta}_{ik_1\cdots k_m}}{b''(\theta_{k_1\cdots k_m})},$$

where q is the number of components of $\boldsymbol{\beta}$ and $\theta_{k_1\cdots k_m}$'s are computed at the updated $\boldsymbol{\beta}$. The rationale of this updating formula is as follows. Suppose the unknown parameters are fixed at their true values, we have

$$
\begin{aligned}
E\left[(Y_i - \mu_{k_1\cdots k_m})^2 \tilde{\Delta}_{ik_1\cdots k_m}\right] &= E\left[\Delta_{ik_1\cdots k_m} E[(Y_i - \mu_{k_1\cdots k_m})^2 | \Delta_{ik_1\cdots k_m}]\right] \\
&= \tau b''(\theta_{k_1\cdots k_m}) E[\Delta_{ik_1\cdots k_m}],
\end{aligned}
$$

and hence,

$$E\left[\sum_{(k_1\cdots k_m)} \sum_{i=1}^{n} \frac{(Y_i - \mu_{k_1\cdots k_m})^2 \tilde{\Delta}_{ik_1\cdots k_m}}{b''(\theta_{k_1\cdots k_m})}\right] = \tau \sum_{i=1}^{n} \sum_{(k_1\cdots k_m)} E[\Delta_{ik_1\cdots k_m}] = n\tau.$$

When the unknown parameters are replaced by their estimates, n is replaced by $n - q$ to compensate the loss of degrees of freedom due to the estimation. Thus the updating formula for τ provides an asymptotically unbiased estimate of τ.

The EM algorithm is summarized as follows.

Initialization: Form the matrix X. Set starting values $\boldsymbol{\beta}^{(0)}, \tau^{(0)}$ and $\boldsymbol{r}^{(0)}$.

E step: At $(\boldsymbol{\beta}^{(0)}, \tau^{(0)}, \boldsymbol{r}^{(0)})$, compute the arrays

$$\{\tilde{\Delta}_{ik_1\cdots k_m}\} = \{E(\Delta_{ik_1\cdots k_m}|\boldsymbol{y}, \boldsymbol{\beta}^{(0)}, \tau^{(0)}, \boldsymbol{r}^{(0)})\},$$

$$\{\tilde{\delta}_{ij}\} = \{E(\delta_{ij}|\boldsymbol{y}, \boldsymbol{\beta}^{(0)}, \tau^{(0)}, \boldsymbol{r}^{(0)})\}, \quad \{\tilde{\Delta}^{[l]}_{k_1\cdots k_m}\}, \quad l = 0, 1, 2.$$

M step: (i) Update β:

    (a) Set $\beta^{\text{OLD}} = \beta^{(0)}$.

    (b) Compute $z(\beta^{\text{OLD}})$ and $W(\beta^{\text{OLD}})$.

    (c) Solve for β^{NEW} in the equation:

$$X'W(\beta^{\text{OLD}})X\beta^{\text{NEW}} = X'W(\beta^{\text{OLD}})z(\beta^{\text{OLD}}).$$

    (d) Check convergence. If convergent, set $\beta^{(1)} = \beta^{\text{NEW}}$ and go to (ii), otherwise set $\beta^{\text{OLD}} = \beta^{\text{NEW}}$ and goto (b).

(ii) Update τ by the following formula evaluated at $\beta^{(1)}$.

$$\hat{\tau} = \frac{1}{n-q} \sum_{(k_1 \cdots k_m)} \frac{\tilde{\Delta}^{[2]}_{k_1 \cdots k_m} - 2\mu_{k_1 \cdots k_m} \tilde{\Delta}^{[1]}_{k_1 \cdots k_m} + \mu^2_{k_1 \cdots k_m} \tilde{\Delta}^{[0]}_{k_1 \cdots k_m}}{b''(\theta_{k_1,\ldots,k_m})}.$$

(iii) Update r: for $j = 1, \ldots, m$, maximizing, with respect to r_j,

$$\sum_{l=1}^{4}[M_{jl} \ln \frac{p(l, r_j\gamma_j)}{1 - p(l, r_j, \gamma_j)} + N_{jl} \ln(1 - p(l, r_j\gamma_j))],$$

where $M_{jl} = \sum_{i:x_{ij}=l} \tilde{\delta}_{ij}$, $N_{jl} = \sum_i I\{x_{ij} = l\}$ and x_{ij} is the code of the genotype class of individual i on interval j.

(iv) Check convergence; if convergent, stop, otherwise set $\beta^{(0)} = \beta^{(1)}$, $\tau^{(0)} = \tau^{(1)}$, $r^{(0)} = r^{(1)}$ and go to E step.

The mixture GLIMs with Bernoulli and Poisson distributions are two important special cases. The former is used for mapping binary traits and the latter is used for mapping count traits. In these two special cases, the dispersion parameter $\tau \equiv 1$. Therefore, in the M step of the EM algorithm, the updating of τ becomes irrelevant and is omitted.

5.4.2 Adaptation for Intercross Design

The EM algorithm for intercross design can be obtained by a simple adaptation of the algorithm developed in the previous sub section. The following are a few modifications in the adaptation. The $\ell_1(r|)$ in (5.24) is replaced by

$$\ell_1(r|) = \sum_{j=1}^{m}\sum_{i=1}^{n}[(1 - \delta_{ij1} - \delta_{ij2}) \ln p_0(x_{ij}, r_j, \gamma_j)$$
$$+\delta_{ij1} \ln p_1(x_{ij}, r_j, \gamma_j) + \delta_{ij2} \ln p_2(x_{ij}, r_j, \gamma_j)].$$

Thus, in (iii) of the M step, to update r,

$$\sum_{i=1}^{n}[(1 - \delta_{ij1} - \delta_{ij2}) \ln p_0(x_{ij}, r_j, \gamma_j)$$
$$+\delta_{ij1} \ln p_1(x_{ij}, r_j, \gamma_j) + \delta_{ij2} \ln p_2(x_{ij}, r_j, \gamma_j)]$$

is maximized with respect to r_j for each j.

The conditional expectation $\tilde{\Delta}_{ik_1\cdots k_m}$ is still computed by formula (5.26) but with the $\sum_{(\nu_1,\ldots,\nu_m)}$ taken over the set

$$\{(\nu_1,\ldots,\nu_m) : \nu_j = (1,0)^\tau \text{ or } (0,1)^\tau, \ j = 1,\ldots,m\},$$

and

$$p_{ik_1\cdots k_m} = \prod_{j=1}^{m} p_0(x_{ij}, r_j, \gamma_j)^{1-k_{1j}-k_{2j}} p_1(x_{ij}, r_j, \gamma_j)^{k_{1j}} p_2(x_{ij}, r_j, \gamma_j)^{k_{2j}}.$$

The conditional expectations $\tilde{\delta}_{ij1}$ and $\tilde{\delta}_{ij2}$ are computed as

$$\tilde{\delta}_{ij1} = \sum_{k_j=(1,0)^\tau} \tilde{\Delta}_{ik_1\cdots k_m}, \quad \tilde{\delta}_{ij2} = \sum_{k_j=(0,1)^\tau} \tilde{\Delta}_{ik_1\cdots k_m},$$

where, in the sums, the other k_l's, $l \neq j$, varies over $(1,0)^\tau$ and $(0,1)^\tau$. The matrix X is again obtained by stacking the $x_{k_1\cdots k_m}$'s one above another in lexicographical order. Symbolically, identify the three vectors $(0,0)^\tau$, $(1,0)^\tau$ and $(0,1)^\tau$ by 0, 1 and 2. The lexicographical order is as follows

$$(00\cdots00),$$
$$(00\cdots01),$$
$$(00\cdots02),$$
$$(00\cdots10),$$
$$(00\cdots11),$$
$$(00\cdots12),$$
$$\cdots$$
$$(22\cdots22).$$

With the above changes, we obtain the EM algorithm for intercross design.

5.4.3 A Special Issue on Binary Traits

A situation called data separation arises in binary models with categorical covariates. A complete data separation refers to the situation that the observed responses in the sample are either all 1 or all 0 in a category classified by categorical covariates. If the responses are almost all 1 or 0, the situation is called a quasi-complete separation. In the context of mixture GLIM, complete or quasi-complete separation is unavoidable as the number of intervals involved becomes large. When complete or quasi-complete separation presents, the MLEs of the coefficients corresponding to the separation categories tend to be either positive infinity (if responses are all 1) or negative infinity (if responses are all 0). Mathematically, this causes a serious bias of the MLE. Computationally, it breaks down the computation algorithm because it causes non-convergence.

A bias reduction method is proposed in [60] to circumvent the difficulty caused by separation in logistic regression models, a special case of GLIM with Bernoulli distribution where the link function g is taken as the canonical link $\ln[\mu/(1-\mu)]$. To deal with the separation problem in the mixture GLIM with Bernoulli distributions, we adapt the bias reduction idea of [60] in the M step of the EM algorithm as follows. Replace the original log likelihood $\ell_2(\boldsymbol{\beta}|\mathcal{T})$ by a penalized log likelihood

$$\ell_{2p}(\boldsymbol{\beta}|\mathcal{T}) = \ell_2(\boldsymbol{\beta}|\mathcal{T}) + 2^{m-2}\ln|X^{'}W(\boldsymbol{\beta})X|,$$

where the term 2^{m-2} reflects the severity of the penalty in accordance with the number of intervals. When $m = 1$, it reduces to the Jeffreys penalty considered in [60]. With the canonical link, the score vector of $\ell_2(\boldsymbol{\beta}|\mathcal{T})$ reduces to

$$\boldsymbol{u}(\boldsymbol{\beta}) = \sum_{(k_1,\ldots,k_m)} \boldsymbol{x}_{k_1\cdots k_m}(\tilde{\Delta}^{[1]}_{k_1\cdots k_m} - \mu_{k_1\cdots k_m}\tilde{\Delta}^{[0]}_{k_1\cdots k_m}).$$

The score vector of $\ell_{2p}(\boldsymbol{\beta}|\mathcal{T})$ is obtained as

$$\boldsymbol{u}^*(\boldsymbol{\beta}) = \boldsymbol{u}(\boldsymbol{\beta}) - 2^{m-1}\sum_{(k_1,\ldots,k_m)} \boldsymbol{x}_{k_1\cdots k_m}h_{k_1\cdots k_m}(\mu_{k_1\cdots k_m} - 1/2),$$

where $h_{k_1\cdots k_m}$'s are the diagonal elements of $W^{1/2}X(X^{'}WX)^{-1}X^{'}W^{1/2}$. Therefore, the modification is easily implemented by replacing $\tilde{\Delta}^{[0]}_{k_1\cdots k_m}$ and $\tilde{\Delta}^{[1]}_{k_1\cdots k_m}$ with $\tilde{\Delta}^{[0]}_{k_1\cdots k_m} + 2^{m-1}h_{k_1\cdots k_m}$ and $\tilde{\Delta}^{[1]}_{k_1\cdots k_m} + 2^{m-2}h_{k_1\cdots k_m}$ respectively. This penalized EM algorithm produces estimates of $\boldsymbol{\beta}$ shrunk towards zero. The estimates are always finite even if there is complete or quasi-complete separation while the MLEs are infinite. In general, the estimates are biased. But, as the sample size n gets larger, the biases diminish.

5.5   Multiple Interval Mapping Procedures

Multiple interval mapping amounts to the selection of a mixture model that best interprets the relationship between the quantitative trait and the QTLs. In this section, we present a stepwise procedure for the model selection with mixture models. Mimicking the stepwise regression in ordinary linear models, the procedure proceeds by adding and deleting QTL effects alternatively according to a certain criterion. The extended BIC (EBIC) discussed in § 2.10 is used as the criterion. Since we consider both main effects and interactions in the mapping procedure, we use the EBIC for interaction models. Let G denote a model. Suppose that the model contains q_M main effects and q_I interactions, the EBIC of model G is given by

$$\mathrm{EBIC}_{\gamma_\mathrm{M}\gamma_\mathrm{I}}(G) = -2\ln L(\hat{\boldsymbol{\theta}}_\mathrm{G}|\boldsymbol{y}) + (q_\mathrm{M}+q_\mathrm{I})\ln n + 2\gamma_\mathrm{M}\ln\binom{M}{q_\mathrm{M}} + 2\gamma_\mathrm{I}\ln\left(\frac{\frac{M(M-1)}{2}}{q_\mathrm{I}}\right),$$

where $\hat{\boldsymbol{\theta}}_G$ is the MLE of the parameter in model G, M is the total number of intervals, $\gamma_M = 1 - \ln n/(2 \ln M)$ and $\gamma_I = 1 - \ln n/(4 \ln M)$. Note that $L(\hat{\boldsymbol{\theta}}_G|\boldsymbol{y})$ is the maximum likelihood based on $\boldsymbol{y}$ which has form (5.19) for backcross design and form (5.21) for intercross design.

Let G_1 and G_2 be two models where G_2 is nested in G_1 and G_1 has exactly one additional term. In the mapping procedure, to evaluate whether or not the additional term should be removed from G_1 (or added to G_2), we compare $\text{EBIC}_{\gamma_M\gamma_I}(G_1)$ and $\text{EBIC}_{\gamma_M\gamma_I}(G_2)$. If $\text{EBIC}_{\gamma_M\gamma_I}(G_2) < \text{EBIC}_{\gamma_M\gamma_I}(G_1)$, then the additional term is removed from G_1, otherwise, it is added to G_2.

To facilitate the description, we adopt the following model representations: an interval, say interval j, with only main QTL effect is represented by δ_j in the model; an interaction between the QTLs in two intervals , say intervals j and l, is represented by $\delta_j\delta_l$. For example, a model containing intervals j and l with only main QTL effects is represented by $\delta_j + \delta_l$, and a model containing intervals j, l and s with interaction between intervals l and s in addition to all the main effects is represented by $\delta_j + \delta_l + \delta_s + \delta_l\delta_s$. By convention, once an interaction is included in a model, the main effects of the constituent intervals are also included. The stepwise multiple interval mapping procedure is as follows.

Step 1: Consider all the models containing only a single interval. Compute the EBIC of the model with the largest maximum likelihood. If the EBIC of this model is smaller than that of the model without any QTL, the interval is selected and continue to the next step; otherwise, stop.

Step 2: Denote the selected model in step 1 by δ_{j_1}. Consider all the models of the form: $\delta_{j_1} + \delta_j + \delta_j\delta_{j_1}$, for all j $(\neq j_1)$. Denote by j_2 the j corresponding to the model of the above form having the largest maximum likelihood. Then evaluate by using EBIC whether or not the term $\delta_{j_1}\delta_{j_2}$ should be removed from the model. If not, continue to the next step; if yes, further evaluate whether or not the term δ_{j_2} can be removed. If yes, stop; otherwise, continue to the next step.

Step m: In general, after $m - 1$ steps, $m - 1$ intervals have been selected, and a model with all the $m-1$ marginal effects and possible interactions among them has been built. Denote this model by $\sum_{r=1}^{m-1} \delta_{j_r} + G_{I,m-1}$ where $G_{I,m-1}$ contains all the selected interaction terms. Now, for each remaining j, consider the models $\sum_{r=1}^{m-1} \delta_{j_r} + G_{I,m-1} + \delta_j + \sum_{r=1}^{m-1} \delta_j\delta_{j_r}$. Let j_m be the index such that the model $G_{m1} = \sum_{r=1}^{m-1} \delta_{j_r} + G_{I,m-1} + \delta_j + \sum_{r=1}^{m-1} \delta_{j_m}\delta_{j_r}$ has the largest maximum likelihood. Then, evaluate whether or not any of the interaction terms in G_{m1} can be removed as follows. First, delete the interaction terms from G_{m1} one at a time, compute the maximum likelihood of the resultant model. Let r^* be the index such that while $\delta_{j_m}\delta_{j_{r^*}}$ is deleted the resultant model has the smallest maximum likelihood. Denote the resultant model by G_{m2}. If $\text{EBIC}_{\gamma_M\gamma_I}(G_{m2}) > \text{EBIC}_{\gamma_M\gamma_I}(G_{m2})$, retain model G_{m1} and go to step

$m + 1$; otherwise, replace G_{m1} by G_{m2} and repeat the same process on G_{m2}. If at any stage, the interaction term under evaluation cannot be removed, go to step $m + 1$. If all the added interactions as well as the new main effect are removed, stop.

When the procedure stops, each selected interval is taken as one that contains a QTL and the location of the QTL is obtained from the estimated QTL position parameter. The procedure emphasizes the selection of intervals instead of particular effects. The consideration of interactions helps to increase the power and accuracy of the mapping.

In the following, we discuss a strategy which reduces the computation amount of the mapping procedure and is useful in some particular cases. The strategy first identifies the intervals that contain QTLs by using a fixed surrogate locus in each interval. After such intervals are identified, the QTL positions together with the QTL effects are re-estimated.

To argue the rationale of the surrogate strategy, consider the case of back-cross design. If an interval is short enough and the two flanking markers are either both homozygous or both heterozygous, the putative QTL must be also either homozygous or heterozygous regardless of its location on the interval. The frequency that both flanking markers are either homozygous or heterozygous is $1 - \gamma$ where γ is the recombination fraction between the two flanking markers of the interval. When γ is small, the frequency is high. This implies that, except in a small portion of the data, the location of the putative QTL has no impact at all on the estimation of the QTL effects. Since only the significance of the QTL effects is of concern in the process of model selection, we can take, in principle, any locus in the interval as a surrogate of the putative QTL. Thus, if the average interval length is short, the strategy can be applied. The optimal choice of the surrogate locus is the middle of the interval since it has the minimum expected distance from a randomly located QTL on the interval.

The mapping procedure with the surrogate strategy remains unchanged. The only modification is that, in the EM algorithm for the computation of the MLEs, the QTL position parameters r are fixed throughout and are not updated in the M step. By using the surrogate strategy, a hypothesis testing measure can also be used as the criterion for adding and deleting effects in the mapping procedure. To evaluate whether or not the additional term in G_1 should be removed, the likelihood ratio test statistic for testing the null model G_2 versus the alternative model G_1, which follows an asymptotic χ^2-distribution with 1 degree of freedom, is compared with a Bonferroni adjusted critical value. If the additional term is a main effect, the adjusted critical value is taken as $\chi_1^2(\alpha/(2M))$. If the additional term is an interaction term, the adjusted critical value taken as $\chi_1^2(\alpha/[M(M-1)])$, where $\chi_1^2(\beta)$ denotes the upper β-quantile of the χ^2-distribution with 1 degree of freedom. These Bonferroni adjusted critical values are used somehow to control the overall false discovery rate at α.

5.6 Example: The Analysis of Radiata Pine Data

The radiata pine data was introduced in § 3.5. The data contains 120 markers in 12 linkage groups with a total length 1676.3 cM. The genetic distances (in units of cM) between markers in the linkage groups are provided in Table 3.9. A portion of the data is given in Table 3.10. The data consists of the marker genotypes of 134 experiment units from a backcross population together with their values of three quantitative traits: the annual brown cone number at eight years of age (CN), diameter of stem at breast height (DBH) and branch quality score ranging from 1 (poorest) to 6 (best) (BS). Here, we consider the trait CN only. The trait CN has a range from 1 to 45 in the data. The mapping of CN was first done in [100] using a multiple interval mapping approach with mixture Gaussian models where the square root transformation of CN is treated as a normal variable.

In this example, we treat the CN as a truncated Poisson random variable with probability density function $P(Y = y) = e^{-\lambda}\lambda^y/\{(1 - e^{-\lambda})y!\}$, $y = 1, 2, \ldots$ We consider the mixture generalized linear model with the above truncated Poisson distribution and the link $\eta = \ln\lambda$. The parameter λ is related to the mean of the truncated Poisson distribution by $\mu = \lambda/(1 - e^{-\lambda})$. The linear predictor η is of the form

$$\eta = \beta_0 + \sum_{j=1}^{m} \beta_j \delta_j + \sum_{j,k} \xi_{jk} \delta_j \delta_k,$$

where δ_j is the dummy variable for QTL j which takes value 1 or 0 according as the genotype is homozygous or heterozygous.

Out of the 134 experiment units, the CN counts of 17 of them are missing. These 17 experiment units are excluded from the analysis. The mapping procedure described in § 3.5 is applied to this data, 5 QTL main effects together with 3 interactions among them are detected. The detected intervals and the estimates of the QTL effects and positions (in terms of recombination fraction between the left marker and the QTL of the detected interval) are given in Table 5.1. The notation $[k, I]$ in the table specifies the chromosome number k and the interval number I.

5.7 Multiple Interval Mapping with Polytomous Traits

A polytomous trait could be ordinal or nominal. An ordinal trait takes categorical values where the categories are ordered. The order of the categories carries information about their differences. For example, the stage of a

TABLE 5.1: Estimated QTL effects and positions for the cone number trait in the radiata pine data (for the notation of intervals, the first digit in the brackets indicates the linkage group, and the second digit indicates the interval).

Intervals	Parameter	Estimate	Standard Deviation
	β_0	2.089	0.102
$[11, 4]$	β_1	-1.271	0.104
	r_1	0.12	0.019
$[11, 5]$	β_2	1.232	0.106
	r_2	0.04	0.022
$[4, 4]$	β_3	0.611	0.113
	r_3	0.07	0.019
$[10, 1]$	β_4	0.331	0.101
	r_4	0.08	0.026
$[10, 9]$	β_5	-0.521	0.110
	r_5	0.05	0.021
$[11, 5] \times [4, 4]$	ξ_{23}	-0.616	0.152
$[11, 4] \times [10, 1]$	ξ_{14}	-0.559	0.147
$[10, 1] \times [10, 9]$	ξ_{45}	0.564	0.149

cancer is an ordinal trait. A nominal trait also takes categorical values but the categories are exchangeble and devoid of structure. For example, the flower color of a plant is a nominal trait. The multiple interval mapping techniques for polytomous traits are essentially the same as those for continuous traits or binary traits discussed in the previous sections. However, since a polytomous trait is represented by an indicator vector, the GLIM with polytomous traits is essentially a multivariate response model. It gives rise to some specific aspects and entails special treatments. The major difficulty with polytomous GLIM is the formulation of the linear structure, which corresponds to the linear predictor in univariate GLIMs, and the link function that connects this structure with the mean vector of the response. In this section, the methods in [52] is applied to develop mixture GLIMs with polytomous traits, and the EM algorithm is tailored to meet the special needs of these models. The mapping strategy for polytomous traits is the same as that given in § 5.5 and hence is omitted. We first consider ordinal traits, which are more common in QTL mapping studies, in detail, and pass the results to nominal traits afterwards.

5.7.1 Mixture Models for Ordinal Traits

We can consider that an ordinal trait is produced through a threshold mechanism by a latent continuous variable referred to as the liability, see [201] and [53]. Let O be the ordinal trait taking values $1, \ldots, K + 1$. These values, which indicate the ordered categories, have no numerical meaning. Let U be the liability variable. Suppose that the categories of the ordinal trait O are determined as follows:

$$O = r, \quad \text{if and only if} \quad \theta_{r-1} < U \le \theta_r, \ r = 1, \dots, K+1, \qquad (5.27)$$

where the θ's are fixed but unknown threshold values satisfying $-\infty = \theta_0 < \theta_1 < \cdots < \theta_K < \theta_{K+1} = +\infty$. It is more convenient to represent O by a vector $\boldsymbol{y} = (y_1, \dots, y_K)^t$ defined below:

$$y_r = \begin{cases} 1, & \text{if } O = r, \\ 0, & \text{otherwise}, \end{cases} \quad r = 1, \dots, K.$$

Let $\pi_r(\boldsymbol{\delta})$ be the probability that $y_r = 1$ conditioning on the QTL genotypes $\boldsymbol{\delta}$. The conditional distribution of $\boldsymbol{y}$ is the multinomial distribution with density function given by

$$p(y_1, \dots, y_K | \boldsymbol{\delta}) = \prod_{r=1}^{K+1} \pi_r(\boldsymbol{\delta})^{y_r}, \qquad (5.28)$$

where $y_r = 1$ or 0, $y_{K+1} = 1 - \sum_{r=1}^K y_r$ and $\pi_{K+1}(\boldsymbol{\delta}) = 1 - \sum_{r=1}^K \pi_r(\boldsymbol{\delta})$. In the form of exponential family, the density function is expressed as

$$
\begin{aligned}
p(y_1, \dots, y_K | \boldsymbol{\delta}) &= \exp\left\{ \sum_{r=1}^K y_r \ln\left(\frac{\pi_r(\boldsymbol{\delta})}{1 - \sum_{r=1}^K \pi_r(\boldsymbol{\delta})} \right) + \ln(1 - \sum_{r=1}^K \pi_r(\boldsymbol{\delta})) \right\} \\
&= \exp\left\{ \boldsymbol{y}^{\mathsf{T}} \boldsymbol{\theta}(\boldsymbol{\delta}) + b(\boldsymbol{\theta}(\boldsymbol{\delta})) \right\},
\end{aligned}
$$

where $\boldsymbol{y} = (y_1, \dots, y_K)^{\mathsf{T}}$, $\boldsymbol{\theta}(\boldsymbol{\delta}) = (\theta_1(\boldsymbol{\delta}), \dots, \theta_K(\boldsymbol{\delta}))$ with $\theta_r(\boldsymbol{\delta}) = \ln\left(\frac{\pi_r(\boldsymbol{\delta})}{1 - \sum_{r=1}^K \pi_r(\boldsymbol{\delta})} \right)$, and $b(\boldsymbol{\theta}(\boldsymbol{\delta})) = -\ln(1 + \sum_{r=1}^K e^{\theta_r(\boldsymbol{\delta})})$.

Assume that the QTLs affect the ordinal trait through the liability variable U in the form of a linear model. For the time being, we focus on the case of backcross design and ignore the epistasis effects among the m QTL and let

$$U = \tilde{\beta}_0 - \sum_{j=1}^m \beta_j \delta_j + \epsilon,$$

where ϵ is a random variable. Without loss of generality, assume that ϵ follows a standard distribution such as a standard normal distribution or a logistic distribution with cumulative distribution function $e^\epsilon / (1 + e^\epsilon)$. Let F be the cumulative distribution function of ϵ. The mean vector of $\boldsymbol{y}$, i.e., $\boldsymbol{\pi} = (\pi_1(\boldsymbol{\delta}), \dots, \pi_K(\boldsymbol{\delta}))^{\mathsf{T}}$, is determined by F as follows.

$$
\begin{aligned}
\pi_1(\boldsymbol{\delta}) + \cdots + \pi_r(\boldsymbol{\delta}) &= P(U \le \theta_r | \boldsymbol{\delta}) \\
&= F(\theta_r - \tilde{\beta}_0 + \sum_{j=1}^m \beta_j \delta_j) = F(\xi_r + \sum_{j=1}^m \beta_j \delta_j), \qquad (5.29)
\end{aligned}
$$

where $\xi_r = \theta_r - \tilde{\beta}_0$. We can take

$$\tilde{\eta}_r = \xi_r + \sum_{j=1}^{m} \beta_j \delta_j, r = 1, \ldots, K, \tag{5.30}$$

as the multiple linear predictors. However, the parameters ξ_r's have the order constraint: $\xi_1 \leq \cdots \leq \xi_K$. This constraint can cause serious computational difficulties. To get rid of this constraint, re-parameterize as follows. Let

$$\xi_1 = \alpha_1,$$
$$\xi_2 = \alpha_1 + e^{\alpha_2},$$
$$\ldots$$
$$\xi_K = \alpha_1 + \sum_{r=2}^{K} e^{\alpha_r}.$$

Or, equivalently,

$$\alpha_1 = \xi_1,$$
$$\alpha_2 = \ln(\xi_2 - \xi_1),$$
$$\ldots$$
$$\alpha_K = \ln(\xi_K - \xi_{K-1}).$$

Thus, we have a one-to-one correspondence between $(\xi_1, \ldots, \xi_K)$ and $(\alpha_1, \ldots, \alpha_K)$. There is no constraint on α_r's. It can be seen from (5.30) that

$$\xi_r - \xi_{r-1} = \tilde{\eta}_r - \tilde{\eta}_{r-1}, \quad r = 2, \ldots, K. \tag{5.31}$$

Since $\tilde{\eta}_r = F^{-1}(\pi_1 + \cdots + \pi_r), r = 1, \ldots, K$, we obtain

$$\begin{aligned}
\alpha_1 + \sum_{j=1}^{m} \beta_j \delta_j &= F^{-1}(\pi_1), \\
\alpha_2 &= \ln(F^{-1}(\pi_1 + \pi_2) - F^{-1}(\pi_1)), \\
&\ldots \\
\alpha_K &= \ln(F^{-1}(\pi_1 + \cdots + \pi_K) - F^{-1}(\pi_1 + \cdots + \pi_{K-1})).
\end{aligned} \tag{5.32}$$

Consider a new multiple linear predictor as

$$\begin{aligned}
\eta_1 &= \alpha_1 + \sum_{j=1}^{m} \beta_j \delta_j, \\
\eta_2 &= \alpha_2, \\
&\ldots \\
\eta_K &= \alpha_K.
\end{aligned}$$

Let $z = (\delta_1, \ldots, \delta_m)^T$ and

$$
Z = \begin{pmatrix}
1 & 0 & \cdots & 0 & z^T \\
0 & 1 & \cdots & 0 & \mathbf{0}^T \\
\cdots & \cdots & \cdots & \cdots & \cdots \\
0 & 0 & \cdots & 1 & \mathbf{0}^T
\end{pmatrix}.
\tag{5.33}
$$

Then, in matrix form, the new multiple linear predictor is expressed as

$$
\boldsymbol{\eta} = Z\boldsymbol{\beta},
\tag{5.34}
$$

where $\boldsymbol{\eta} = (\eta_1, \ldots, \eta_K)^T$ and $\boldsymbol{\beta} = (\alpha_1, \ldots, \alpha_K, \beta_1, \ldots, \beta_m)^T$. From (5.32), we have the multiple link function $\boldsymbol{g}(\boldsymbol{\pi}) = (g_1(\boldsymbol{\pi}), \ldots, g_K(\boldsymbol{\pi}))^T$ given by

$$
\eta_1 = g_1(\boldsymbol{\pi}) = F^{-1}(\pi_1),
$$

$$
\eta_r = g_r(\boldsymbol{\pi}) = \ln[F^{-1}(\sum_{l=1}^{r} \pi_l) - F^{-1}(\sum_{l=1}^{r-1} \pi_l)],
\tag{5.35}
$$

$$
r = 2, \ldots, K.
$$

Equivalently, we have the inverse link:

$$
\begin{aligned}
\pi_1 &= F(\eta_1), \\
\pi_2 &= F(e^{\eta_2} + \eta_1) - F(\eta_1), \\
\pi_3 &= F(e^{\eta_3} + e^{\eta_2} + \eta_1) - F(e^{\eta_2} + \eta_1), \\
\pi_r &= F(e^{\eta_r} + \cdots + e^{\eta_2} + \eta_1) - F(e^{\eta_{r-1}} + \cdots + e^{\eta_2} + \eta_1), \quad (5.36) \\
r &= 4, \ldots, K.
\end{aligned}
$$

The explicit form of the multiple link function can be obtained once F is specified. In the following, we provide a few examples for the multiple link function.

1. F is taken as the logistic distribution:

$$
F(\epsilon) = \frac{e^\epsilon}{1 + e^\epsilon}.
$$

The multiple link function is given by

$$
g_1(\boldsymbol{\pi}) = \ln \frac{\pi_1}{1 - \pi_1},
$$

$$
g_r(\boldsymbol{\pi}) = \ln \left(\ln \frac{\sum_{l=1}^{r} \pi_l}{1 - \sum_{l=1}^{r} \pi_l} - \ln \frac{\sum_{l=1}^{r-1} \pi_l}{1 - \sum_{l=1}^{r-1} \pi_l} \right),
$$

$$
r = 2, \ldots, K.
$$

2. F is taken as extreme-minimal-value distribution:

$$
F(\epsilon) = 1 - e^{-e^\epsilon}.
$$

The multiple link function is given by

$$g_1(\boldsymbol{\pi}) = \ln(-\ln(1 - \pi_1)),$$

$$g_r(\boldsymbol{\pi}) = \ln[\ln(-\ln(1 - \sum_{l=1}^{r} \pi_l)) - \ln(-\ln(1 - \sum_{l=1}^{r-1} \pi_l))],$$

$$r = 2, \dots, K.$$

3. F is taken as the extreme-maximal-value distribution:

$$F(\epsilon) = e^{-e^{-\epsilon}}.$$

The multiple link function is given by

$$g_1(\boldsymbol{\pi}) = -\ln(-\ln(\pi_1)),$$

$$g_r(\boldsymbol{\pi}) = \ln[-\ln(-\ln(\sum_{l=1}^{r} \pi_l)) + \ln(-\ln(\sum_{l=1}^{r-1} \pi_l))],$$

$$r = 2, \dots, K.$$

To summarize, we have derived a multivariate generalized linear model for the ordinal trait vector $\boldsymbol{y}$: given QTL genotypes $\boldsymbol{\delta}$, $\boldsymbol{y}$ follows the multinomial distribution (5.28), the QTLs affect the trait through the multiple linear predictor (5.34), and the multiple linear predictor is related to the mean vector of $\boldsymbol{y}$ by the multiple link function (5.35).

Following the same argument as in § 5.1, with observations $\{\boldsymbol{y}_i, (x_{i1}, \dots, x_{im}) : i = 1, \dots, n\}$, we arrive at the following likelihood functions:

$$
\begin{aligned}
L(\boldsymbol{\theta}|\boldsymbol{y}) &= \prod_{i=1}^{n} \sum_{(k_1,\dots,k_m)} \prod_{j=1}^{m} p(x_{ij}, r_j, \gamma_j)^{k_j} [1 - p(x_{ij}, r_j, \gamma_j)]^{1-k_j} \\
&\quad \times \prod_{r=1}^{K+1} \pi_r^{y_{ir}}(\boldsymbol{k}), \quad (5.37)
\end{aligned}
$$

$$
\begin{aligned}
L(\boldsymbol{\theta}|\mathcal{T}) &= \prod_{i=1}^{n} \prod_{j=1}^{m} p(x_{ij}, r_j, \gamma_j)^{\delta_{ij}} [1 - p(x_{ij}, r_j, \gamma_j)]^{1-\delta_{ij}} \\
&\quad \times \prod_{r=1}^{K+1} \pi_r^{y_{ir}}(\boldsymbol{\delta}_i). \quad (5.38)
\end{aligned}
$$

In the derivation above, we confined to the backcross design and assumed no epistasis QTL effects. The results are easily extended to the general case as follows. To include epistasis effects, we only need to add some product terms $\delta_j \delta_k$ to the sum $\sum_{j=1}^{m} \beta_j \delta_j$. For the case of intercross design, we only need to replace the products of Bernoulli densities by the products of trinomial densities in the above likelihood functions.

5.7.2 EM Algorithm for Mixture Polytomous Models

The EM algorithm for mixture polytomous models is similar to that for mixture GLIM (univariate) given in § 5.4.1. The algorithm is based on the likelihood function:

$$L(\boldsymbol{\theta}|\mathcal{T}) = \prod_{i=1}^{n} p_{i\boldsymbol{\delta}_i} \prod_{r=1}^{K+1} \pi_r^{y_{ir}}(\boldsymbol{\delta}_i), \qquad (5.39)$$

where $p_{i\boldsymbol{\delta}_i} = \prod_{j=1}^{m} p(x_{ij}, r_j, \gamma_j)^{\delta_{ij}} [1 - p(x_{ij}, r_j, \gamma_j)]^{1-\delta_{ij}}$ in the case of back-cross design, and $\prod_{j=1}^{m} p_0(x_{ij}, r_j, \gamma_j)^{1-\delta_{ij1}-\delta_{ij2}} p_1(x_{ij}, r_j, \gamma_j)^{\delta_{ij1}} p_2(x_{ij}, r_j, \gamma_j)^{\delta_{ij2}}$ in the case of intercross design. The log of this likelihood is again decomposed into two components:

$$\ell(\boldsymbol{\theta}|\mathcal{T}) = \ell_1(\boldsymbol{r}) + \ell_2(\boldsymbol{\beta})$$

where

$$\ell_1(\boldsymbol{r}) = \sum_{i=1}^{n} \ln p_{i\boldsymbol{\delta}_i}$$

which is linear in the QTL genotype functions and dependent of only QTL position parameters, and

$$\ell_2(\boldsymbol{\beta}) = \sum_{i=1}^{n} \left[\sum_{r=1}^{K} y_{ir} \ln \frac{\pi_r(\boldsymbol{\delta}_i)}{1 - \sum_{s=1}^{K} \pi_r(\boldsymbol{\delta}_i)} + \ln(1 - \sum_{s=1}^{K} \pi_r(\boldsymbol{\delta}_i)) \right].$$

As in § 5.4.1, we need a treatment on $\ell_2(\boldsymbol{\beta})$ for the EM algorithm. Let $\Delta_{ik_1 \cdots k_m}$ be defined the same as in §5.4.1. Let

$$V_{k_1 \ldots k_m} = \sum_{i=1}^{n} \Delta_{ik_1 \ldots k_m}, \quad U_{rk_1 \ldots k_m} = \sum_{i=1}^{n} y_{ir} \Delta_{ik_1 \ldots k_m},$$

$$\pi_{rk_1 \cdots k_m} = \pi_r((k_1, \cdots, k_m)).$$

We express $\ell_2(\boldsymbol{\beta})$ in the following form:

$$\ell_2(\boldsymbol{\beta})$$
$$= \sum_{(k_1 \cdots k_m)} \left[\sum_{r=1}^{K} U_{rk_1 \ldots k_m} \ln \frac{\pi_{rk_1 \cdots k_m}}{1 - \sum_{s=1}^{K} \pi_{sk_1 \cdots k_m}} + V_{k_1 \ldots k_m} \ln(1 - \sum_{r=1}^{K} \pi_{rk_1 \cdots k_m}) \right].$$

In order to describe the EM algorithm, we introduce more notations in the following. Let $Z_{k_1 \cdots k_m}$ be the matrix obtained from the matrix Z defined in (5.33) by replacing the vector $\boldsymbol{z}$ with $(k_1, \ldots, k_m)^{\tau}$. Then construct a big matrix $\boldsymbol{Z}$ by stacking $Z_{k_1 \cdots k_m}$'s one above another in lexicographical order. We can derive from (5.36) the matrix of the first derivatives of $\boldsymbol{\pi}$ with respect to $\boldsymbol{\eta}$:

$$D = \begin{pmatrix} \frac{\partial \pi_1}{\partial \eta_1} & \cdots & \frac{\partial \pi_1}{\partial \eta_K} \\ \cdots & \cdots & \cdots \\ \frac{\partial \pi_K}{\partial \eta_1} & \cdots & \frac{\partial \pi_K}{\partial \eta_K} \end{pmatrix}.$$

Let $D_{k_1\cdots k_m}$ be the matrix D evaluated at $\boldsymbol{\eta} = Z_{k_1\cdots k_m}\boldsymbol{\beta}$. Let

$$\Omega_{k_1\cdots k_m} = \text{Diag}(\pi_{1k_1\cdots k_m}^{-1}, \ldots, \pi_{Kk_1\cdots k_m}^{-1}) + (1 - \sum_{r=1}^{K} \pi_{rk_1\cdots k_m})^{-1}\mathbf{11}^T,$$

where $\text{Diag}(\cdots)$ denotes a diagonal matrix with diagonal elements given by its arguments and $\mathbf{1}$ is a vector of all 1's. Define the weight matrix

$$W_{k_1\cdots k_m} = \tilde{V}_{k_1\cdots k_m} D_{k_1\cdots k_m}^T \Omega_{k_1\cdots k_m} D_{k_1\cdots k_m}.$$

Let $\boldsymbol{W}$ be the diagonal block matrix with diagonal blocks $W_{k_1\cdots k_m}$ arranged in lexicographical order. Let

$$\boldsymbol{z}_{k_1\cdots k_m} = Z_{k_1\cdots k_m}\boldsymbol{\beta} + D_{k_1\cdots k_m}^{-1}(\tilde{V}_{k_1\cdots k_m}^{-1}\tilde{\boldsymbol{U}}_{k_1\cdots k_m} - \boldsymbol{\pi}_{k_1\cdots k_m}),$$

where

$$\tilde{\boldsymbol{U}}_{k_1\cdots k_m} = (\tilde{U}_{1k_1\cdots k_m}, \ldots, \tilde{U}_{Kk_1\cdots k_m})^\tau, \quad \boldsymbol{\pi}_{k_1\cdots k_m} = (\pi_{1k_1\cdots k_m}, \ldots, \pi_{Kk_1\cdots k_m})^\tau.$$

Denote by $\boldsymbol{z}$ the vector obtained by concatenating $\boldsymbol{z}_{k_1\cdots k_m}$'s in lexicographical order.

The EM algorithm for mixture polytomous GLIM is tailored as follows.

Initialization: Form the matrix $\boldsymbol{Z}$. Set starting values $\boldsymbol{\beta}^{(0)}$ and $\boldsymbol{r}^{(0)}$.

E step: At $(\boldsymbol{\beta}^{(0)}, \boldsymbol{r}^{(0)})$, compute the arrays $\{\tilde{\Delta}_{ik_1\cdots k_m}\}$ by the formula

$$\tilde{\Delta}_{ik_1\cdots k_m} = \frac{p_{ik_1\cdots k_m} \prod_{r=1}^{K+1} \pi_{rk_1\cdots k_m}^{(0)y_{ir}}}{\sum_{(\nu_1,\ldots,\nu_m)} p_{i\nu_1\cdots\nu_m} \prod_{r=1}^{K+1} \pi_{r\nu_1\cdots\nu_m}^{(0)y_{ir}}}.$$

M step: (i) Update $\boldsymbol{\beta}$:

    (a) Set $\boldsymbol{\beta}^{\text{OLD}} = \boldsymbol{\beta}^{(0)}$.

    (b) Compute $\boldsymbol{z}(\boldsymbol{\beta}^{\text{OLD}})$ and $\boldsymbol{W}(\boldsymbol{\beta}^{\text{OLD}})$.

    (c) Solve for $\boldsymbol{\beta}^{\text{NEW}}$ in the equation:

$$\boldsymbol{Z}^\tau \boldsymbol{W}(\boldsymbol{\beta}^{\text{OLD}})\boldsymbol{Z}\boldsymbol{\beta}^{\text{NEW}} = \boldsymbol{Z}^\tau \boldsymbol{W}(\boldsymbol{\beta}^{\text{OLD}})\boldsymbol{z}(\boldsymbol{\beta}^{\text{OLD}}).$$

    (d) Check convergence. If convergent, set $\boldsymbol{\beta}^{(1)} = \boldsymbol{\beta}^{\text{NEW}}$ and go to (ii), otherwise set $\boldsymbol{\beta}^{\text{OLD}} = \boldsymbol{\beta}^{\text{NEW}}$ and goto (b).

  (ii) Update $\boldsymbol{r}^{(0)}$ to $\boldsymbol{r}^{(1)}$ by maximizing $\ell_1(\boldsymbol{r})$.

  (iii) Check convergence, if convergent, stop, otherwise set $\boldsymbol{\beta}^{(0)} = \boldsymbol{\beta}^{(1)}$, $\boldsymbol{r}^{(0)} = \boldsymbol{r}^{(1)}$ and go to E step.

The step (ii) of the M step above is the same as step (iii) of the M step of the EM algorithm in §5.4.

5.7.3 Multiple Interval Mapping with Nominal Traits

We consider nominal traits in this sub section. Nominal traits cannot be treated in the same way as the ordinal traits. While an ordinal trait can be determined by a single latent continuous variable with an ordered sequence of threshold values, there is no such mechanism for a nominal trait. The categories of a nominal trait do not have a meaningful order. Instead, we can consider that the categories of a nominal trait are determined by different biological forces. Each biological force competes with the others. If a particular force beats all the others then a certain particular category is realized. This mechanism is called the principle of maximum random utility in [52]. In what follows, we use this mechanism to formulate the mixture GLIM for nominal traits.

Let $\{U_r : r = 1, \ldots, K + 1\}$ represent the biological forces. The trait phenotype takes category r if U_r beats all the other forces, i.e., if $U_r > U_s$ for all $s \neq r$. Suppose that the competing forces are governed by the QTLs as follows:

$$U_r = u_r + \epsilon_r, \quad r = 1, \ldots, K + 1,$$

where

$$u_r = \alpha_{r0} + \sum_{j=1}^{m} \alpha_{rj}\delta_j.$$

Here, for the ease of description, we focus on the case of backcross design and ignore epstasis effects. The ϵ_r's are assumed to be independent identically distributed random errors with a common distribution function G and density function g. Under the above assumption, the probabilities in the multinomial density function (5.28) are determined as follows:

$$
\begin{aligned}
\pi_r(\boldsymbol{\delta}) &= P(U_r \geq U_1, \ldots, U_r \geq U_{K+1}) \\
&= P(\epsilon_1 \leq u_r - u_1 + \epsilon_r, \ldots, \epsilon_{K+1} \leq u_r - u_{K+1} + \epsilon_r) \\
&= \int \left[\prod_{s \neq r} G(u_r - u_s + \epsilon) \right] g(\epsilon)d\epsilon, \\
& r = 1, \ldots, K.
\end{aligned}
$$

Now, consider the linear structure:

$$
\begin{aligned}
\eta_r &= u_r - u_{K+1} = \beta_{r0} + \sum_{j=1}^{m} \beta_{rj}\delta_j, \\
& r = 1, \ldots, K,
\end{aligned}
\tag{5.40}
$$

where $\beta_{rj} = \alpha_{rj} - \alpha_{K+1j}$, $j = 0, \ldots, m$. Let $\boldsymbol{z} = (1, \delta_1, \ldots, \delta_m)^{\tau}$, $\boldsymbol{\beta}_r =$

$(\beta_{r0}, \beta_{r1}, \ldots, \beta_{rm})^\tau$ and $\boldsymbol{\beta} = (\boldsymbol{\beta}_1^\tau, \ldots, \boldsymbol{\beta}_K^\tau)^\tau$. Define

$$
\boldsymbol{Z} = \begin{pmatrix}
\boldsymbol{z}^\tau & \boldsymbol{0}^\tau & \cdots & \boldsymbol{0}^\tau & \boldsymbol{0}^\tau \\
\boldsymbol{0}^\tau & \boldsymbol{z}^\tau & \cdots & \boldsymbol{0}^\tau & \boldsymbol{0}^\tau \\
\cdots & \cdots & \cdots & \cdots & \cdots \\
\boldsymbol{0}^\tau & \boldsymbol{0}^\tau & \cdots & \boldsymbol{z}^\tau & \boldsymbol{0}^\tau \\
\boldsymbol{0}^\tau & \boldsymbol{0}^\tau & \cdots & \boldsymbol{0}^\tau & \boldsymbol{z}^\tau
\end{pmatrix}.
$$

Then we can express $\boldsymbol{\eta} = (\eta_1, \ldots, \eta_K)^\tau$ as

$$
\boldsymbol{\eta} = \boldsymbol{Z}\boldsymbol{\beta}. \tag{5.41}
$$

We take this linear structure as the multiple linear predictor in the GLIM. In terms of η_r's,

$$
\pi_r(\boldsymbol{\delta}) = \int \left[\prod_{s \neq r} G(\eta_r - \eta_s + \epsilon) \right] g(\epsilon) d\epsilon, \tag{5.42}
$$

$$
r = 1, \ldots, K.
$$

When G is specified, we can solve the η_r's from (5.42) to obtain the multiple link function. In particular, if G is taken to be the extreme-value distribution function: $G(\epsilon) = e^{-e^{-\epsilon}}$, it can be derived that

$$
\pi_r = \frac{e^{\eta_r}}{1 + \sum_{s=1}^{K} e^{\eta_s}}, \quad r = 1, \ldots, K.
$$

By solving the above equations for the η_r's, we obtain the link function as

$$
\eta_r = \ln \frac{\pi_r}{1 - \sum_{s=1}^{K} \pi_s}, \quad r = 1, \ldots, K.
$$

The mixture GLIM for nominal traits is the same as that for ordinal traits, except that the multiple linear predictor and the multiple link function are determined as above. The EM algorithm also remains unchanged except that the content of matrix $\boldsymbol{Z}$ is different.

5.8 Multiple Interval Mapping of Traits with Spike Distributions

In survival analysis where the trait of concern is the survival time, experimental units are usually censored. Especially, in animal studies, the experimental units are censored at the end of the experiment. In such situations, the trait distribution displays a spike; that is, there is a positive point mass in the distribution, otherwise, the distribution is continuous. For example, in an animal study on mouse infection of *Listeria monocytogenes* bacterium [16], among 116 infected F2 (CB6F2/ByJ) mice, a large proportion survived past the 240-hour time point, which is considered to be recovered, and the other mice displayed a wide range of survival times with a mean of 106 hours. The distribution of the survival time has a continuous part between 0 and 240 and, in addition, has a positive mass at point 240. For more examples, see [91] [199]. We refer to the distribution that displays a spike as a spike distribution. Traits with spike distributions need special treatments. In this section, we focus on the multiple interval mapping approach developed in [118]. Some single-QTL methods are briefly discussed in § 5.8.5.

5.8.1 The Mixture Models for Traits with Spike Distributions

A spike distribution can be considered as a mixture of a continuous distribution and a point mass distribution. Without loss of generality, assume the point mass is at zero. The density function of a random variable Y with a spike distribution is given by

$$(1 - \pi)I\{y = 0\} + \pi I\{y \neq 0\}f(y),$$

where π is the proportion that Y has a continuous distribution and $f(y)$ is the density function of that distribution. Let $z = I\{y \neq 0\}$. The density of the spike distribution can be expressed as

$$(1 - \pi)^{1-z}[\pi f(y)]^z. \tag{5.43}$$

For a trait with a spike distribution, assume its continuous part follows an exponential family distribution:

$$f(y) = \exp\{\frac{\theta y - b(\theta)}{\tau}\}.$$

Given the QTL genotype variables $\boldsymbol{\delta}$, the density of the trait distribution is given by

$$[1 - \pi(\boldsymbol{\delta})]^{1-z}[\pi(\boldsymbol{\delta}) \exp\left\{\frac{\theta(\boldsymbol{\delta})y - b(\theta(\boldsymbol{\delta}))}{\tau}\right\}]^z. \tag{5.44}$$

By the same argument as in the previous sections, the joint density of Y and $\boldsymbol{\delta}$ is given by

$$f(\boldsymbol{y}, \boldsymbol{\delta}|\boldsymbol{\theta}) = p(\boldsymbol{x}, \boldsymbol{r}, \boldsymbol{\delta})[1 - \pi(\boldsymbol{\delta})]^{1-z}\left[\pi(\boldsymbol{\delta}) \exp\left\{\frac{\theta(\boldsymbol{\delta})y - b(\theta(\boldsymbol{\delta}))}{\tau}\right\}\right]^z, \tag{5.45}$$

where $p(\boldsymbol{x}, \boldsymbol{r}, \boldsymbol{\delta})$ is defined the same as in the previous sections. The marginal density of $\boldsymbol{y}$ is obtained by summing up the product over all possible values of the $\boldsymbol{\delta}$.

Let

$$\eta_\mu(\boldsymbol{\delta}) = \beta_{\mu 0} + \sum_{j=1}^{m} \beta_{\mu j} \delta_j,$$

and

$$\eta_\pi(\boldsymbol{\delta}) = \beta_{\pi 0} + \sum_{j=1}^{m} \beta_{\pi j} \delta_j.$$

Suppose that the QTLs affect μ, the mean of the continuous part, through the linear predictor η_μ and affect the proportion π through the linear predictor η_π. If epistasis effects are considered, product terms are added to the linear predictors. Let g_1 and g_2 be two link functions such that

$$\eta_\mu = g_1(\mu), \quad \eta_\pi = g_2(\pi).$$

A common choice for g_2 is the logistic link $g_2(\pi) = \log\left(\frac{\pi}{1-\pi}\right)$.

With the above assumptions, we arrive at the mixture GLIM for the trait Y. Given observations $\{y_i, x_{i1}, \ldots, x_{im} : i = 1, \ldots, n\}$, the incomplete and complete likelihood functions are given as follows:

$$
\begin{aligned}
L(\boldsymbol{\theta}|\boldsymbol{y}) &= \prod_{i=1}^{n} \sum_{(k_1,\ldots,k_m)} p(\boldsymbol{x}_i, \boldsymbol{r}, \boldsymbol{k})[1 + e^{\eta_\pi(\boldsymbol{k})}]^{-1+z_i} \\
&\quad \times \left[\frac{e^{\eta_\pi(\boldsymbol{k})}}{1 + e^{\eta_\pi(\boldsymbol{k})}} \exp\{[y_i\theta(\boldsymbol{k}) - b(\theta(\boldsymbol{k})]/\tau\} \right]^{z_i}, \quad (5.46)
\end{aligned}
$$

$$
\begin{aligned}
L(\boldsymbol{\theta}|\mathcal{T}) &= \prod_{i=1}^{n} p(\boldsymbol{x}_i, \boldsymbol{r}, \boldsymbol{\delta}_i)[1 + e^{\eta_\pi(\boldsymbol{\delta}_i)}]^{-1+z_i} \\
&\quad \times \left[\frac{e^{\eta_\pi(\boldsymbol{\delta}_i)}}{1 + e^{\eta_\pi(\boldsymbol{\delta}_i)}} \exp\{[y_i\theta(\boldsymbol{k}) - b(\theta(\boldsymbol{k})]/\tau\} \right]^{z_i}. \quad (5.47)
\end{aligned}
$$

5.8.2 EM Algorithm for Traits with Spike Distributions

By modifying the the general EM algorithm in § 5.3, we obtain the EM algorithm for traits with spike distributions. The log of (5.47) is decomposed into three components:

$$\ell(\boldsymbol{\theta}|\mathcal{T}) = \ell_0(\boldsymbol{r}) + \frac{1}{\tau}\ell_1(\boldsymbol{\beta}_\mu) + \ell_2(\boldsymbol{\beta}_\pi).$$

The components are given by

$$\ell_0(\boldsymbol{r}) = \sum_{i=1}^{n} \ln(p(\boldsymbol{x}_i, \boldsymbol{r}, \boldsymbol{\delta}_i)),$$

$$\ell_1(\boldsymbol{\beta}_\mu) = \sum_{i=1}^{n} z_i[(y_i\theta_\mu(\boldsymbol{\delta}_i, \boldsymbol{\beta}_\mu) - b_\mu(\theta_\mu(\boldsymbol{\delta}_i, \boldsymbol{\beta}_\mu)))],$$

$$\ell_2(\boldsymbol{\beta}_\pi) = \sum_{i=1}^{n} [z_i\theta_\pi(\boldsymbol{\delta}_i, \boldsymbol{\beta}_\pi) - b_\pi(\theta_\pi(\boldsymbol{\delta}_i, \boldsymbol{\beta}_\pi))],$$

where $\theta_\pi(\boldsymbol{\delta}_i) = \ln\left(\frac{\pi(\boldsymbol{\delta}_i,\boldsymbol{\beta}_\pi)}{1-\pi(\boldsymbol{\delta}_i,\boldsymbol{\beta}_\pi)}\right)$ and $b_\pi(\theta_\pi) = \ln(1 + \exp(\theta_\pi))$. Let $\Delta_{ik_1\cdots k_m}$ be defined the same as before. Let $\theta_{\mu k_1\cdots k_m} = \theta_\mu((k_1,\ldots,k_m), \boldsymbol{\beta}_\mu)$ and $\theta_{\pi k_1\cdots k_m} = \theta_\pi((k_1,\ldots,k_m), \boldsymbol{\beta}_\pi)$. Then $\ell_1(\boldsymbol{\beta}_\mu)$ and $\ell_2(\boldsymbol{\beta}_\pi)$ are expressed as

$$\ell_1(\boldsymbol{\beta}_\mu) = \sum_{(k_1\cdots k_m)} [\theta_{\mu k_1\cdots k_m} \sum_{i=1}^{n} z_iy_i\Delta_{ik_1\cdots k_m} - b_\mu(\theta_{\mu k_1\cdots k_m}) \sum_{i=1}^{n} z_i\Delta_{ik_1\cdots k_m}],$$

$$\ell_2(\boldsymbol{\beta}_\pi) = \sum_{(k_1\cdots k_m)} [\theta_{\pi k_1\cdots k_m} \sum_{i=1}^{n} z_i\Delta_{ik_1\cdots k_m} - b_\pi(\theta_{\pi k_1\cdots k_m}) \sum_{i=1}^{n} \Delta_{ik_1\cdots k_m}].$$

We now introduce some notations similar to those in the general EM algorithm. Let

$$\Delta_{k_1\cdots k_m}^{[z]} = \sum_{i=1}^{n} z_i\Delta_{ik_1\cdots k_m}$$

and

$$w_{\mu k_1\cdots k_m} = \frac{\Delta_{k_1\cdots k_m}^{[z]}}{[g_1'(\mu_{k_1\cdots k_m})]^2 b_\mu''(\theta_{\mu k_1\cdots k_m})}.$$

Denote by $W_\mu(\boldsymbol{\beta}_\mu)$ the diagonal matrix with diagonal elements $w_{\mu k_1\cdots k_m}$. Let

$$\Delta_{k_1\cdots k_m}^{[zl]} = \sum_{i=1}^{n} z_iy_i^l\Delta_{ik_1\cdots k_m}, \quad l = 1, 2,$$

and

$$v_{\mu k_1\cdots k_m} = g_1'(\mu_{k_1\cdots k_m})\left(\frac{\Delta_{k_1\cdots k_m}^{[z1]}}{\Delta_{k_1\cdots k_m}^{[z]}} - \mu_{k_1\cdots k_m}\right).$$

Define $\boldsymbol{z}_\mu(\boldsymbol{\beta}_\mu) = X_\mu\boldsymbol{\beta}_\mu + \boldsymbol{v}_\mu(\boldsymbol{\beta}_\mu)$, where $\boldsymbol{v}_\mu(\boldsymbol{\beta}_\mu)$ is the vector with components $v_(\mu, k_1\cdots k_m)$, and X_μ is the matrix obtained by stacking the vectors $\boldsymbol{x}_{\mu k_1\cdots k_m}$ one above another in lexicographical order. Here $\boldsymbol{x}_{\mu k_1\cdots k_m}$ is the vector of the covariable terms in the linear predictor $\eta_\mu(\boldsymbol{k})$. Similarly, define $W_\pi(\boldsymbol{\beta}_\pi)$ and $\boldsymbol{z}_\pi(\boldsymbol{\beta}_\pi)$ by replacing $g_1, b_\mu, \boldsymbol{\beta}_\mu, \Delta^{[z]}, \Delta^{[z1]}$ and X_μ with $g_2, b_\pi, \boldsymbol{\beta}_\pi, \Delta^{[0]}, \Delta^{[z]}$ and X_π respectively. Here X_π is the matrix obtained in the same way as X_μ

from the row vectors $x_{\pi k_1 \cdots k_m}$ which is vector of the covariable terms in the linear predictor $\eta_{\pi}(k)$.

The modified EM algorithm is as follows.

Initialization: Form the matrix X. Set starting values $\beta_{\mu}^{(0)}, \beta_{\pi}^{(0)}, \tau^{(0)}$ and $r^{(0)}$.

E step: At $\beta_{\mu}^{(0)}, \beta_{\pi}^{(0)}, \tau^{(0)}, r^{(0)})$, compute the arrays

$$\{\tilde{\Delta}_{ik_1 \cdots k_m}\} = \{E(\Delta_{ik_1 \cdots k_m}|y, \beta^{(0)}, \tau^{(0)}, r^{(0)})\},$$
$$\{\tilde{\delta}_{ij}\} = \{E(\delta_{ij}|y, \beta^{(0)}, \tau^{(0)}, r^{(0)})\}, \quad \{\tilde{\Delta}_{k_1 \cdots k_m}^{[l]}\}, \ l = 0, 1, 2.$$

M step: (i) Update β_{μ}:

    (ia) Set $\beta_{\mu}^{\mathrm{OLD}} = \beta_{\mu}^{(0)}$.

    (ib) Compute $z_{\mu}(\beta_{\mu}^{\mathrm{OLD}})$ and $W_{\mu}(\beta_{\mu}^{\mathrm{OLD}})$.

    (ic) Solve for $\beta_{\mu}^{\mathrm{NEW}}$ in the equation:

$$X_{\mu}' W_{\mu}(\beta_{\mu}^{\mathrm{OLD}}) X_{\mu} \beta_{\mu}^{\mathrm{NEW}} = X_{\mu}' W_{\mu}(\beta_{\mu}^{\mathrm{OLD}}) z_{\mu}(\beta_{\mu}^{\mathrm{OLD}}).$$

    (id) Check convergence. If convergent, set $\beta_{\mu}^{(1)} = \beta_{\mu}^{\mathrm{NEW}}$ and goto (ii), otherwise set $\beta_{\mu}^{\mathrm{OLD}} = \beta_{\mu}^{\mathrm{NEW}}$ and goto (ib).

(ii) Update β_{π}:

    (iia) Set $\beta_{\pi}^{\mathrm{OLD}} = \beta_{\pi}^{(0)}$.

    (iib) Compute $z_{\pi}(\beta_{\pi}^{\mathrm{OLD}})$ and $W_{\pi}(\beta_{\pi}^{\mathrm{OLD}})$.

    (iic) Solve for $\beta_{\pi}^{\mathrm{NEW}}$ in the equation:

$$X_{\pi}' W_{\pi}(\beta_{\pi}^{\mathrm{OLD}}) X_{\pi} \beta_{\pi}^{\mathrm{NEW}} = X_{\pi}' W_{\pi}(\beta_{\pi}^{\mathrm{OLD}}) z_{\pi}(\beta_{\pi}^{\mathrm{OLD}}).$$

    (iid) Check convergence. If convergent, set $\beta_{\pi}^{(1)} = \beta_{\pi}^{\mathrm{NEW}}$ and go to (iii), otherwise set $\beta_{\pi}^{\mathrm{OLD}} = \beta_{\pi}^{\mathrm{NEW}}$ and go to (iib).

(iii) Update τ by the following formula evaluated at $\beta_{\mu}^{(1)}$.

$$\hat{\tau} = \frac{1}{n_{\mu} - q} \sum_{(k_1 \cdots k_m)} \frac{\tilde{\Delta}_{k_1 \cdots k_m}^{[z2]} - 2\mu_{k_1 \cdots k_m} \tilde{\Delta}_{k_1 \cdots k_m}^{[z1]} + \mu_{k_1 \cdots k_m}^2 \tilde{\Delta}_{k_1 \cdots k_m}^{[z]}}{b_{\mu}''(\theta_{k_1, \ldots, k_m})}.$$

(iv) Update r by maximizing

$$\sum_{i=1}^{n} \ln p(x_i, r, \tilde{\delta}_i).$$

(v) Check convergence, if convergent, stop, otherwise set $\beta_{\mu}^{(0)} = \beta_{\mu}^{(1)}$, $\beta_{\pi}^{(0)} = \beta_{\pi}^{(1)}$, $\tau^{(0)} = \tau^{(1)}$, $r^{(0)} = r^{(1)}$ and go to E step.

5.8.3 The Mapping Strategy

The mapping procedure in § 5.5 is modified for the mapping of traits with spike distributions in this sub section. For the representation of the models, we follow the convention in § 5.5 with linear predictors η_μ and η_π. The QTL genotype variable with a superscript μ or π indicates that the variable is involved in η_μ or η_π. For example, $\delta_1^\mu + \delta_2^\mu + \delta_1^\pi$ represents the model with linear predictor $\eta_\mu = \beta_{\mu 0} + \beta_{\mu 1}\delta_1 + \beta_{\mu 2}\delta_2$ and $\eta_\pi = \beta_{\pi 0} + \beta_{\pi 1}\delta_1$. The modified procedure is as follows.

Step 1: Consider all the models of the form $\delta_j^\mu + \delta_j^\pi$. For the model with the largest maximum likelihood, evaluate whether or not either of δ_j^μ and δ_j^π can be removed using EBIC. If at least one of them cannot be removed, continue to the next step; otherwise, stop.

Step m: Denote the model selected after $m-1$ steps by $M_{m-1} = \sum_{r \in M_\mu} \delta_r^\mu + G_{I,m-1}^\mu + \sum_{s \in M_\pi} \delta_s^\pi + G_{I,m-1}^\pi$ where M_μ and M_π are the index sets of the δ_j's in η_μ and η_π respectively, and $G_{I,m-1}^\mu$ and $G_{I,m-1}^\pi$ are the sums of interaction terms in η_μ and η_π respectively. Now, for each remaining $j \notin M_\mu \cup M_\pi$, consider the models $M_{m-1} + \delta_j^\mu + \sum_{r \in M_\mu} \delta_r \delta_j + \delta_j^\pi + \sum_{s \in M_\pi} \delta_s \delta_j$. Let j_m be the index such that the above model has the largest maximum likelihood. Then, evaluate whether or not any of the new interaction terms can be removed by deleting them one at a time in the same way as in the procedure in § 5.5. If at a stage, an interaction term involved in η_μ cannot be removed, then evaluate whether or not any new terms in η_π can be removed and then continue to the next step; or if an interaction term involved in η_π cannot be removed, then evaluate whether or not any new terms in η_μ can be removed and then continue to the next step; If all the new interactions as well as the new main effects are removed, stop.

The EBIC for the spike distribution model is also modified. Let $q_{\mu M}$, $q_{\mu I}$ and $q_{\pi M}$, $q_{\pi I}$ be the numbers of main effects and interactions in η_μ and η_π respectively. Let q_M and q_I be the total number of distinct main effects and interactions in both η_μ and η_π. Note that they are not the sum of the corresponding numbers in η_μ and η_π. Let $n_\mu = \sum_{i=1}^n z_i$ and $n_\pi = n - n_\mu$. The modified version is given below:

$$\text{EBIC}(G) = -2 \ln L(\hat{\boldsymbol{\theta}}_G | \boldsymbol{y}) + (q_{\mu M} + q_{\mu I}) \ln n_\mu + (q_{\pi M} + q_{\pi I}) \ln n_\pi$$
$$+ 2\gamma_M \ln \binom{M}{q_M} + 2\gamma_I \ln \binom{\frac{M(M-1)}{2}}{q_I}.$$

5.8.4 Example: The Analysis of Mouse *Listeria monocytogenes* Infection Data

The mouse *Listeria monocytogenes* infection data was mentioned at the beginning of this section. The data consists of the time to death (in units of hours) following infection with *L. monocytogenes* of 116 F_2 mice from an intercross between the BALB/cByJ and C57BL/6ByJ strains and the mice's genotypes at 133 markers on 20 chromosomes. The data was analyzed in [118] using the methods described in this section. The analysis is presented in the following.

In the mixture GLIM, the exponential family distribution f for the continous part is taken as the normal distribution (which serves as a satisfactory approximation to the uncensored survival time in this example). In the two linear predictors, only the main effects of the QTL are assumed; that is,

$$\eta_\mu(\delta) = \beta_{\mu 0} + \sum_{j=1}^m \beta_{\mu j}\delta_j, \quad \text{and} \quad \eta_\pi(\delta) = \beta_{\pi 0} + \sum_{j=1}^m \beta_{\pi j}\delta_j.$$

The link function for the continuous part is taken as the identity link, i.e., $\eta_\mu = g_1(\mu) = \mu$. The link for the binary part is taken as the logistic link $g_2(\pi) = \log\left(\frac{\pi}{1-\pi}\right)$.

When there are no interaction terms in both linear predictors, the procedure described in § 5.8.3 reduces to a simple forward procedure. In the forward procedure used in [118], a QTL is considered simultaneously in both η_μ and η_π; that is, a term δ_j is either present in or absent from both η_μ and η_π, which is slightly different from that described in § 5.8.3.

In what follows, the notation $[k, d]$ denotes a locus on chromosome k with a genetic distance d cM from the left end of the chromosome. At step 1 of the forward procedure, locus [13, 27] is selected with an EBIC value 153.02. At step 2, loci [13, 26.5] and [5, 28] are selected with an EBIC value 141.11, and so on. The loci selected together with the EBIC value at each step of the whole procedure are given in Table 5.2. The detected loci in the last step of the procedure are: [1, 81], [2, 3.5], [5, 29.0], [6, 13.0], [8, 10.0], [13, 26.5] and [13, 13.05]. The positions of the loci are slightly different from the intermediate steps because they are re-estimated at each step.

5.8.5   Single-QTL Methods and Comparisons

A few single-QTL methods are studied in [19]. In this sub section, we provide a brief description of those methods and summarize the findings from the literature on the comparison among the single-QTL methods as well as the comparison between the single-QTL model methods and the multiple interval mapping approach.

The trait data can be considered consisting of two parts: a binary part, i.e, the z_i's, and the continuous part, i.e., the y_i's which are non-zeros (by the convention of § 5.8.1). The first single-QTL method is a naive one that separately analyzes the binary part and the continuous part. Single interval mapping approaches are used for the analyses. For the continuous part, the distribution of the trait is assumed Gaussian, the method discussed in § 4.2.1 is applied. For the binary part, the mixture models in § 4.2.1 are adapted by

TABLE 5.2: The loci included and the corresponding EBIC value at each step of the multiple interval mapping procedure in the mouse *Listeria monocytogenes* infection example.

Step	Loci included	EBIC
1	[13, 27]	153.02
2	[13, 26.5] [5, 28]	141.11
3	[13, 26.5] [5, 28] [1, 81]	138.78
4	[13, 26.5] [5, 28] [1, 81] [6, 14]	136.88
5	[13, 26.5] [5, 28] [1, 81] [6, 14] [2, 4]	136.74
6	[13, 26.5] [5, 29] [1, 81] [6, 14] [2, 3.5] [8, 8.5]	128.31
7	[13, 26.5] [5, 29] [1, 81] [6, 13] [2, 3.5] [8, 10] [13, 13.05]	124.48

replacing the normal density functions with the Bernoulli density:

$$\pi_\delta^{1-z_i}(1 - \pi_\delta)^{z_i}.$$

The second single-QTL method is referred to as the two-part model method in [19]. It is essentially a special case of the multiple interval mapping model when the number of interval is reduced to one. Since only one interval is involved, there is no need for the GLIM structure; it is simply a mixture of three components of the form:

$$L(\boldsymbol{\theta}) = \prod_{i=1}^{n} \sum_{j=0}^{2} p_j(x_i, r, \gamma)(1 - \pi_j)^{1-z_i}[\pi_j f(y_i; \mu_j, \sigma^2)]^{z_i},$$

where $\boldsymbol{\theta} = (\pi_0, \pi_1, \pi_2, \mu_1, \mu_1, \mu_2, \sigma^2)^\tau$. The LOD score given by $\log_{10}[L(\hat{\boldsymbol{\theta}})/L(\hat{\boldsymbol{\theta}}_0)]$, where $\hat{\boldsymbol{\theta}}$ and $\hat{\boldsymbol{\theta}}_0$ are the MLE of $\boldsymbol{\theta}$ under full and null hypothesis respectively, is taken as the test statistic at the fixed position determined by r and the interval.

The third single-QTL method studied in [19] is a mixture model version of the Kruskal-Wallis test approach proposed in [108]. The approach is as follows. Rank the trait values y_i, $i = 1, \ldots, n$. Let R_i be the rank of individual i. If there are ties, each individual in a group of ties receives the average rank of the group. For $j = 0, 1, 2$, corresponding to QTL genotype qq, Qq and QQ, let $S_j = \sum_{i=1}^{n} p_j(x_i, r, \gamma)R_i$. For simplicity, we denote $p_j(x_i, r, \gamma)$ by p_{ij}. The test statistic is given by

$$H = \sum_{j=0}^{2} \left(\frac{n - \sum_{i=1}^{n} p_{ij}}{n} \right) \left[\frac{(S_j - E_{0j})^2}{V_{0j}} \right],$$

where E_{0j} and V_{0j} are the mean and variance of S_j under the null hypothesis that $\mu_0 = \mu_1 = \mu_2$. It is derived that

$$H = \frac{12}{n(n+1)} \sum_{j=1}^{2} \frac{(n - \sum_{i=1}^{n} p_{ij})(\sum_{i=1}^{n} p_{ij})^2}{n \sum_{i=1}^{n} p_{ij}^2 - (\sum_{i=1}^{n} p_{ij})^2} \left(\frac{S_j}{\sum_{i=1}^{n} p_{ij}} - \frac{n+1}{2} \right)^2.$$

The permutation method discussed in § 4.4.3 is used to determine the threshold value in the above single-QTL model approaches.

A comparison among the three single-QTL model methods is made by simulation studies in [19]. It is found that (i) if the QTL affects the probabilities π_j's only, or the conditional means μ_j only, the separate analysis over-performs both the two-part model and the non-parametric approaches, (ii) if the QTL affects both the π_j's and the μ_j's, the non-parametric approach over-performs the separate analysis and the two-part model approach. However, the two-part model approach is robust in the sense that, in all situations, it is competitive to the best and better than the other method.

The single-QTL methods suffer the general drawbacks mentioned at the beginning of this chapter. Simulation studies are carried out in [118] to compare the single-QTL two-part model method and the multiple interval mapping approach. It is revealed that the multiple interval mapping approach is much more efficient than the single-QTL two-part model method in terms of positive discovery rate (PDR) and false discovery rate (FDR). Unless the heritability is very low, the multiple interval mapping approach has a much higher PDR and a comparable or lower FDR than the two-part model method.

Chapter 6

QTL Mapping with Dense Markers

With the advance of biotechnology, genotyping of a huge number of densely distributed genetic markers becomes cheap and fast. Single nucleotide polymorphisms (SNPs), which are tens or hundreds of thousands in number, are typical genetic markers. Since the markers are densely distributed, they present in the vicinity of any loci such as QTLs. When a marker is close to a QTL, the recombination fraction r between them is close to zero. Thus the genotypes of the marker and the QTL are simultaneously either homozygous or heterozygous with a negligibly small chance of exception. Thus the markers in the vicinity of a QTL can be treated as surrogates of that QTL. The mapping of QTL can then be done through the mapping of those surrogate markers.

When using markers as surrogates of QTLs, the mixture models considered in Chapter 5 simplify to non-mixture linear and generalized linear models. But these linear and generalized linear models differ from the traditional models in an important aspect: in these models, the number of covariates (the markers) is huge and is usually much larger than the number of observations (experiment units). Such models are referred to as small-n-large-p models in the statistical literature, where n stands for the number of observations and p for the number of covariates. The problem of QTL mapping boils down to the problem of feature selection in small-n-large-p models; that is, the selection of the markers that fall into close vicinities of QTLs. A brief overview on feature selection in small-n-large-p models is given in § 2.9.

In this chapter, we consider the method of feature selection for QTL mapping with dense markers. In § 6.1, we discuss the strategy and method of feature selection for QTL mapping without the consideration of epstatic effects. In § 6.2, we deal with the case that epistatic effects are taken into account.

6.1 Feature Selection Methods for QTL Mapping

In practical feature selection, when the number of features, p, is too large, a preliminary screening is commonly adopted to reduce the unnecessary huge amount of computation. We first discuss methods of preliminary screening for

QTL mapping in § 6.1.1. In § 6.1.2, we consider QTL mapping using penalized likelihood methods and in § 6.1.4 we consider QTL mapping using sequential methods.

In the sparse high-dimensional models for QTL mapping, the response variable Y is the quantitative trait value and the covariate x_j is a variable associated with the jth marker. The covariate x_j could be the genotype, the expression level, and so on. In most of studies, x_j is the dummy variable determined by the genotype of the jth marker as follows. Let mm, mM and MM denote the genotypes of marker j. For backcross and RIL designs,

$$x_j = \begin{cases} 1, & \text{if the genotype at marker } j \text{ is } MM, \\ 0, & \text{otherwise.} \end{cases}$$

For intercross design,

$$x_j = \begin{cases} 2, & \text{if the genotype at marker } j \text{ is } MM, \\ 1, & \text{if the genotype at marker } j \text{ is } mM, \\ 0, & \text{otherwise.} \end{cases}$$

In the sparse high-dimensional models, this single covariate is used to represent the genotype of marker j in all the designs. As we have discussed in § 3.3, in general, the effect of a QTL is decomposed as an additive effect and a dominant effect, which are represented by two features — an additive feature as defined above and a dominant feature, see (3.5). By using only x_j, the additive feature, to represent marker j, we ignore the dominant feature when the design is intercross. This rarely affects the efficiency of the feature selection procedures since the dominant effect is either zero (if the genetic mode of the marker is additive) or small (in the case of non-additive modes) and hence plays very little role in mapping QTLs. However, the computation amount is greatly reduced since the total number of features (including both additive and dominant ones) is reduced by half.

6.1.1 Preliminary Screening

A simple screening approach called sure independence screening (SIS) [56] is introduced in this sub section. The SIS uses single-feature models to rank the features in terms of their association with the response variable. Denote by $\boldsymbol{y} = (y_1, \ldots, y_n)^\tau$ the vector of Y-values and by $\boldsymbol{x}_j = (x_{1j}, \ldots, x_{nj})^\tau$ the vector of x_j-values observed on n experimental units. Let the observations be standardized so that, for each variable, the sample mean is 0 and the sample standard deviation is 1. The same notation is used to denote the standardized vectors.

Let $X = (\boldsymbol{x}_1, \ldots, \boldsymbol{x}_p)$. In the case of Gaussian traits, a p-vector $\boldsymbol{\omega} = (\omega_1, \ldots, \omega_p)^\tau$ is computed as

$$\boldsymbol{\omega} = X\boldsymbol{y}.$$

This is equivalent to fitting the following single-feature models:

$$y_i = \beta_j x_{ij} + \epsilon_i, \ i = 1, \ldots, n; j = 1, \ldots, p.$$

There is no intercept term in the models because of the standardization. The component ω_j of $\boldsymbol{\omega}$ is in fact the fitted β_j for feature j.

The components of $\boldsymbol{\omega}$ are ordered according to the magnitude of their absolute values. Then, for a specified number d, the features with the first d largest absolute values are retained for further study and the others are screened out. The number d is specified as $d = [\xi n]$ for a $\xi \in (0, 1)$ in [56], where $[x]$ denotes the largest integer less than or equal to x. The screening therefore reduces the dimensionality of the features from the original p to d which is smaller than the sample size n. The screening procedure is called sure independence screening (SIS) because it is shown in [56] that, under certain conditions, it has a so-called sure screening property. Denote the set of the exact relevant features by s_0 as before. Let s^* be the set of feature retained by the screening procedure. The sure screening property refers to the property below:

$$P(s_0 \subset s^*) \to 1, \text{as } n \to \infty.$$

The sure screening property is an asymptotic one. The rate of the above convergence could be low. However, if we do not reduce the dimensionality so drastically below n, the above probability can be greatly increased. In fact, as a preliminary screening, it is not necessary to reduce the dimensionality below n. As long as the computational capacity allows, we can specify d as $d = [kn]$ for a k that is much larger than 1. Specifically, d can be set at the level of thousands.

In the case of non-Gaussian traits, a single-feature generalized linear model is fitted for each feature. Specifically, for feature j, the following log likelihood function is maximized:

$$\ell_j = \sum_{i=1}^{n} [y_i \theta_{ij} - b(\theta_{ij})],$$

where θ_{ij} is related to $\eta_{ij} = \beta_0 + \beta_j x_{ij}$ through the functions:

$$\mu_{ij} = b'(\theta_{ij}), \eta_{ij} = g(\mu_{ij}).$$

Here, $g(\cdot)$ is the link function of the generalized linear model. For more details of the generalized linear model and the computation algorithm, see § 2.5.

Let $\hat{\ell}_j$ denote the maximum over β_0 and β_j. The $\hat{\ell}_j$'s then play the role of ω_j's in the linear models. The $\hat{\ell}_j$'s are ordered according to their magnitude and the features with the first d largest $\hat{\ell}_j$'s are retained.

There is an alternative way to determine the retained features. Instead of specifying the number of retained features d, we can specify a level α for the p-value of the test for testing $H_0 : \beta_j = 0$, for each marker. Let $\hat{\beta}_j$ denote the MLE of β and s.d.$(\hat{\beta}_j)$ the estimate of its standard deviation. The test statistic is given by

$$t_j = \frac{\hat{\beta}_j}{\text{s.d.}(\hat{\beta}_j)}.$$

Under the null hypothesis, the test statistic has an asymptotic standard normal distribution. The p-value of the statistic is thus

$$p_j = 2P(Z \geq |t_j|),$$

where Z is the standard normal random variable. If $p_j < \alpha$, feature j is retained; otherwise, feature j is discarded. Usually, people set $\alpha = 0.05$.

For Gaussian traits, t_j is of the explicit form in terms of $\boldsymbol{y}$ and $\boldsymbol{x}_j$ as follows:

$$t_j = \frac{\boldsymbol{x}_j^\top \boldsymbol{y}}{\hat{\sigma} \sqrt{\boldsymbol{x}_j^\top \boldsymbol{x}_j}},$$

where

$$\hat{\sigma}^2 = \frac{1}{n-1} \left\| \left(I - \frac{\boldsymbol{x}_j \boldsymbol{x}_j^\top}{\boldsymbol{x}_j^\top \boldsymbol{x}_j} \right) \boldsymbol{y} \right\|_2^2.$$

For non-Gaussian traits, there is no explicit form of t_j in terms of $\boldsymbol{y}$ and $\boldsymbol{x}_j$. It must be computed using the algorithm given in § 2.5.

6.1.2 QTL Mapping by Penalized Likelihood Methods

The penalized likelihood method is based on the following penalized likelihood function:

$$\ell_p(\lambda, \boldsymbol{\beta}) = -2 \ln L_n(\boldsymbol{y}, X\boldsymbol{\beta}) + \sum_{j=1}^{p} p_\lambda(|\beta_j|), \qquad (6.1)$$

where $L_n(\boldsymbol{y}, X\boldsymbol{\beta})$ is the likelihood function and $p_\lambda(\cdot)$ is a penalty function regulated by a parameter λ. As mentioned in § 2.9, various penalty functions have been proposed for the penalized likelihood method in the literature. The Lasso penalty (i.e., the L_1 norm penalty) is the most popular because it is convex and easy to compute. However, it introduces too much shrinkage for large regression coefficients and lacks the desired oracle property except under very restricted unrealistic conditions. The SCAD penalty and MCP penalty have been shown to have the oracle property under milder conditions and have demonstrated their superior performance to the other penalties in many published simulation studies. Therefore, we only consider these two penalties in the penalized likelihood method for QTL mapping.

The SCAD penalty is defined through its derivative given below:

$$p_\lambda'(|\beta|) = \lambda \left\{ I(|\beta| \leq \lambda) + \frac{(a\lambda - |\beta|)_+}{(a-1)\lambda} I(|\beta| > \lambda) \right\}$$

for some choice of $a > 2$. The SCAD penalty is the same as the L_1 norm for β near zero, is a constant when β is relatively far away from zero, the straight line of the L_1 norm and the constant are connected by a smooth spline in between.

The MCP penalty is defined as follows:

$$p_\lambda(|\beta|) = \begin{cases} \lambda|\beta|(1 - \frac{|\beta|}{2\gamma\lambda}), & \text{if } |\beta| < \gamma\lambda, \\ \frac{\gamma\lambda^2}{2}, & \text{if } |\beta| \geq r\gamma\lambda, \end{cases}$$

for some $\gamma > 0$.

The mechanism of the penalized likelihood method for QTL mapping is as follows. For a fixed value of the regulating parameter λ, by minimizing the penalized likelihood function (6.1), only a certain number (corresponding to the fixed λ value) of x_j's have non-zero estimated coefficients. Roughly speaking, the x_j's with non-zero estimated coefficients make more contribution to the likelihood function and hence are more important than those with zero estimated coefficients. The value of λ determines in a certain sense the relative importance of the x_j's. The larger the value of λ, the smaller the number of x_j's having non-zero estimated coefficients. When λ is extremely large, none of the x_j's has a non-zero estimated coefficient. In the other extreme case that λ is taken as zero, all the x_j's have non-zero estimated coefficients. By considering different values of λ between the two extreme cases, we obtain different sets of x_j's with different relative importance. Then we can choose a particular set using a certain criterion. The chosen set of x_j's are considered as the identified QTLs.

In implementation, first, a set of regularly spaced points in a certain range of λ is determined, then, at each point, the penalized likelihood function (6.1) is minimized to yield a set of x_j's having non-zero estimated coefficients, finally, each of such sets are evaluated by a model selection criterion. There are two types of criteria which are commonly used. The first one is a d-fold cross-validation (CV), usually, d is taken as 5 or 10. The second one is a BIC-type criterion such as the EBIC discussed in § 2.10. If the purpose of the mapping is to build a QTL model for prediction (or classification), the CV criterion is usually used. If the purpose is to eventually detect QTLs, the EBIC criterion is more appropriate.

The computation for the penalized likelihood approach with either the SCAD or MCP penalty function can be carried out by using a publicly available R language package `ncvreg` [18] which is available from the CRAN website `http://cran.r-project.org`.

6.1.3 Example: Analysis of CGEMS Prostate Cancer Data

The prostate cancer Phase 1A data in the Prostate, Lung, Colon and Ovarian (PLCO) cancer screening trial can be obtained from The CGEMS data portal of National Cancer Institute, USA (`http://cgems.cancer.gov`). The CGEMS data protal provides public access to the summary results of approximately 550,000 SNPs genotyped in the CGEMS prostate cancer whole genome scan.

The prostate cancer Phase 1A data in PLCO contains 294,179 autosomal

SNPs on 1,111 controls and 1,148 cases (673 cases are aggressive, Gleason ≥ 7 or stage $\geq$ III; 475 cases are non-aggressive, Gleason < 7 and stage $<$ III). In this example, all the cases are grouped together without distinguishing aggressive and non-aggressive ones, and the logistic regression model (the GLIM with binary response and logistic link function) is used to model the case-control data. The data was analyzed in [222] by using the penalized likelihood approach with both main-effect models and interaction models. The analysis with main-effect models is presented in the following.

The pre-screening was done in two steps. In the first step, the SIS approach discussed in § 6.1.1 is used to screen the 294,179 SNPs; that is, for each SNP, a single-SNP logistic model is fitted and the p-value of the significance test of the SNP effect is obtained. Those SNPs with a p-value bigger than 0.05 are discarded. There are 17,387 SNPs which have a p-value less than 0.05 and are retained.

In the second step, a screen approach called the tournament procedure [31] is applied to simultaneously screen the SNPs. In general, the tournament procedure is as follows. Let S be the index set of all features under consideration. Partition S into groups of size n_G, where n_G is chosen such that the minimization of the penalized likelihood with n_G features can be efficiently carried out. Suppose the partition is $S = \cup_k s_k$. For each s_k, minimize

$$-2 \log L(\boldsymbol{\beta}(s_k)|s_k) + \lambda \sum_{j \in s_k} |\beta_j|$$

by tuning the value of λ to retain n_k features. This can be effectively done using the R package `glmpath` [143] while setting the maximum steps at n_k. If $\sum_j n_j > n_G$, repeat the above process with all retained features; otherwise, apply the L_1-penalized logistic model to the retained features to reduce the number to n_M, a pre-specified number for the retained features. Applying the tournament procedure, the 17,387 SNPs are randomly partitioned into groups of size 1,000, except one group of size 1,387, and 100 SNPs are selected from each group. A second round of screening is applied to the selected 1,700 SNPs out of which 100 SNPs are retained.

Taking into account the issue of separation in logistic regression models addressed in §5.4.3, the following penalized likelihood is used:

$$l_p(\boldsymbol{\beta}(s^*)|\lambda) = -2 \log L(\boldsymbol{\beta}(s^*)|s^*) - \log |I(\boldsymbol{\beta}(s^*))| + \sum_{j \in s^*} p_\lambda(|\beta_j|),$$

where s^* is the index set of the retained SNPs, p_λ is the SCAD penalty and $I(\boldsymbol{\beta}(s^*))$ is the Fisher information matrix. The EBIC discussed in § 2.10 is used for the choice of the penalty parameter λ. The penalized likelihood approach is applied to the 100 retained SNPs. The EBIC for main-effect models with $\gamma = 1 - \ln n/(2 \ln p) \approx 0.6$ is used for the final model selection. The reference of the detected SNPs and their located chromosomes are given in Table 6.1

TABLE 6.1: The reference of the detected SNPs and their located chromosomes in the analysis of the CGEMS prostate cancer data.

Chromosome	SNP reference
1	rs1721525
1	rs10518441
2	rs4521097
6	rs727056
6	rs1885693
8	rs7837688
7	rs7384464
10	rs3134883
10	rs1887244
13	rs4941462

Among the detected SNPs, the one on chromosome 8, rs7837688, falls into region 8q24. Many prostate cancer studies are focused on this region. It has been reported in a number of studies that rs1447295, one of the 4 tightly linked SNPs in the "locus 1" region of 8q24, is associated with prostate cancer, and it has been established as a benchmark for prostate cancer association studies. Though rs1447295 is not directly detected in the above analysis, the detected one falling into 8q24, i.e., rs7837688, is highly correlated with rs1447295 ($r^2 = 0.9$) and is more significant than rs1447295 based on single-SNP models. These two SNPs, which are in the same recombination block, are also physically close. Therefore, they can be considered essentially the same.

In the analysis with interaction models in [222], the following three interaction terms are detected: rs1885693×rs12537363, rs7837688×rs2256142, rs1721525×rs2243988. It is interesting to note that no main-effect terms are detected and that the three SNPs: rs1885693, rs7837688, rs1721525, which are involved in the detected interaction terms are detected by the analysis with main-effect models. Another fact to notice is that, in the analysis with interaction models, the EBIC for interaction models (2.12) with $\gamma = 1 - \ln n/(4 \ln p) \approx 0.8$ is used. This γ applies to both interaction and main-effect terms. The phenomena that only interaction terms are detected in this example can be explained partially by the use of the above form of EBIC; see the discussion in § 2.10. Further explanations are provided in § 6.2.1.

6.1.4 QTL Mapping by Sequential Methods

The common nature of sequential methods is that features are selected in a step-by-step forward procedure to produce a sequence of subsets of features: $s_{*1} \subset s_{*2} \subset \cdots \subset s_{*k} \subset \cdots$, and a model selection criterion is used to decide the final selected set. In this section, we consider sequential methods for both Gaussian and non-Gaussian traits.

6.1.4.1 Sequential Methods for Gaussian Traits

A Gaussian trait is modeled by a sparse high-dimensional linear model as follows:

$$\boldsymbol{y} = \sum_{j=1}^{p} \beta_j \boldsymbol{x}_j + \boldsymbol{\epsilon} = X\boldsymbol{\beta} + \boldsymbol{\epsilon}.$$

As mentioned in § 2.9, for sparse high-dimensional linear models, the sequential methods include the traditional forward regression (FR), the least angle regression (LAR) and orthogonal matching pursuit (OMP). Denote by s_{*k} the set of features selected up to step k. Let

$$g_1(j) = \frac{|\boldsymbol{x}_j^\tau[I - H(s_{*k})]\boldsymbol{y}|}{\sqrt{\boldsymbol{x}_j^\tau[I - H(s_{*k})]\boldsymbol{x}_j}} = \frac{|\boldsymbol{x}_j^\tau(\boldsymbol{y} - X(s_{*k})\hat{\boldsymbol{\beta}}(s_{*k})|}{\sqrt{\boldsymbol{x}_j^\tau[I - H(s_{*k})]\boldsymbol{x}_j}},$$

$$g_2(j) = |\boldsymbol{x}_j^\tau[I - H(s_{*k})]\boldsymbol{y}| = |\boldsymbol{x}_j^\tau(\boldsymbol{y} - X(s_{*k})\hat{\boldsymbol{\beta}}(s_{*k})|$$

$$g_3(j) = |\boldsymbol{x}_j^\tau[\boldsymbol{y} - X(s_{*k})\tilde{\boldsymbol{\beta}}(s_{*k})]|,$$

where $X(s_{*k})$ is the matrix formed by the $\boldsymbol{x}_j$'s with $j \in s_{*k}$, $H(s_{*k})$ is the projection matrix of $X(s_{*k})$, $\hat{\boldsymbol{\beta}}(s_{*k})$ and $\tilde{\boldsymbol{\beta}}(s_{*k})$ are, respectively, the least squares estimate and a shrunk estimate of $\boldsymbol{\beta}(s_{*k})$. To select features at step $k + 1$, FR, OMP and LAR maximize $g_1(j)$, $g_2(j)$ and $g_3(j)$ respectively. The OMP selects the next feature that has the highest correlation with the current residual but the FR selects the next feature that has the highest inflated correlation with the current residual. The correlation is inflated by a factor $[\boldsymbol{x}_j^\tau[I - H(s_{*k})]\boldsymbol{x}_j]^{-1/2}$. The more correlated the $\boldsymbol{x}_j$ is with the features in s_{*k}, the larger the inflating factor. If two features have the same absolute correlation with the current residual, the FR will select the one that is more correlated with the features in s_{*k}. If one feature has a lower correlation with the current residual but is more correlated with the features in s_{*k} than another feature, it might turn out that this feature has a higher inflated correlation and is selected by FR. This is a disadvantage of FR when there are high spurious correlations among the features which is the case in QTL mapping with dense markers. The LAR also selects the next feature that has the highest correlation with the current residual. But, the current residual is obtained from a shrunk estimate of $\boldsymbol{\beta}(s_{*k})$. In the shrunk estimate, the effects on $\boldsymbol{y}$ of the features in s_{*k} are not fully counted. This leaves more chance for those features that have high spurious correlations with the features in s_{*k} to be selected in subsequent steps. This is a potential disadvantage for the identification of relevant features.

Because of the advantage of OMP over FR and LAR for the identification of relevant features in sparse high-dimesional linear models, we only consider OMP for QTL mapping of Gaussian traits. The OMP is closely related to the sequential Lasso (SLasso) proposed in [124]. Conceptually, after the set s_{*k} has been selected, the SLasso selects the next features at step $k + 1$ by minimizing the following partially penalized likelihood:

$$l_{k+1}(\boldsymbol{\beta}) = -2 \ln L_n(\boldsymbol{y}, X\boldsymbol{\beta}) + \lambda_{k+1} \sum_{j \notin s_{*k}} |\beta_j|,$$

where no penalty is imposed on the β_j's for $j \in s_{*k}$ and λ_{k+1} is the largest value of the penalty parameter such that at least one of the β_j's, $j \notin s_{*k}$, will be estimated non-zero. If at each step there is only one feature which can be selected by OMP or a partial positive cone condition holds, see [124], the OMP and the SLasso are equivalent. In the framework of SLasso, the OMP is shown to have the following properties under certain mild conditions. First, with probability converging to 1, the OMP procedure exhausts all relevant features before any irrelevant features can be selected; that is, if we knew the number of relevant features, p_0, and stop at the step when we have selected exactly p_0 features, the selected set will be exactly the set of relevant features s_0. Second, if the EBIC is used as the model selection criterion then, with probability converging to 1, uniformly for all k such that $|s_{*k}| < p_0$, $\text{EBIC}(s_{*k}) > \text{EBIC}(s_{*k+1})$ and $\text{EBIC}(s) > \text{EBIC}(s_0)$ for all s such that $p_0 < |s| \leq k_0 p_0$ for some fixed $k_0 > 1$. These properties imply that, asymptotically, the set of features selected by OMP together with EBIC will be exactly the set of relevant features with probability 1.

The sequential procedure using OMP together with EBIC for QTL mapping of Gaussian traits is described in the following algorithm.

Sequential Algorithm for Mapping Gaussian Traits

- Initial Step: Standardize $\boldsymbol{y}$, $\boldsymbol{x}_j$, $j = 1, \ldots, p$, such that $\boldsymbol{y}^\tau \mathbf{1} = 0$, $\boldsymbol{x}_j^\tau \mathbf{1} = 0$ and $\boldsymbol{x}_j^\tau \boldsymbol{x}_j = n$. Compute $\boldsymbol{x}_j^\tau \boldsymbol{y}$ for $j \in S$. The set of selected features at this step is given by

$$s_{*1} = \{j : |\boldsymbol{x}_j^\tau \boldsymbol{y}| = \max_{l \in S} |\boldsymbol{x}_l^\tau \boldsymbol{y}|\}.$$

  Compute $I - H(s_{*1})$ and $\text{EBIC}(s_{*1})$.

- General Step: For $k \geq 1$, compute $\boldsymbol{x}_j^\tau \tilde{\boldsymbol{y}}$ for $j \in s_{*k}^c$, where $\tilde{\boldsymbol{y}} = [I - H(s_{*k})]\boldsymbol{y}$. Let

$$s_{\text{TEMP}} = \{j : |\boldsymbol{x}_j^\tau \tilde{\boldsymbol{y}}| = \max_{l \in s_{*k}^c} |\boldsymbol{x}_l^\tau \tilde{\boldsymbol{y}}|\}.$$

  Update s_{*k} to s_{*k+1} as $s_{*k+1} = s_{*k} \cup s_{\text{TEMP}}$. Compute $\text{EBIC}(s_{*k+1})$. If $\text{EBIC}(s_{*k+1}) > \text{EBIC}(s_{*k})$, stop and take s_{*k} as the set of selected markers; otherwise, compute $I - H(s_{*k+1})$ and continue.

The EBIC for s_{*k}, $k = 1, 2, \ldots$, in the above algorithm is given by

$$\text{EBIC}(s_{*k}) = n \ln \left(\frac{\|[I - H(s_{*k})]\boldsymbol{y}\|_2^2}{n} \right) + |s_{*k}| \ln n + 2(1 - \frac{\ln n}{2 \ln p}) \ln \left(\frac{p}{|s_{*k}|} \right).$$

The matrix $I - H(s_{*k+1})$ can be updated from $I - H(s_{*k})$ recursively. Suppose there are K features in s_{TEMP} with indices $\{j_l : l = 1, \ldots, K\}$ at step $k+1$. Denote by $J_l = \{j_1, \ldots, j_l\}$ for $l = 1, \ldots, K$. Let $J_0 = \phi$. The recursive formula is given by

$$I - H(s_{*k} \cup J_l) = [I - H(s_{*k} \cup J_{l-1})] \left\{ I - \frac{\boldsymbol{x}_{j_l} \boldsymbol{x}_{j_l}^{\tau} [I - H(s_{*k} \cup J_{l-1})]}{\boldsymbol{x}_{j_l}^{\tau} [I - H(s_{*k} \cup J_{l-1})] \boldsymbol{x}_{j_l}} \right\}. \quad (6.2)$$

A comprehensive simulation study in [124] demonstrated that, compared with other methods available in the literature, the above sequential method has higher or comparable positive selection rates and lower false selection rates. In other words, it is more accurate in detecting the QTLs in QTL mapping.

6.1.4.2 Sequential Methods for Non-Gaussian Traits

A non-Gaussian trait is modeled by a sparse high-dimensional generalized linear model. With observations $(y_i, x_{i1}, \ldots, x_{ip}), i = 1, \ldots, n$, the log likelihood function is given by

$$\ln L_n(\boldsymbol{y}, X\boldsymbol{\beta}) = \sum_{i=1}^{n} [y_i \theta_{ij} - b(\theta_{ij})],$$

where θ_{ij} is related to $\eta_{ij} = \beta_0 + \sum_{j=1}^{p} \beta_j x_{ij}$ through the link function g as follows:

$$\eta_{ij} = g(b^{'}(\theta_{ij})).$$

Because of the relationship between OMP and SLasso, the OMP can be extended to generalized linear models as follows. Again, denote by s_{*k} the set of features selected up to step k. At step $k+1$, we minimize the partially penalized likelihood

$$l_{k+1}(\boldsymbol{\beta}) = -2\ln L_n(\boldsymbol{y}, X\boldsymbol{\beta}) + \lambda_{k+1} \sum_{j \in s_{*k}^c} |\beta_j|, \quad (6.3)$$

where λ_{k+1} is the largest value of the penalty parameter such that at least one of the β_j's with $j \in s_{*k}^c$ is estimated non-zero. The $\boldsymbol{x}_j$'s having non-zero coefficients are then selected at step $k+1$. It is shown in [126] that the $\boldsymbol{x}_j$'s having non-zero coefficients maximize the absolute value of the partial marginal score defined below. Let $\ell(X\boldsymbol{\beta}) = \ln L(\boldsymbol{y}, X\boldsymbol{\beta})$. For any $s \subset \mathcal{S}$, let $\tilde{\ell}(\boldsymbol{\beta}_{s^c})$ denote the profile log likelihood function of $\boldsymbol{\beta}_{s^c}$ given by

$$\tilde{\ell}(\boldsymbol{\beta}_{s^c}) = \max_{\boldsymbol{\beta}_s} \ell(X_s \boldsymbol{\beta}_s + X_{s^c} \boldsymbol{\beta}_{s^c}) = \ell(X_s \tilde{\boldsymbol{\beta}}_s(\boldsymbol{\beta}_{s^c}) + X_{s^c} \boldsymbol{\beta}_{s^c}),$$

i.e., $\tilde{\ell}(\boldsymbol{\beta}_{s^c})$ is the maximum of $\ell(X\boldsymbol{\beta})$ over $\boldsymbol{\beta}_s$ while $\boldsymbol{\beta}_{s^c}$ is kept fixed. Note that $\tilde{\boldsymbol{\beta}}_s(\boldsymbol{\beta}_{s^c})$ is the maximizer which depends on $\boldsymbol{\beta}_{s^c}$. For $j \in s^c$, let

$$\psi(\boldsymbol{x}_j|s) = \left. \frac{\partial \tilde{\ell}(\boldsymbol{\beta}_{s^c})}{\partial \beta_j} \right|_{\boldsymbol{\beta}_{s^c} = \boldsymbol{0}}. \quad (6.4)$$

The $\psi(\boldsymbol{x}_j|s)$ is referred to as the profile marginal score of $\boldsymbol{x}_j$. The $\boldsymbol{x}_j$'s for $j \in s_{*k}^c$ have non-zero coefficients in the minimization of (6.3) if and only if

$$|\psi(\boldsymbol{x}_j|s_{*k})| = \max_{l \in s_{*k}^c} |\psi(\boldsymbol{x}_l|s_{*k})|.$$

It is also shown in [126] that $\psi(\boldsymbol{x}_j|s)$ is given by

$$\psi(\boldsymbol{x}_j|s) = \sum_{i=1}^n (y_i - \hat{\mu}_{is})\hat{\nu}_{is}x_{ij},$$

where $\hat{\mu}_{is} = b'(\theta(\boldsymbol{x}_{(i)s}^\tau \hat{\boldsymbol{\beta}}_s))$, $\boldsymbol{x}_{(i)}$ being the ith row vector of X, $\hat{\nu}_{is}$ is the partial derivative $\frac{\partial \theta(\eta_i)}{\partial \eta_i}\Big|_{\boldsymbol{\beta}_s = \hat{\boldsymbol{\beta}}_s} = [b''(\theta(X_s\hat{\boldsymbol{\beta}}_s))g'(\hat{\mu}_{is})]^{-1}$, and $\hat{\boldsymbol{\beta}}_s$ is the MLE of $\boldsymbol{\beta}_s$ in the GLIM consisting of only $\boldsymbol{x}_l$'s with $l \in s$. In matrix form,

$$\psi(\boldsymbol{x}_j|s) = \boldsymbol{x}_j^\tau \hat{V}_s(\boldsymbol{y} - \hat{\mu}_s),$$

where $\hat{V}_s = \text{Diag}(\hat{\nu}_{1s}, \ldots, \hat{\nu}_{ns})$ and $\hat{\mu}_s = (\hat{\mu}_{1s}, \ldots, \hat{\mu}_{ns})^\tau$.

The sequential method for mapping non-Gaussian traits is described in the following algorithm.

Sequential Algorithm for Mapping non-Gaussian Traits

- Initial Step: Standardize $\boldsymbol{y}$, $\boldsymbol{x}_j$, $j = 1, \ldots, p$, such that $\boldsymbol{y}^\tau \mathbf{1} = 0, \boldsymbol{x}_j^\tau \mathbf{1} = 0$ and $\boldsymbol{x}_j^\tau \boldsymbol{x}_j = n$. Compute $\boldsymbol{x}_j^\tau \boldsymbol{y}$ for $j \in S$. The set of selected features at this step is given by

$$s_{*1} = \{j : |\boldsymbol{x}_j^\tau \boldsymbol{y}| = \max_{l \in S} |\boldsymbol{x}_l^\tau \boldsymbol{y}|\}.$$

Compute $\hat{\mu}_{s_{*1}}$, $\hat{V}_{s_{*1}}$ and $\text{EBIC}(s_{*1})$.

- General Step: For $k \geq 1$, compute $\psi(\boldsymbol{x}_j|s_{*k})$ for $j \in s_{*k}^c$. Let

$$s_{\text{TEMP}} = \{j : |\psi(\boldsymbol{x}_j|s_{*k})| = \max_{l \in s_{*k}^c} |\psi(\boldsymbol{x}_l|s_{*k})|\}.$$

Update s_{*k} to s_{*k+1} as $s_{*k+1} = s_{*k} \cup s_{\text{TEMP}}$. Compute $\text{EBIC}(s_{*k+1})$. If $\text{EBIC}(s_{*k+1}) > \text{EBIC}(s_{*k})$, stop and take s_{*k} as the set of selected markers; otherwise, compute $\hat{\mu}_{s_{*k+1}}$, $\hat{V}_{s_{*k+1}}$ and continue.

In the remainder of this subsection, we give the explicit forms of $\hat{\mu}_{s_{*k}}$ and $\hat{V}_{s_{*k}}$ for binary traits with Bernoulli distribution and count traits with Poisson distribution.

For binary traits with Bernoulli distribution, the log likelihood function with markers in s_{*k} is as follows:

$$\ell(X_{s_{*k}}\boldsymbol{\beta}_{s_{*k}}) = \sum_{i=1}^n [y_i\theta_i - \ln(1 + e^{\theta_i})],$$

where $\theta_i = \ln \frac{\mu_i}{1-\mu_i}$ which is related to $\eta_i = \beta_0 + \sum_{j \in s_{*k}} \beta_j x_{ij}$ by $\eta_i = g(\mu_i)$. Here g is the link function. To reduce the bias caused by possible data separation, the $\hat{\boldsymbol{\beta}}_{s_{*k}}$ is obtained by maximizing the following function:

$$\ell(X_{s_{*k}} \boldsymbol{\beta}_{s_{*k}}) + \frac{1}{2} \ln |X_{s_{*k}}^{\tau} W(\boldsymbol{\beta}_{s_{*k}}) X_{s_{*k}}|,$$

where $W(\boldsymbol{\beta}_{s_{*k}}) = \mathrm{Diag}(w_1(\boldsymbol{\beta}_{s_{*k}}), \ldots, w_n(\boldsymbol{\beta}_{s_{*k}}))$ with $w_i(\boldsymbol{\beta}_{s_{*k}}) = [\mu_i(1 - \mu_i)(g'(\mu_i))^2]^{-1}$. Then $\hat{\mu}_{is} = h(\hat{\beta}_0 + \sum_{j \in s_{*k}} \hat{\beta}_j x_{ij})$ where h is the inverse function of the link function g, and $\hat{\nu}_{is} = [\hat{\mu}_{is}(1 - \hat{\mu}_{is})g'(\hat{\mu}_{is})]^{-1}$. When g is the canonical link $g(\mu) = \ln \frac{\mu}{1-\mu}$, $\hat{\nu}_{is} \equiv 1$ and $\hat{\mu}_{is} = \frac{e^{\hat{\beta}_0 + \sum_{j \in s_{*k}} \hat{\beta}_j x_{ij}}}{1 + e^{\hat{\beta}_0 + \sum_{j \in s_{*k}} \hat{\beta}_j x_{ij}}}$.

For count traits with Poisson distribution, the log likelihood function with markers in s_{*k} is as follows:

$$\ell(X_{s_{*k}} \boldsymbol{\beta}_{s_{*k}}) = \sum_{i=1}^{n} (y_i \theta_i - e^{\theta_i}),$$

where $\theta_i = \ln \mu_i$. The $\hat{\boldsymbol{\beta}}_{s_{*k}}$ is obtained by using the algorithm given in § 2.5. Again, $\hat{\mu}_{is} = h(\hat{\beta}_0 + \sum_{j \in s_{*k}} \hat{\beta}_j x_{ij})$. The $\hat{\nu}_{is}$ is computed as $[\hat{\mu}_{is} g'(\hat{\mu}_{is})]^{-1}$. When g is the canonical link $g(\mu) = \ln \mu$, $\hat{\nu}_{is} \equiv 1$ and $\hat{\mu}_{is} = e^{\hat{\beta}_0 + \sum_{j \in s_{*k}} \hat{\beta}_j x_{ij}}$.

6.1.4.3   Example: Analysis of F_2 Rat Data

The data, which was reported in [158], consists of the expression levels of over 31,042 different probes from 120 F_2 male rats generated from an intercross experiment. A cross of SR/JrHsd male rats and SHRSP female rats was performed to generate F_1 and the F_1 rats were intercrossed to generate the F_2 rats. The probes that were not expressed in the eye or that lacked sufficient variation were excluded. A probe was considered expressed if its maximum expression value observed among the 120 F2 rats was greater than the 25th percentile of the entire set of RMA (robust multi-chip averaging) expression values. A probe was considered "sufficiently variable" if it exhibited at least 2-fold variation in expression level among the 120 F2 rats. A total of 18,976 probes that met these criteria were retained. Among the 18,976 probes, there is one, 1389163_at, from gene TRIM32. This gene was found to cause Bardet-Biedl syndrom [36]. Of interest is to find the probes among the remaining 18, 975 probes that are most related to TRIM32. Therefore, the response variable is taken as the expression level of probe 1389163_at. The remaining 18, 975 probes are taken as the covariates.

The expression levels are standardized to have mean 0 and standard deviation 1 in the analysis. The probes are first screened according to their variances and the top 3,000 probes with the largest variances are retained for further selection. The sequential method described in § 6.1.4.1 is applied to these 3,000 probes. Two probes, 1383110_at and 1392692_at, are eventually selected by the sequential procedure. These two probes have been detected separately by using other methods reported in the literature, see [90], [104], [92], [54] and [175].

The above analysis was originally carried out in [124].

6.2 QTL Mapping with Consideration of QTL Epistatic Effects

In the previous sections, we implicitly assumed that the QTLs act on the quantitative trait independently, i.e., no epistatic effects among the QTLs are assumed. However, for most quantitative traits which are affected by multiple QTLs, the QTLs do interact with each other. To deal with QTLs with epistatic effects, special treatments are needed. This section is devoted to the mapping procedure with special treatment for epistatic effects.

6.2.1 Special Natures of Interaction Models for QTL Mapping

The QTL epistatic effects are called interaction effects in statistical terminology. The existence of interaction between two QTLs implies that the effects of one QTL are different when the other has a different genotype. The interaction has different natures: it either enhances the total effect or reduces the total effect. The nature of the interaction effect on a quantitative trait with two QTLs under a backcross design is illustrated in Figure 6.1. The panel (a) of Figure 6.1 shows the case of no interaction; panel (b) shows the case where the total effect is enhanced; panel (c) shows the case where the total effect is reduced; panel (d) shows an extreme case where the effects of the first QTL are exactly opposite when the second QTL has different genotypes.

In the presence of interaction, the correct model for the trait with two QTLs is the following interaction model:

$$y = \alpha + \beta_1 x_1 + \beta_2 x_2 + \gamma x_1 x_2 + \epsilon, \tag{6.5}$$

where x_j takes value 0 when QTL j has a homozygous genotype and 1 when it has a heterozygous genotype, $j = 1, 2$. When x_1 is fixed at value 0, the difference between the effects of the two genotypes of QTL 2 is β_2. When x_1 is fixed at value 1, the difference between the effects of the two genotypes of QTL 2 is $\beta_2 + \gamma$. The interaction enhances or reduces the total effect accordingly as γ is positive or negative. However, if the trait is wrongly modeled by the following additive model:

$$y = \alpha + \beta_1 x_1 + \beta_2 x_2 + \epsilon, \tag{6.6}$$

instead of representing the effect of QTL 2 when x_1 is fixed at value 0, β_2

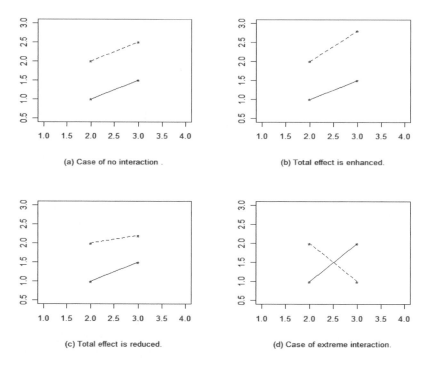

FIGURE 6.1: The nature of interaction effect of two QTLs on a quantitative trait with backcross designs.

represents the average effect of QTL 2 over the two genotypes of QTL 1. In the extreme case in panel (d), the average effect is 0, though the effect of QTL 2 is significant at each fixed genotype of QTL 1. Thus, if model (6.6) is used for the mapping, there is no chance for the two QTLs to be detected. Therefore, we are obliged to consider the interaction model (6.5).

In general, in the QTL mapping for Gaussion traits, we consider the interaction model given below:

$$y_i = \alpha + \sum_{j=1}^{p} \beta_j x_j + \sum_{j \neq k} \gamma_{jk} x_j x_k + \epsilon_i, \quad i = 1, \ldots, n. \tag{6.7}$$

For non-Gaussian traits, we consider a generalized linear model with the linear predictor given by

$$\eta = \alpha + \sum_{j=1}^{p} \beta_j x_j + \sum_{j \neq k} \gamma_{jk} x_j x_k.$$

For the sake of convenience, we refer to the single variables x_j's as main-effect features, the products $x_j x_k$'s as interaction features, and the associated coefficients as their effects. Note that the effect of a main-effect feature is not the complete effect of the corresponding marker (or QTL); it is only the effect of the marker when all the other markers have their values fixed at 0, as argued in the previous paragraph.

The approaches considered in the previous sections cannot be directly applied to interaction models. It has been observed in [222] and reported by other researchers as well that, by using those approaches for the interaction model, the main-effect features are hardly detectable. As has been seen in § 6.1.3, in the application of the penalized likelihood method to the analysis of the CGEMS prostate cancer data, only interaction features are detected. Though those detected interaction features involve the SNPs detected by using main-effect models, the main-effect features corresponding to those SNPs are not detected. To explain this phenomenon, let us consider the two-QTL interaction model with backcross design again. Suppose that for each QTL, one allele is the wild allele in the population and the other is the minor one which is responsible for the increase or decrease of the quantitative trait from its normal level. In two situations, the main-effect feature is not detectable while the interaction feature is. The first situation is that the minor allele of one QTL can only be activated in the presence of the minor allele of the other QTL. The second situation is that the minor allele has a small effect without the presence of the minor allele of the other QTL but in the presence of the minor allele of the other QTL its effect can be greatly enhanced. In these situations, there is an intrinsic cause for the phenomenon. There is another cause which is not intrinsic. In model (6.7), suppose that markers j and k, for some particular j and k, represent two real QTLs which have an interaction effect. Suppose that the minor allele of QTL j has an effect in the absence of the minor allele of QTL k smaller than that in the presence of the minor allele of QTL k, that is, in terms of the parameters, β_j is smaller than γ_{jk}. Since the number of markers is large, it causes spurious correlations among the markers in the data. If there are a lot of x_l's that are correlated with x_k in the data due to the spurious correlation, then, by fitting model (6.7), all the coefficients γ_{jl}'s corresponding to those x_l's will appear more significant than β_j. If they are judged by the same standard, those spurious interaction features rather than the true main-effect feature will be identified. Because of the above phenomenon, in the QTL mapping with model (6.7), the mapping procedure needs to be modified. This is handled in the next sub section.

6.2.2 The Mapping Procedure with Interaction Models

In QTL mapping with interaction models, the main-effect features and interaction features should be treated differently. If p is so large that the data exceeds the capacity of the computation facility and a preliminary screening needs to be called, the main-effect features and the interaction features should be screened separately. That is, the method described in § 6.1.1 is first applied to the p main-effect features to retain a given number, say N_M, of of them, and then the method is applied to the $p(p-1)/2$ interaction features to retain a given number, say N_{I}, of them. The retained main-effect and interaction natures are subjected to further selection. The penalized likelihood approach in § 6.1.2 is not applicable for interaction models. While generating the candidate sets of features by fixing the regularization parameter λ at different values, it cannot distinguish between main-effect features and interaction features and all the features are treated on an equal footing. Only the sequential methods in § 6.1.4 are appropriate. But some adaption is required. In the following, we present the adapted sequential method for QTL mapping with interaction models.

The major adaptation is that, at each step, the main-effect features and the interaction features are assessed separately first, then the most significant main-effect feature and the most significant interaction feature are assessed by the $\mathrm{EBIC}_{\gamma_{\mathrm{M}}\gamma_{\mathrm{I}}}$ given in § 2.10, the one that reduces the $\mathrm{EBIC}_{\gamma_{\mathrm{M}}\gamma_{\mathrm{I}}}$ more is selected at the step. The computation algorithm is given below. In the computation algorithm, we don't distinguish Gaussian and non-Gaussian traits since the profile marginal score $\psi(\boldsymbol{x}_j|s_{*k})$ reduces to $\boldsymbol{x}_j^{\tau}[I - H(s_{*k})]\boldsymbol{y}$ when the traits are Gaussian. For the sake of convenience, we use single indexed vectors $\boldsymbol{z}_j$'s to denote interaction feature vectors. The main-effect feature vectors are still denoted by $\boldsymbol{x}_j$'s. Let $S_{\mathrm{M}} = \{1, \cdots, p\}$ and $S_{\mathrm{I}} = \{p+1, \cdots, p(p+1)/2\}$.

Sequential Algorithm for QTL Mapping with Interactions

- Initial Step: Standardize $\boldsymbol{y}$, $\boldsymbol{x}_j$ and $\boldsymbol{z}_j$ such that $\boldsymbol{y}^{\tau}\boldsymbol{1} = 0, \boldsymbol{x}_j^{\tau}\boldsymbol{1} = 0$, $\boldsymbol{z}_j^{\tau}\boldsymbol{1} = 0$, $\boldsymbol{x}_j^{\tau}\boldsymbol{x}_j = n$, and $\boldsymbol{z}_j^{\tau}\boldsymbol{z}_j = n$. Compute $\boldsymbol{x}_j^{\tau}\boldsymbol{y}$ for $j \in S_{\mathrm{M}}$ and $\boldsymbol{z}_j^{\tau}\boldsymbol{y}$ for $j \in S_{\mathrm{I}}$. Let

$$s_{\mathrm{TEMP}}^{\mathrm{M}} = \{j : |\boldsymbol{x}_j^{\tau}\boldsymbol{y}| = \max_{l \in S_{\mathrm{M}}} |\boldsymbol{x}_l^{\tau}\boldsymbol{y}|\},$$

$$s_{\mathrm{TEMP}}^{\mathrm{I}} = \{j : |\boldsymbol{z}_j^{\tau}\boldsymbol{y}| = \max_{l \in S_{\mathrm{I}}} |\boldsymbol{z}_l^{\tau}\boldsymbol{y}|\}.$$

Compute $\mathrm{EBIC}_{\gamma_{\mathrm{M}}\gamma_{\mathrm{I}}}(s_{\mathrm{TEMP}}^{\mathrm{M}})$ and $\mathrm{EBIC}_{\gamma_{\mathrm{M}}\gamma_{\mathrm{I}}}(s_{\mathrm{TEMP}}^{\mathrm{I}})$. If $\mathrm{EBIC}_{\gamma_{\mathrm{M}}\gamma_{\mathrm{I}}}(s_{\mathrm{TEMP}}^{\mathrm{M}}) < \mathrm{EBIC}_{\gamma_{\mathrm{M}}\gamma_{\mathrm{I}}}(s_{\mathrm{TEMP}}^{\mathrm{I}})$, let $s_{*1}^{\mathrm{M}} = s_{\mathrm{TEMP}}^{\mathrm{M}}$ and $s_{*1}^{\mathrm{I}} = \phi$, the set of selected features is given by $s_{*1} = s_{*1}^{\mathrm{M}}$; otherwise, let $s_{*1}^{\mathrm{I}} = s_{\mathrm{TEMP}}^{\mathrm{I}}$ and $s_{*1}^{\mathrm{M}} = \phi$, the set of selected features is given by $s_{*1} = s_{*1}^{\mathrm{I}}$.

- General Step: For $k \geq 1$, compute $\psi(\boldsymbol{x}_j|s_{*k})$ for $j \in S_{\mathrm{M}}/s_{*k}^{\mathrm{M}}$ and $\psi(\boldsymbol{z}_j|s_{*k})$ for $j \in S_{\mathrm{I}}/s_{*k}^{\mathrm{I}}$. Let

$$s_{\mathrm{TEMP}}^{\mathrm{M}} = \{j : |\psi(\boldsymbol{x}_j|s_{*k})| = \max_{l \in S_{\mathrm{M}}/s_{*k}^{\mathrm{M}}} |\psi(\boldsymbol{x}_l|s_{*k})|\},$$

$$s_{\mathrm{TEMP}}^{\mathrm{I}} = \{j : |\psi(\boldsymbol{z}_j|s_{*k})| = \max_{l \in S_{\mathrm{I}}/s_{*k}^{\mathrm{I}}} |\psi(\boldsymbol{z}_l|s_{*k})|\}.$$

Compute $\mathrm{EBIC}_{\gamma_{\mathrm{M}}\gamma_{\mathrm{I}}}(s_{*k} \cup s_{\mathrm{TEMP}}^{\mathrm{M}})$ and $\mathrm{EBIC}_{\gamma_{\mathrm{M}}\gamma_{\mathrm{I}}}(s_{*k} \cup s_{\mathrm{TEMP}}^{\mathrm{I}})$. If $\mathrm{EBIC}_{\gamma_{\mathrm{M}}\gamma_{\mathrm{I}}}(s_{*k} \cup s_{\mathrm{TEMP}}^{\mathrm{M}}) < \mathrm{EBIC}_{\gamma_{\mathrm{M}}\gamma_{\mathrm{I}}}(s_{*k} \cup s_{\mathrm{TEMP}}^{\mathrm{I}})$, update $s_{*k+1}^{\mathrm{M}} =$

$s_{*k}^{\mathrm{M}} \cup s_{\mathrm{TEMP}}^{\mathrm{M}}$, $s_{*k+1}^{\mathrm{I}} = s_{*k}^{\mathrm{I}}$ and $s_{*k+1} = s_{*k} \cup s_{\mathrm{TEMP}}^{\mathrm{M}}$; otherwise, update $s_{*k+1}^{\mathrm{I}} = s_{*k}^{\mathrm{I}} \cup s_{\mathrm{TEMP}}^{\mathrm{I}}$, $s_{*k+1}^{\mathrm{M}} = s_{*k}^{\mathrm{M}}$ and $s_{*k+1} = s_{*k} \cup s_{\mathrm{TEMP}}^{\mathrm{I}}$. If $\mathrm{EBIC}_{\gamma_{\mathrm{M}}\gamma_{\mathrm{I}}}(s_{*k+1}) \leq \mathrm{EBIC}_{\gamma_{\mathrm{M}}\gamma_{\mathrm{I}}}(s_{*k})$, continue; otherwise, stop.

6.2.3 Example: Mapping QTL for Locomotor Activation and Anxiety with F_2 Mouse Model

Anxiety-related disorders, which are common mental health problems affecting humans, are affected by a genetic component, as shown in epidemiological studies. It is also revealed that the genetic mechanism behind the anxiety-related disorders is shared among species [76]. It is a common practice in genetic study to use animal models for the investigation of the genetic mechanism.

A mouse model is used in [5] for mapping QTL of locomotor activation and anxiety with F_2 progeny of two phenotypically similar inbred mouse strains: C57BL/6J and C58/J. An open-field assay test was conducted for 196 female and 166 male F_2 mice. The following six measures were considered in the open-field assay test: total distance traveled (in cm), ambulatory episodes (number of times animal breaks user defined number of beams before coming to rest), percent time resting, average velocity (in centimeters per second), number of rearings and percent time spent in center of arena. The data consists of the measurements on the six measures for each of the 362 mice together with their genotypes at 211 SNP markers with an average SNP density of one marker per 11 mega base pairs (Mb).

The sequential method described in the previous sub section is applied to the above data. First, the R package Imputation is used to impute some missing values in the genotypes of the data set. Then, with the complete data, the sequential method is applied to map QTL for each of the six traits (the six behavioral measures). With 211 SNPs, there are a total of 211 main-effect features and 22,155 interaction features. The sequential method is applied directly to these features without preliminary screening. A total of 10 main-effect features and 3 interaction features are identified for the six measures. Some of the features affect multiple measures, some of them affect only one measure. The details of these features including their location in the genome and the measures affected are given in Table 6.2.

TABLE 6.2: The locations and affected behavioral measures of the features detected by the sequential method

	Chr	Location(Mb)	Effect	Interaction
Percent Time in Center	13	89.444	main	
	2	178.315	interaction	Chr13:22.251
	13	22.251	interaction	Chr2:178.315
Total Distance	8	57.724	main	
	17	56.801	main	
	6	102.455	interaction	Chr12:20.058
	12	20.058	interaction	Chr6:102.445
Total Rearing	2	153.094	main	
Ambulatory Episodes	8	68.129	main	
	17	56.801	main	
	6	102.455	interaction	Chr12:20.058
	12	20.058	interaction	Chr6:102.455
Average Velocity	8	89.447	main	
Percent Resting	2	97.379	main	
	7	63.356	main	
	8	89.447	main	

Chapter 7

Bayesian Approach to QTL Mapping

The methods discussed in Chapter 4 and Chapter 5 are in the vein of the frequentist school. The inference on the QTL location and effect is essentially based on the likelihood function of the unknowns under a formulated model. The unknowns are treated as fixed parameters. The Bayesian methods which we are discussing in this chapter are in a different vein. In the Bayesian approach, all the unknowns are considered as random variables. The inference is based on the posterior distribution of the unknowns which summarizes all the available information including those provided by the prior distribution and by the data through the likelihood function; see § 2.7.

The Bayesian approach has long been used for QTL mapping in both inbred and outbred animal and plant populations, as well as in humans. The earlier application of Bayesian approach is limited to simple genetic models such as single-QTL models with a single marker or multiple markers; see [87], [88], [179], [180] and [190]. Bayesian methods for more complicated models were subsequently developed. Methods for fixed number of QTLs were proposed in [152] and [153]. Methods for unknown variable number of QTLs were developed in [163] for inbred line cross population and in [164] for outbred offsprings. Methods for models with QTL interactions were studied in [213] and [212]. Methods for binary traits are considered in [211].

In this chapter, we discuss various Bayesian methods proposed in the literature. In § 7.1, a general framework for Bayesian QTL mapping is presented. In § 7.2, we deal with the case of inbred line crosses. In § 7.3, we consider the case of outbred crosses. In § 7.4, the Bayesian methods for mapping ordinal polytomous traits including binary traits are treated.

7.1 The Framework of the Bayesian Approach to QTL Mapping

Throughout this chapter, the following notations are used:

- n: the number of experiment subjects;

- $\boldsymbol{y} = (y_1, \ldots, y_n)^\tau$: the vector of trait values, where y_i is the trait value of subject i;

- $\boldsymbol{d}_{\text{M}}$: the vector of marker locations in terms of genetic distance relative to a fixed location on a chromosome;

- $\boldsymbol{M} = (\boldsymbol{M}_1, \ldots, \boldsymbol{M}_n)^\tau$: the matrix of marker genotypes, where $\boldsymbol{M}_i$ is the vector of marker genotypes of subject i;

- m: the number of QTLs;

- $\boldsymbol{d}_{\text{Q}}$: the vector of QTL locations in terms of genetic distance relative to a fixed location on a chromosome;

- $\boldsymbol{Q} = (\boldsymbol{Q}_1, \ldots, \boldsymbol{Q}_n)^\tau$: the matrix of QTL genotypes, where $\boldsymbol{Q}_i$ is the vector of QTL genotypes of subject i;

- $\boldsymbol{\beta}$: the vector of QTL effect parameters;

- σ^2: the variance of the conditional trait distribution.

The general framework of the Bayesian approach is described in this section. For the clarity of the description, we assume there is no missing values in the marker genotypes for the time being. The case of missing marker genotypes will be considered when we come to specific mapping procedures.

A Bayesian model for QTL mapping consists of two components: (i) the conditional distribution of the trait given the unknowns and (ii) the prior distribution of the unknowns. The unknowns include m, $\boldsymbol{d}_{\text{Q}}$, $\boldsymbol{Q}$, $\boldsymbol{\beta}$ and σ^2. Suppose that the conditional distribution of the trait given the unknowns has a density function

$$\phi(\boldsymbol{y}|m, \boldsymbol{d}_{\text{Q}}, \boldsymbol{Q}, \boldsymbol{\beta}, \sigma^2) = \phi(\boldsymbol{y}|m, \boldsymbol{Q}, \boldsymbol{\beta}, \sigma^2).$$

Note that the QTL locations do not affect the distribution of the trait. The common assumption on the trait distribution is a normal linear model for Gaussian traits and a generalized linear model for non-Gaussian traits. Let $p(m, \boldsymbol{d}_{\text{Q}}, \boldsymbol{Q}, \boldsymbol{\beta}, \sigma^2)$ denote the density of the prior distribution of the unknowns. The prior density can be factorized as

$$p(m, \boldsymbol{d}_{\text{Q}}, \boldsymbol{Q}, \boldsymbol{\beta}, \sigma^2) = p(\boldsymbol{\beta}|m)p(\boldsymbol{Q}|\boldsymbol{d}_{\text{Q}}, \boldsymbol{d}_{\text{M}}, \boldsymbol{M})p(\boldsymbol{d}_{\text{Q}}|m)p(m)p(\sigma^2).$$

Since, given the QTL genotypes, the traits of the experiment subjects are independent, $\phi(\boldsymbol{y}|m, \boldsymbol{Q}, \boldsymbol{\beta}, \sigma^2)$ can be further factorized as

$$\phi(\boldsymbol{y}|m, \boldsymbol{Q}, \boldsymbol{\beta}, \sigma^2) = \prod_{i=1}^{n} \phi(y_i|m, \boldsymbol{Q}_i, \boldsymbol{\beta}, \sigma^2).$$

The details of the conditional and prior distributions will be specified later when we consider particular mapping procedures. With the specified conditional distribution and the prior distribution, we arrive at the density of the joint distribution of the trait and the unknowns

$$p(m, \boldsymbol{d}_{\mathrm{Q}}, \boldsymbol{Q}, \boldsymbol{\beta}, \sigma^2, \boldsymbol{y})$$

$$= p(\boldsymbol{\beta}|m)p(\boldsymbol{d}_{\mathrm{Q}}|m)p(m)p(\sigma^2)p(\boldsymbol{Q}|\boldsymbol{d}_{\mathrm{Q}}, \boldsymbol{d}_{\mathrm{M}}, \boldsymbol{M}) \prod_{i=1}^{n} \phi(y_i|m, \boldsymbol{Q}_i, \boldsymbol{\beta}, \sigma^2).$$

Thus the density of the posterior distribution of the unknowns given $\boldsymbol{y}$ is as follows:

$$\pi(m, \boldsymbol{d}_{\mathrm{Q}}, \boldsymbol{Q}, \boldsymbol{\beta}, \sigma^2|\boldsymbol{y}) \tag{7.1}$$

$$\propto p(\boldsymbol{\beta}|m)p(\boldsymbol{d}_{\mathrm{Q}}|m)p(m)p(\sigma^2)p(\boldsymbol{Q}|\boldsymbol{d}_{\mathrm{Q}}, \boldsymbol{d}_{\mathrm{M}}, \boldsymbol{M}) \prod_{i=1}^{n} \phi(y_i|m, \boldsymbol{Q}_i, \boldsymbol{\beta}, \sigma^2).$$

The posterior density above is specified up to a normalizing constant which is not necessarily required in the Bayesian analysis.

The inference on QTLs is based on the above posterior distribution. However, the posterior density (7.1) is usually too complicated. It is even impossible to compute simple characteristics such as means. The Markov Chain Monte Carlo (MCMC) discussed in § 2.8 is applied to generate random samples from the posterior distribution. The inference is then made by certain empirical procedures based on the generated random samples. The details will be given when we consider various particular QTL mapping Bayesian models in the subsequent sections.

7.2 Bayesian QTL Mapping with Inbred Line Cross Data

The Bayesian approach for QTL mapping with experimental crosses are treated in this section. We focus on the Gaussian traits. The approach can be easily extended to non-Gaussian traits. We first deal with in § 7.2.1 the case where no epistatic effect is taken into account. Two methods which follow the principles in [153] and [163] are developed. In the first method, the number of QTLs is taken fixed and the true number of QTLs is determined by Bayes factors. In the second method, the number of QTLs are taken as a random variable. In the original work of [163] , the mapping is focused on only one chromosome, selected markers from other chomosomes are used as background adjustment for the QTL effects. However, in the method presented in this section, the mapping is over the whole genome or a whole mapping region not necessarily consisting of only one chromsome. The case of models with QTL epstatic effects is tackled in § 7.2.2.

7.2.1 Bayesian Models without Epistatic Effects

First, we specify the details of the components in the posterior density:

$$\pi(m, \boldsymbol{d}_{\text{Q}}, \boldsymbol{Q}, \boldsymbol{\beta}, \sigma^2 | \boldsymbol{y})$$

$$\propto \quad p(\boldsymbol{\beta}|m)p(\boldsymbol{d}_{\text{Q}}|m)p(m)p(\sigma^2)p(\boldsymbol{Q}|\boldsymbol{d}_{\text{Q}}, \boldsymbol{d}_{\text{M}}, \boldsymbol{M}) \prod_{i=1}^{n} \phi(y_i|m, \boldsymbol{Q}_i, \boldsymbol{\beta}, \sigma^2).$$

To consider the conditional distribution of the trait given all the other unknowns, we use the same dummy variables defined in § 4.2.1 to represent the QTL genotypes. The marker genotypes are represented similarly when they arise. Given the number and the genotypes of the QTLs, the Gaussian traits are modeled as

$$y_i = \beta_0 + \sum_{j=1}^{m} \beta_j \delta_{ij} + \epsilon_i, \ i = 1, \ldots, n, \tag{7.2}$$

in the case of backcross and RIL designs, and

$$y_i = \beta_0 + \sum_{j=1}^{m} (\beta_{j1}\delta_{ij1} + \beta_{j2}\delta_{ij2}) + \epsilon_i, \ i = 1, \ldots, n, \tag{7.3}$$

in the case of intercross designs, where ϵ_i are i.i.d. normal random variables with mean zero and variance σ^2. In what follows, we concentrate on the backcross designs unless otherwise mentioned. The extension to RIL and intercross designs is trivial. The linear model (7.2) specifies the conditional density $\phi(y_i|m, \boldsymbol{Q}_i, \boldsymbol{\beta}, \sigma^2)$ in (7.1):

$$\phi(y_i|m, \boldsymbol{Q}_i, \boldsymbol{\beta}, \sigma^2) = \frac{1}{\sqrt{2\pi\sigma^2}} \exp\{-\frac{(y_i - \eta_i)^2}{2\sigma^2}\},$$

where $\boldsymbol{Q}_i = (\delta_{i1}, \ldots, \delta_{im})^{\tau}$, $\boldsymbol{\beta} = (\beta_0, \beta_1, \ldots, \beta_m)^{\tau}$ and $\eta_i = \beta_0 + \sum_{j=1}^{m} \beta_j \delta_{ij}$.

Suppose that there is at most one QTL between two adjacent markers, which is the common assumption in multiple interval mapping discussed in the previous chapter. In the case of experimental crosses, given the marker genotypes, the QTL genotypes are independent between experiment subjects and between QTLs, furthermore, the distribution of the genotypes of a QTL is determined by the genotypes of its two flanking markers. Thus the prior $p(\boldsymbol{Q}|\boldsymbol{d}_{\text{Q}}, \boldsymbol{d}_{\text{M}}, \boldsymbol{M})$ in (7.1) can be factorized as

$$p(\boldsymbol{Q}|\boldsymbol{d}_{\text{Q}}, \boldsymbol{d}_{\text{M}}, \boldsymbol{M}) = \prod_{i=1}^{n} p(\boldsymbol{Q}_i|\boldsymbol{d}_{\text{Q}}, \boldsymbol{d}_{\text{M}}, \boldsymbol{M}_i)$$

$$= \prod_{i=1}^{n} \prod_{j=1}^{m} p(\delta_{ij}|d_{\text{Q}j}, d_{\text{M}j-}, d_{\text{M}j+}, M_{ij-}, M_{ij+}),$$

where M_{ij-} and M_{ij+} are the genotypes of the two markers flanking QTL j, d_{Mj-} and d_{Mj+} are their locations. Let $r(d)$ be the Haldane mapping function, the recombination fractions between the two flanking markers and between the left flanking marker and the QTL are determined respectively as $\gamma_j = r(d_{Mj+} - d_{Mj-})$ and $r_j = r(d_{Qj} - d_{Mj-})$. Given the genotype combination (M_{ij-}, M_{ij+}), $p(\delta_{ij}|d_{Qj}, d_{Mj-}, d_{Mj+}, M_{ij-}, M_{ij+})$ is given in Table 3.4.

A conjugate prior is usually assumed for the QTL effect vector $\boldsymbol{\beta}$. Furthermore, the components of $\boldsymbol{\beta}$ are assumed independent. Thus the prior on $\boldsymbol{\beta}$ is factorized as

$$p(\boldsymbol{\beta}|m) = \prod_{j=1}^{m} p(\beta_j|m).$$

For Gaussian traits, the $p(\beta_j|m)$'s are assumed as identical normal densities with two superparameters μ_β, the super-mean, and σ_β^2, the super-variance. For the error variance σ^2, an inverse gamma prior is assumed.

If there is no prior knowledge available on the locations of the QTLs, the QTL locations, i.e., the components of $\boldsymbol{d}_Q$, are assumed independently uniformly distributed over the mapping region, subject to at most one QTL between any two adjacent markers. The prior density is then given by

$$p(\boldsymbol{d}_Q|m) = p(d_{Q1}) \prod_{j=2}^{m} p(d_{Qj}|d_{Q1} \cdots d_{Qj-1}),$$

where $p(d_{Q1})$ is the density function of the uniform distribution over the whole mapping region, and $p(d_{Qj}|d_{Q1} \cdots d_{Qj-1})$ is the density function of the uniform distribution across the marker intervals unoccupied by $d_{Q1} \cdots d_{Qj-1}$. If some prior knowledge is available from other studies, that knowledge should be incorporated into the prior, and more mass should be put on the regions where more QTLs are suggested by those studies.

When m is considered as a random variable, the prior on m is taken as a truncated Poisson distribution with a pre-determined maximum number of QTLs, $m_{\max}$, and intensity λ, though other priors such as a discrete uniform distribution can be considered as well.

The MCMC procedure for simulating the posterior distribution (7.1) and the method for making inference are different when m is treated as fixed and when m is taken as a random variable. We distinguish these two situations in the following.

7.2.1.1 The Number of QTLs Treated as Fixed

When m, though unknown, is treated as fixed, the MCMC procedure using a hybrid of the Gibbs sampler and the Metropolis-Hastings algorithm is applied to simulate the posterior distribution (7.1) for each fixed possible number m. The inference on the true value of m is made by using Bayes factors of the models with different m values. The final inference on QTL locations and effects are then based on the model selected by the Bayes factors.

For each fixed m, we need to simulate the Markov chain:

$$(\boldsymbol{d}_{\mathsf{Q}}^{(0)}, \boldsymbol{Q}^{(0)}, \boldsymbol{\beta}^{(0)}, \sigma^{2(0)}), (\boldsymbol{d}_{\mathsf{Q}}^{(1)}, \boldsymbol{Q}^{(1)}, \boldsymbol{\beta}^{(1)}, \sigma^{2(1)}), \cdots,$$
$$(\boldsymbol{d}_{\mathsf{Q}}^{(t)}, \boldsymbol{Q}^{(t)}, \boldsymbol{\beta}^{(t)}, \sigma^{2(t)}), \cdots, (\boldsymbol{d}_{\mathsf{Q}}^{(N)}, \boldsymbol{Q}^{(N)}, \boldsymbol{\beta}^{(N)}, \sigma^{2(N)}).$$

The hybrid procedure is to update each component of $(\boldsymbol{d}_{\mathsf{Q}}^{(t)}, \boldsymbol{Q}^{(t)}, \boldsymbol{\beta}^{(t)}, \sigma^{2(t)})$, in turn, by using their full conditional posterior distribution, i.e., the conditional distribution given all the other components. For those full conditional distributions which have the form of a standard distribution, the updates are directly generated from the standard distribution. For those which do not have a standard form, the Metropolis-Hastings algorithm is used to generate the updates.

By the specification of Bayesian model in this section, all the components of $(\boldsymbol{d}_{\mathsf{Q}}^{(t)}, \boldsymbol{Q}^{(t)}, \boldsymbol{\beta}^{(t)}, \sigma^{2(t)})$ except $\boldsymbol{d}_{\mathsf{Q}}^{(t)}$ have a standard full conditional distribution. These full conditional distributions are given below:

The full conditional distribution of β_0:

$$\beta_0 \sim N\left(\frac{\sum_{i=1}^{n}(y_i - \eta_i + \beta_0)}{\frac{\sigma^2}{\sigma_\beta^2} + n}, \frac{1}{\frac{1}{\sigma_\beta^2} + \frac{n}{\sigma^2}}\right) \tag{7.4}$$

where $\eta_i = \beta_0 + \sum_{l=1}^{m} \beta_l \delta_{il}$.

The full conditional distribution of β_j, $j = 1, \ldots, m$:

$$\beta_j \sim N\left(\frac{\sum_{i=1}^{n} \delta_{ij}(y_i - \eta_i + \beta_j \delta_{ij})}{\frac{\sigma^2}{\sigma_\beta^2} + \sum_{i=1}^{n} \delta_{ij}^2}, \frac{1}{\frac{1}{\sigma_\beta^2} + \frac{\sum_{i=1}^{n} \delta_{ij}^2}{\sigma^2}}\right). \tag{7.5}$$

The full conditional distribution of σ^2:

$$\sigma^2 \sim IG\left(3 + \frac{n}{2}, \left[3 + \frac{1}{2}\sum_{i=1}^{n}(y_i - \beta_0 - \sum_{j=1}^{m}\beta_j\delta_{ij})^2\right]^{-1}\right). \tag{7.6}$$

The full conditional distribution of δ_{ij}:

$$
\begin{aligned}
\pi_{ij} &= P(\delta_{ij} = 1 || \boldsymbol{d}_{\mathsf{Q}}, \boldsymbol{Q}/\{\delta_{ij}\}, \boldsymbol{\beta}, \sigma^2, \boldsymbol{y}) \\
&= \frac{p(\delta_{ij} = 1 | d_{\mathsf{Q}j})\phi(y_i | \boldsymbol{\beta}, \sigma^2, \boldsymbol{Q}_i, \delta_{ij} = 1)}{\sum_{\delta=0}^{1} p(\delta_{ij} = \delta | d_{\mathsf{Q}j})\phi(y_i | \boldsymbol{\beta}, \sigma^2, \boldsymbol{Q}_i, \delta_{ij} = \delta)}
\end{aligned} \tag{7.7}
$$

where $p(\delta_{ij} = \delta | d_{\mathsf{Q}j}) = p(\delta_{ij} | d_{\mathsf{Q}j}, d_{\mathsf{M}j-}, d_{\mathsf{M}j+}, M_{ij-}, M_{ij+})$.

In the full conditional distributions above, the superparameter μ_β has been taken as zero.

The detail of the hybrid MCMC procedure is described in the following. Let D be the total length of the mapping region. If there is only one chromosome, d_{Qj} denotes the genetic distance from the left end of the chromosome. In general, let the chromosomes be artificially connected in any given order, the value d_Q at any location of a chromosome is taken as the genetic distance of this location from the left end of the chromosome plus the total length of the chromosomes preceding this chromosome. Without loss of generality, assume that $0 < d_{Q1} < d_{Q2} < \cdots < d_{Qm} < D$. The MCMC procedure starts with an arbitrary state $(\boldsymbol{d}_Q^{(0)}, \boldsymbol{Q}^{(0)}, \boldsymbol{\beta}^{(0)}, \sigma^{2(0)})$ which falls into the domain of positive posterior density. Given the current state $(\boldsymbol{d}_Q^{(t)}, \boldsymbol{Q}^{(t)}, \boldsymbol{\beta}^{(t)}, \sigma^{2(t)})$, the components are updated as follows.

MCMC algorithm I:

Updating $\boldsymbol{d}_Q^{(t)}$ and $\boldsymbol{Q}^{(t)}$: The QTL locations and genotypes are simultaneously updated one QTL at a time by the Metropolis-Hastings algorithm. Let $a > 0$ be a fixed number. For each j and any d_Q in between $d_{Qj-1}^{(t)}$ and $d_{Qj+1}^{(t)}$, define the interval

$$U_j(d_Q) = [\max(d_{Qj-1}^{(t)}, d_Q - a), \min(d_{Qj+1}^{(t)}, d_Q + a)].$$

Let $u_j(d_Q, d)$, $d \in U_j(d_Q)$, denote the density function of the uniform distribution over the interval $U_j(d_Q)$. For updating the location and genotypes of the jth QTL, a proposal location d^* is generated from $u_j(d_{Qj}^{(t)}, d)$. The proposed location is accepted with probability

$$\alpha(d_{Qj}^{(t)}, d^*) = \min\left\{ \frac{\pi(d^*|\boldsymbol{d}_{Q-j}^{(t)}, \boldsymbol{Q}^{(t)}, \boldsymbol{\beta}^{(t)}, \sigma^{2(t)}) u_j(d^*, d_{Qj}^{(t)})}{\pi(d_{Qj}^{(t)}|\boldsymbol{d}_{Q-j}^{(t)}, \boldsymbol{Q}^{(t)}, \boldsymbol{\beta}^{(t)}, \sigma^{2(t)}) u_j(d_{Qj}^{(t)}, d^*)}, 1 \right\},$$

where $\boldsymbol{d}_{Q-j}^{(t)} = \{\boldsymbol{d}_{Ql}^{(t)} : 1 \leq l \leq m, l \neq j\}$, $\pi(d|\boldsymbol{d}_{Q-j}^{(t)}, \boldsymbol{Q}^{(t)}, \boldsymbol{\beta}^{(t)}, \sigma^{2(t)})$ is the full conditional posterior density of d_{Qj} up to a constant, in fact,

$$\pi(d|\boldsymbol{d}_{Q-j}^{(t)}, \boldsymbol{Q}^{(t)}, \boldsymbol{\beta}^{(t)}, \sigma^{2(t)}) \propto p(\boldsymbol{d}_Q^{(t)}|m) p(\boldsymbol{Q}^{(t)}|\boldsymbol{d}_Q^{(t)}, \boldsymbol{d}_M, \boldsymbol{M}),$$

where the jth component of $\boldsymbol{d}_Q^{(t)}$ in the right hand side is replaced by d. If the proposed location is accepted, the genotypes at the new location are generated by using the full conditional posterior probability (7.7), otherwise, both the location and the genotypes of the QTL remain unchanged.

Updating $\boldsymbol{\beta}^{(t)}$ and $\sigma^{2(t)}$: The components of $\boldsymbol{\beta}^{(t)}$ and $\sigma^{2(t)}$ are updated by directly sampling from their full conditional posterior distribution given in (7.4), (7.5) and (7.6). For each of the full conditional distributions, it is conditioned on the most updated values of the other components, as described in the general algorithm of the Gibbs sampler.

For a fixed m, the QTL locations are estimated from the extracted marginal sequence $\{\boldsymbol{d}_{\scriptscriptstyle Q}^{(t)}, t = 1, 2, \ldots, N\}$, as

$$\hat{d}_{\scriptscriptstyle Qj} = \frac{1}{N} \sum_{t=1}^{N} d_{\scriptscriptstyle Qj}^{(t)}, j = 1, \ldots, m.$$

The marginal posterior densities of the QTL effects are estimated from the extracted marginal sequence $\{\boldsymbol{\beta}^{(t)}, t = 1, 2, \ldots, N\}$, by using standard density estimation methods such as kernel density estimation. For each component of $\boldsymbol{\beta}$, a density estimator called the Rao-Blackwell estimator is given by

$$\pi(\beta_j) = \frac{1}{N} \sum_{t=1}^{N} \pi(\beta_j | \boldsymbol{d}_{\scriptscriptstyle Q}^{(t)}, \boldsymbol{Q}^{(t)}, \boldsymbol{\beta}_{-j}^{(t)}, \sigma^{2(t)}),$$

where $\pi(\beta_j | \boldsymbol{d}_{\scriptscriptstyle Q}^{(t)}, \boldsymbol{Q}^{(t)}, \boldsymbol{\beta}_{-j}^{(t)}, \sigma^{2(t)})$ is the full conditional posterior density of β_j given all the other unknowns. Credible intervals for the QTL effects can then be computed from these density estimators. For more details, the reader is referred to [153] and the references therein.

The Bayes factors among models with different m values are used to select the final model, see § 2.7.4. By using Bayes factors to select the final m value is equivalent to comparing the marginal probability density of $\boldsymbol{y}$, $p(\boldsymbol{y}|m)$, among all m values where

$$p(\boldsymbol{y}|m) = \int \phi(\boldsymbol{y}|m, \boldsymbol{Q}, \boldsymbol{\beta}, \sigma^2) p(m, \boldsymbol{d}_{\scriptscriptstyle Q}, \boldsymbol{Q}, \boldsymbol{\beta}, \sigma^2) d\boldsymbol{d}_{\scriptscriptstyle Q} d\boldsymbol{Q} d\boldsymbol{\beta} d\sigma^2.$$

The final selected $m = \hat{m}$ satisfies

$$p(\boldsymbol{y}|\hat{m}) = \max_{m} p(\boldsymbol{y}|m).$$

The marginal probability density $p(\boldsymbol{y}|m)$ is in general not possible to compute analytically. But they can be approximated by using the MCMC samples. Among other approximations, the following one is given in [153]:

$$\hat{p}(\boldsymbol{y}|m) = \left[\frac{1}{N} \sum_{t=1}^{N} \frac{h(\boldsymbol{d}_{\scriptscriptstyle Q}^{(t)})}{\phi(\boldsymbol{y}|m, \boldsymbol{Q}^{(t)}, \boldsymbol{\beta}^{(t)}, \sigma^{2(t)}) p(\boldsymbol{d}_{\scriptscriptstyle Q}^{(t)}|m)} \right]^{-1}, \qquad (7.8)$$

where $h(\boldsymbol{d}_{\scriptscriptstyle Q})$ is a normal density restricted to $0 < d_{\scriptscriptstyle Q1} < d_{\scriptscriptstyle Q2} < \cdots < d_{\scriptscriptstyle Qm} < D$.

7.2.1.2 The Number of QTLs Treated as a Random Variable

When the number of QTLs is treated as a random variable, the dimension of the unknowns is not fixed and varies as the number of QTLs changes. The state of the Markov chain $(m^{(t)}, \boldsymbol{d}_{\scriptscriptstyle Q}^{(t)}, \boldsymbol{Q}^{(t)}, \boldsymbol{\beta}^{(t)}, \sigma^{2(t)}), t = 0, 1, 2, \ldots$, jumps among spaces of different dimensions. The reversible jump MCMC is required for the simulation of the Markov chain. Different methods for the inference on the true number of QTLs and their locations are also in order. In this section, the specific aspects of the reversible jump MCMC are addressed, and the empirical procedures for the inference are described.

In the reversible jump MCMC, from time t to time $t + 1$, a move type is determined first and the state is then updated accordingly. As suggested in [163], three move types are considered: (i) stay in the current space, (ii) move to a space with one more QTL (add one QTL) and (iii) move to a space with one less QTL (delete one QTL). Define the proposal probabilities of the move types (i), (ii) and (iii), respectively, as p_0, p_a and p_d such that $p_a = cI\{m < m_{\max}\}$, $p_d = cI\{m > 0\}$ and $p_0 = 1 - p_a - p_d$, where $0 < c < 0.5$.

At the initial step, initial values of $(m, \boldsymbol{d}_Q, \boldsymbol{Q}, \boldsymbol{\beta}, \sigma^2)$ are determined. The initial values of $m, \boldsymbol{d}_Q, \boldsymbol{\beta}$ and σ^2 can be either specified by the analyst based on his or her prior knowledge or generated from their prior distributions. Once m and $\boldsymbol{d}_Q$ are specified, the initial values of $\boldsymbol{Q}$ are generated from their prior distributions. At step $t + 1$, the state $(m^{(t)}, \boldsymbol{d}_Q^{(t)}, \boldsymbol{Q}^{(t)}, \boldsymbol{\beta}^{(t)}, \sigma^{2(t)})$ is updated as follows.

MCMC algorithm II

Determination of move types: Draw a random variable u from the uniform distribution over the unit interval $[0, 1]$. If $m^{(t)} = 0$, take move type (ii) and (i) according as $0 < u < c$ and $c \leq u < 1$. If $m^{(t)} = m_{\max}$, take move type (iii) and (i) according as $0 < u < c$ and $c \leq u < 1$. If $0 < m^{(t)} < m_{\max}$, take move type (ii), (iii) and (i) according as $0 < u < c$, $c \leq u < 2c$ and $2c \leq u < 1$.

Type (i) update: When the move type is (i), update $(\boldsymbol{d}_Q^{(t)}, \boldsymbol{Q}^{(t)}, \boldsymbol{\beta}^{(t)}, \sigma^{2(t)})$ in the same way as in the hybrid MCMC with fixed number of QTLs.

Type (ii) update: When the move type is (ii), the following sub steps are taken.

(iia): $m^{(t)}$ is updated to $m^{(t+1)} = m^{(t)} + 1$. The location of the new QTL, $d_{Qm^{(t+1)}}$, is proposed from the uniform distribution over the genome. The genotype of individual i at the new QTL is proposed from $p(\delta_{im^{(t+1)}} | d_{Qm^{(t+1)}}, d_{Mm^{(t+1)}-}, d_{Mm^{(t+1)}+}, M_{im^{(t+1)}-}, M_{im^{(t+1)}+})$. The regression coefficient of $\delta_{im^{(t+1)}}$ is proposed from its prior distribution. Obtain $(\tilde{\boldsymbol{d}}_Q^{(t)}, \tilde{\boldsymbol{Q}}^{(t)}, \tilde{\boldsymbol{\beta}}^{(t)})$ by augmenting $(\boldsymbol{d}_Q^{(t)}, \boldsymbol{Q}^{(t)}, \boldsymbol{\beta}^{(t)})$ with the proposed components.

(iib): The whole proposal is accepted with probability

$$\alpha_{II} = \min\left\{\frac{\phi(\boldsymbol{y} | \tilde{\boldsymbol{d}}_Q^{(t)}, \tilde{\boldsymbol{Q}}^{(t)}, \tilde{\boldsymbol{\beta}}^{(t)}, \sigma^{2(t)})}{\phi(\boldsymbol{y} | \boldsymbol{d}_Q^{(t)}, \boldsymbol{Q}^{(t)}, \boldsymbol{\beta}^{(t)}, \sigma^{2(t)})} \frac{\lambda p_d}{[m^{(t+1)}]^2 p_a}, 1\right\},$$

where λ is the intensity parameter of the Poisson prior distribution of m. Note that α_{II} is of the form (2.9) where the Jacobian is 1 since the derivative matrix is the identity matrix.

(iic): When the whole proposal is accepted, update $(\tilde{d}_{\scriptscriptstyle Q}^{(t)}, \tilde{Q}^{(t)}, \tilde{\beta}^{(t)}, \sigma^{2(t)})$ to $(d_{\scriptscriptstyle Q}^{(t+1)}, Q^{(t+1)}, \beta^{(t+1)}, \sigma^{2(t+1)})$ by the same procedure as in the hybrid MCMC with fixed number of QTLs.

Type (iii) update: When the move type is (iii), the following sub steps are taken.

(iiia): $m^{(t)}$ is updated to $m^{(t+1)} = m^{(t)} - 1$. Randomly delete one of the existing QTLs. Obtain $(\tilde{d}_{\scriptscriptstyle Q}^{(t)}, \tilde{Q}^{(t)}, \tilde{\beta}^{(t)})$ by deleting the components corresponding to the deleted QTL from $(d_{\scriptscriptstyle Q}^{(t)}, Q^{(t)}, \beta^{(t)})$.

(iiib): The whole proposal is accepted with probability

$$\alpha_{III} = \min \left\{ \frac{\phi(y|\tilde{d}_{\scriptscriptstyle Q}^{(t)}, \tilde{Q}^{(t)}, \tilde{\beta}^{(t)}, \sigma^{2(t)})}{\phi(y|d_{\scriptscriptstyle Q}^{(t)}, Q^{(t)}, \beta^{(t)}, \sigma^{2(t)})} \frac{[m^{(t)}]^2 p_a}{\lambda p_d}, 1 \right\}.$$

Note that α_{III} is of the form (2.10).

(iiic): When the whole proposal is accepted, update $(\tilde{d}_{\scriptscriptstyle Q}^{(t)}, \tilde{Q}^{(t)}, \tilde{\beta}^{(t)}, \sigma^{2(t)})$ to $(d_{\scriptscriptstyle Q}^{(t+1)}, Q^{(t+1)}, \beta^{(t+1)}, \sigma^{2(t+1)})$ by the same procedure as in the hybrid MCMC with fixed number of QTLs.

After the Markov sequence $(m^{(t)}, d_{\scriptscriptstyle Q}^{(t)}, Q^{(t)}, \beta^{(t)}, \sigma^{2(t)}), t = 0, 1, 2, \ldots$, is simulated by the reversible jump MCMC, the information about the QTL locations and effects is summarized by the empirical QTL intensity function and the empirical cumulative distribution functions of QTL effects [163], which are described in the following.

Let the whole mapping region be divided into equal length intervals: $\Delta_1, \Delta_2, \ldots, \Delta_K$. The length of each interval is then D/K, D being the total length of the mapping region. The interval length, hence K, is determined by the mapping resolution required by the analyst. Let N be the number of MCMC cycles. The empirical QTL intensity over interval Δ_k is defined as

$$\hat{\lambda}_k = \frac{K}{ND} \sum_{t=1}^{N} \sum_{j=1}^{m^{(t)}} I\{d_{\scriptscriptstyle Qj}^{(t)} \in \Delta_k\}.$$

The empirical QTL intensity function is defined as

$$\hat{\lambda}(d) = \sum_{k=1}^{K} \hat{\lambda}_k I\{d \in \Delta_k\}. \tag{7.9}$$

The empirical cumulative distribution function of QTL effects is defined for each interval Δ_k as follows.

$$\hat{D}_k(s) = \frac{\sum_{t=1}^{N} \sum_{j=1}^{m^{(t)}} I\{d_{qj}^{(t)} \in \Delta_k, \beta_j \le s\}.}{\sum_{t=1}^{N} \sum_{j=1}^{m^{(t)}} I\{d_{qj}^{(t)} \in \Delta_k\}.} \tag{7.10}$$

Finally, the true number of QTLs and their locations are inferred from the peaks of the empirical QTL intensity function. The magnitude of the QTL effects is inferred from the empirical cumulative distribution functions at the peaks.

7.2.1.3   Example: Mapping Loci Controlling Flowering Time in *Brassica napus*

In [153], the Bayesian method with fixed number of QTL is illustrated using a data set obtained from an experiment reported in [59]. The objective of the experiment is to identify genomic regions associated with vernalization requirement and flowering in a doubled haploid (DH) population of *Brassica napus*. In the experiment, a single plant of the *Brassica napus* CV. Major used as a female is crossed to a doubled haploid line derived from CV. Stellar. One hundred and four F_1-derived DH lines, the F_1 hybrid and progeny from self-pollination of the parents Major and Stellar are evaluated in the field. Three treatments are given to the plants in the experiment: no vernalization, 4 weeks vernalization and 8 weeks vernalization. For each treatment, the data set consists of the days-to-flowering and vernalization requirement (0 or 1) of the 104 progeny, together with their genotypes at 132 RELP markers (plus a few other loci) on 22 linkage groups covering a total 1016 cM of the *Brassica napus* genome. Only the flowering data for 104 progeny from 8 weeks vernalization treatment and genotypes of 10 markers from linkage group 9 is considered in [153] for the illustration.

The following model for the number of days to flowering for the ith DH line is assumed in the Bayesian approach:

$$y_i = \mu + \sum_{j=1}^{s} \alpha_j \delta_{ij} + \epsilon_i,$$

where y_i is log of the number of days to flower, and δ_{ij} is the dummy variable for the genotype defined as

$$\delta_{ij} = \begin{cases} 1, & \text{if the genotype is } S/S; \\ -1, & \text{if the genotype is } M/M. \end{cases}$$

The allele S is from Stellar and M from Major. In the F_1-derived DH lines, there is no heterozygous genotype. The ϵ_i's are assumed to have independent $N(0, \sigma^2)$ distribution.

The prior distributions for the model parameters are assumed as follows. The parameters are assumed independent. The priors for the overall mean μ

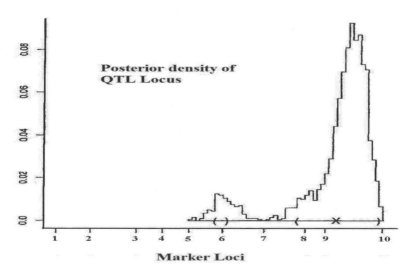

FIGURE 7.1: Posterior probability of QTL locus of the single-QTL model (reproduced from Figure 2 in [153]).

and the α_j's are assumed as independent $N(0, 10)$ distributions. The phenotypic variance σ^2 is assumed to have an inverse gamma prior. The QTL location d_{qj}'s are assumed to have uniform prior along the entire linkage group 9 such that $0 < d_{q1} < \cdots < d_{qs} < D_m$.

The Bayesian analysis is carried out with a single QTL model, a two-QTL model and a three-QTL model. For each analysis, 400,000 cycles of the Markov chain are run and samples are taken from every 200 cycles without initial burn-in, which results in a MCMC sample of size 2000. The number of inter-sample cycles is determined from an autocorrelation analysis. It is found that when the number is taken as 100, the autocorrelation of the MCMC sample is significant even at log 30, but when the number is taken as 200, the autocorrelation becomes negligible.

For the single QTL model, the starting value for the single putative locus is taken as $d_{q1}^{(0)} = 24cM$. From the 2000 MCMC samples, the single QTL is estimated to be at $d_{q1} = \frac{1}{2000} \sum_{m=1}^{2000} d_{q1}^{(m)} == 71.9cM$ between markers 9 and 10. The estimated effect was $\alpha_1 = \frac{1}{2000} \sum_{m=1}^{2000} \alpha_1^{(m)} = -0.165$. The posterior density of d_{q1} is depicted in Figure 7.1. The posterior density displays two modes: a high mode between markers 9 and 10, and a smaller one between markers 5 and 6. This suggests a two-QTL model.

For the two-QTL model, the starting values for the two loci are $d_{q1}^{(0)} = 24cM$ and $d_{q2}^{(0)} = 59cM$. The first locus is estimated at $d_{q1} = 42.2cM$ between markers 5 and 6, and the second locus at $d_{q2} = 76.4cM$ between markers 9 and 10. The effect α_2 of locus d_{q2} is nearly twice that of α_1 at locus d_{q1}. The

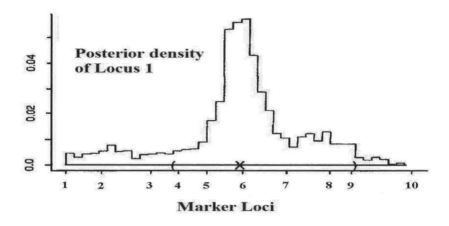

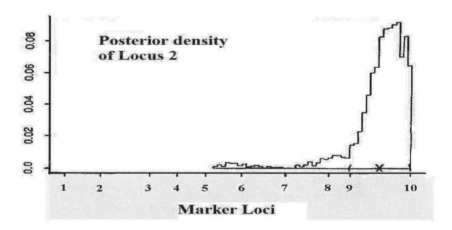

FIGURE 7.2: Posterior probability of QTL loci of the two-QTL model (reproduced from Figure 7 in [153]).

locus d_{q2} having the larger effect corresponds to the high mode in the single QTL model. The posterior densities of d_{q1} and d_{q2} are shown in Figure 7.2.

For the three-QTL model, the starting values for the loci are $d_{q1}^{(0)} = 24cM$, $d_{q2}^{(0)} = 42cM$ and $d_{q3}^{(0)} = 59cM$. The posterior densities of d_{q1}, d_{q2} and d_{q3} are depicted in Figure 7.3. The mode of the posterior density of d_{q1} corresponds to the mode of the posterior density of the first locus of the two-QTL model. The mode of the posterior density of d_{q3} corresponds to the mode of the posterior density of the second locus of the two-QTL model. The posterior density of d_{q2} has two modes which are near the modes of d_{q1} and d_{q3}. This is an indication that only two QTL are supported by the data.

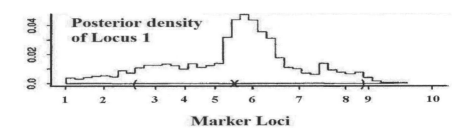

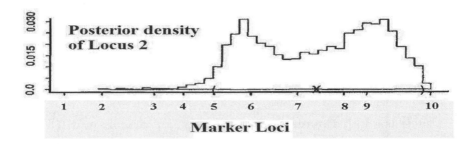

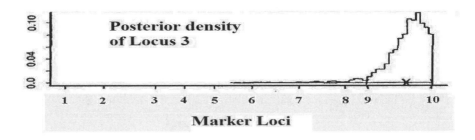

FIGURE 7.3: Posterior probability of QTL loci of the three-QTL model (reproduced from Figure 8 in [153]).

To select the final model, Bayes factors comparing single-QTL with two-QTL model and comparing two-QTL with three-QTL model are computed respectively using the marginal densities computed by (7.8) as 0.12 and 1.1. Thus the two-QTL model is favored by the data. The estimated parameters for the three models and their Bayes factors are summarized in Table 7.1. For more details of the analysis based on the MCMC samples, the reader is referred to [153].

TABLE 7.1: The estimated parameters of the three Bayesian models and their Bayes factors (BF) (numbers in the parentheses are Monte Carlo standard errors)

Parameter	Model 1	Model 2	Model 3
μ	3.060 (0.0010)	3.061 (0.005)	3.060 (0.021)
σ^2	0.081 (0.027)	0.078 (0.010)	0.080 (0.004)
Locus		42.2 (0.026)	38.6 (0.019)
Effect		−0.066 (0.047)	−0.041 (0.060)
Locus			57.8 (0.035)
Effect			−0.044 (0.067)
Locus	71.9 (0.045)	76.4 (0.070)	77.9 (0.016)
Effect	−0.165 (0.034)	−0.128 (0.044)	−0.111 (0.050)
BF	0.12 (1 vs. 2)	1.1 (2 vs. 3)	

7.2.2 Bayesian Models with Epistatic Effects

In dealing with epistatic effects, we consider, for backcross or RIL design, the conditional model of the Gaussian traits as follows:

$$y_i = \beta_0 + \sum_{j=1}^{m} \beta_j \delta_{ij} + \sum_{j>k} \beta_{jk} \delta_{ij} \delta_{ik} + \epsilon_i, \ i = 1, \ldots, n, \qquad (7.11)$$

where ϵ_i are i.i.d. normal random variables with mean zero and variance σ^2.

In the model for backcross or RIL designs, there is only one possible interaction term between any two QTLs since there is only one main-effect term for each QTL. When considering models for intercross designs, there are two main-effect terms for each QTL and hence there are four possible interaction terms between any two QTLs. Apart from this difference, all the other aspects of the models are the same. Hence, again, we concentrate on the model for backcross designs.

The model with epistatic effects is much more complicated than that without epistatic effects. In the model without epistatic effects, a QTL presents if and only if its main-effect term presents. However, in the model with epistatic effects, a QTL can present in different forms: it might only have the main-effect term, it might only have some interaction terms without the main-effect term, it might have both the main-effect term and some interaction terms. Furthermore, when it has an epistatic effect, not all the interaction terms are necessarily existent. In short, QTL j exists if and only if one the coefficients $\beta_j, \beta_{jk}, \beta_{kj}$, is non-zero. In the model without epistatic effects, adding or deleting a QTL is equivalent to adding or deleting the corresponding main-effect term. However, in the model with epistatic effects, when adding a QTL, it is uncertain which effect terms should be added; when deleting an effect term, the involved QTL is not necessarily deleted. In order to tackle this complexity, the following alternative form of (7.11) is considered in [212]:

$$y_i = \beta_0 + \sum_{j=1}^{m} \gamma_j \beta_j \delta_{ij} + \sum_{j>k} \gamma_{jk} \beta_{jk} \delta_{ij} \delta_{ik} + \epsilon_i, \quad i = 1, \ldots, n, \qquad (7.12)$$

where γ_j's and γ_{jk}'s are effect indicators taking values 0 or 1, subject to the constraint that, for any given j, at least one of the $\gamma_j, \gamma_{jk}, k \neq j$, is 1. The development of the method in this section follows the work in [212].

The Bayesian model with epistatic effects has the same structure as the general model (7.1) except that there is an additional component of the unknowns, i.e., the set of the effect indicators $\gamma = \{\gamma_j, \gamma_{jk} : j, k = 1, \ldots, m, k < j, \}$. Let β denote the set of all effect parameters including both the main effects and interaction effects, i.e., $\beta = \{\beta_0, \beta_j, \beta_{jk} : j, k = 1, \ldots, m, k < j, \}$. The posterior density of the model is then given by

$$\pi(m, d_q, Q, \beta, \gamma, \sigma^2 | y) \qquad (7.13)$$

$$\propto \quad p(\beta|m)p(d_q|m)p(\gamma|m)p(m)p(\sigma^2)p(Q|d_q, d_M, M) \prod_{i=1}^{n} \phi(y_i|m, Q_i, \beta, \gamma, \sigma^2).$$

Apart from $p(\gamma|m)$, all the other priors and the conditional distribution of y are assumed the same as in § 7.2.1, except that, in the specification of $\phi(y_i|m, Q_i, \beta, \gamma, \sigma^2)$,

$$\eta_i = \beta_0 + \sum_{j=1}^{m} \gamma_j \beta_j \delta_{ij} + \sum_{j>k} \gamma_{jk} \beta_{jk} \delta_{ij} \delta_{ik}$$

in the current context. The prior on β becomes

$$p(\beta|m) = \prod_{j=0}^{m} p(\beta_j|m) \prod_{j>k}^{m} p(\beta_{jk}|m).$$

For the prior $p(\gamma|m)$, assume that the components of γ are i.i.d. as Bernoulli distribution with probability of success $\frac{1}{2}$, i.e.,

$$p(\gamma|m) = \left(\frac{1}{2}\right)^{m(m+1)/2}.$$

As in the case of the model without epistatic effects, the model with epistatic effects can be analyzed by either treating m, the number of QTL, as an unknown fixed number or as a random variable. When m is treated as an unknown fixed number, the inference procedure is the same as that for the model without epistatic effects. However, instead of the Metropolis-Hastings algorithm, a reversible jump MCMC procedure must be developed, which can be done easily. We will not go into the details of the fixed-m method. In the following, we only consider the method that treats m as a random variable.

In the reversible jump MCMC procedure for generating samples from the posterior distribution (7.13), more move types than those considered in § 7.2.1.2 must be invoked. In general, any reasonable move types can be incorporated into the MCMC procedure. We consider here the four move types proposed in [212]. In the following description, by the configuration of a state of the reversible jump Markov chain, we refer to the number of QTLs and the number and types of the QTL effects at that state.

Type a: Keep the configuration of the current state, update all the other parameters;

Type b: Change the configuration of the current state by keeping the same number of QTL but changing the effect indicators, then update all the other parameters with the updated configuration;

Type c: Change the configuration of the current state by adding a new QTL with main-effect term or interaction terms with existing QTLs, or deleting an existing QTL, then update all the other parameters with the updated configuration;

Type d: Change the configuration of the current state by adding two new QTLs with only the interaction term between them, or deleting two existing QTLs, then update all the other parameters with the updated configuration.

In what follows we present an adapted version of the reversible jump MCMC algorithm developed in [212]. Let the proposal probabilities for the move types be p_a, p_b, p_c, p_d, which are all bigger than zero, such that $p_a + p_b + p_c + p_d = 1$. Let the initial values of $m^{(0)}, d_Q^{(0)}, Q^{(0)}, \beta^{(0)}$ and $\sigma^{2(0)}$ be determined in the same way as in § 7.2.1.2. Let the initial values for $\gamma^{(0)}$ be determined from their prior distribution. We first give a brief description of the algorithm and then provide the details.

MCMC algorithm III

For $t = 0, 1, 2, \ldots,$

Step 1: Draw a random variable u from the uniform distribution over the unit interval $[0, 1]$ to determine move types.

$$\text{Take move type} \begin{cases} a, & \text{if } 0 < u \leq p_a; \\ b, & \text{if } p_a < u \leq p_a + p_b; \\ c, & \text{if } p_a + p_b < u \leq p_a + p_b + p_c; \\ d, & \text{if } p_a + p_b + p_c < u \leq 1. \end{cases}$$

Step 2: Propose an new configuration according to the move type determined in Step 1. If the proposed configuration is accepted go to Step 3, otherwise go back to Step 1.

Step 3: Update $(\boldsymbol{d}_{\mathrm{Q}}^{(t)}, \boldsymbol{Q}^{(t)}, \boldsymbol{\beta}^{(t)}, \sigma^{2(t)})$ with the accepted configuration in Step 2 in the same way as in the hybrid MCMC of § 7.2.1.1.

Step 4: Update t to $t+1$, go to Step 1.

For move type **a**, the current configuration is always accepted, it directly goes to Step 3. We now elaborate the details for the updating of state configuration with the other three move types.

Updating configuration with move Type b: Randomly choose one of the $m^{(t)}(m^{(t)}+1)/2$ effect indicators. Let $\gamma_*^{(t)}$ be the chosen indicator. Then update $\gamma_*^{(t)}$ to $\gamma_*^{(t+1)} = 1 - \gamma_*^{(t)}$ while the other elements of $\boldsymbol{\gamma}^{(t)}$ remain unchanged.

- If $\gamma_*^{(t)} = 0$, then add the effect β_* corresponding to this indicator. The added effect is generated from a proposal density $q_+(\beta_*)$. One choice of the proposal density is the full conditional posterior density given by the normal distribution $N(\mu_a^*, \sigma_*^2)$ where

$$
\mu_a^* = \frac{\mu_\beta/\sigma_\beta^2 + \sum_{i=1}^n \delta_{i*}^{(t)}(y_i - \eta_i^{(t)})/\sigma^{2(t)}}{1/\sigma_\beta^2 + \sum_{i=1}^n [\delta_{i*}^{(t)}]^2/\sigma^{2(t)}},
$$

$$
\sigma_*^2 = \frac{1}{1/\sigma_\beta^2 + \sum_{i=1}^n [\delta_{i*}^{(t)}]^2/\sigma^{2(t)}}.
$$

  Here $\delta_{i*}^{(t)}$ is a single QTL genotype indicator if γ_* is a main-effect indicator, a product of two QTL genotype indicators if γ_* is an interaction indicator. Let $\tilde{\boldsymbol{\beta}}^{(t)}$ be $\boldsymbol{\beta}^{(t)}$ augmented by adding β_*. This step is accepted with probability $\alpha_a = \min(r_a, 1)$ where

$$
r_a = \frac{\phi(\boldsymbol{y}|\boldsymbol{d}_{\mathrm{Q}}^{(t)}, \boldsymbol{Q}^{(t)}, \boldsymbol{\gamma}^{(t+1)}, \tilde{\boldsymbol{\beta}}^{(t)}, \sigma^{2(t)})}{\phi(\boldsymbol{y}|\boldsymbol{d}_{\mathrm{Q}}^{(t)}, \boldsymbol{Q}^{(t)}, \boldsymbol{\gamma}^{(t)}, \boldsymbol{\beta}^{(t)}, \sigma^{2(t)})} \frac{p(\beta_*|m^{(t)})}{q_+(\beta_*)}.
$$

- If $\gamma_*^{(t)} = 1$, then remove the effect corresponding to this indicator. Denote by $\tilde{\boldsymbol{\beta}}^{(t)}$ the $\boldsymbol{\beta}^{(t)}$ with β_* deleted. This step is accepted with probability $\alpha_d = \min(r_d, 1)$ where

$$
r_d = \frac{\phi(\boldsymbol{y}|\boldsymbol{d}_{\mathrm{Q}}^{(t)}, \boldsymbol{Q}^{(t)}, \boldsymbol{\gamma}^{(t+1)}, \tilde{\boldsymbol{\beta}}^{(t)}, \sigma^{2(t)})}{\phi(\boldsymbol{y}|\boldsymbol{d}_{\mathrm{Q}}^{(t)}, \boldsymbol{Q}^{(t)}, \boldsymbol{\gamma}^{(t)}, \boldsymbol{\beta}^{(t)}, \sigma^{2(t)})} \frac{q_-(\beta_*)}{p(\beta_*|m^{(t)})}.
$$

  Here $q_-(\beta_*)$ is the density of the normal distribution $N(\mu_d^*, \sigma_*^2)$ where

$$
\mu_d^* = \frac{\mu_\beta/\sigma_\beta^2 + \sum_{i=1}^n \delta_{i*}^{(t)}(y_i - \eta_i^{(t)} + \beta_*^{(t)}\delta_{i*}^{(t)})/\sigma^{2(t)}}{1/\sigma_\beta^2 + \sum_{i=1}^n [\delta_{i*}^{(t)}]^2/\sigma^{2(t)}}.
$$

Updating configuration with move Type c: Determine whether to add a new QTL or to delete an existing QTL with equal probability.

If a new QTL is to be added, then

- $m^{(t)}$ is updated to $m^{(t+1)} = m^{(t)} + 1$. The location of the new QTL, $d_{Qm^{(t+1)}}$, is proposed from the uniform distribution over the unoccupied intervals on the mapping region.

- The genotype of individual i at the new QTL is proposed from $p(\delta_{im^{(t+1)}}|d_{Qm^{(t+1)}}, d_{Mm^{(t+1)}-}, d_{Mm^{(t+1)}+}, M_{im^{(t+1)}-}, M_{im^{(t+1)}+})$.

- Choose an integer k with equal probabilities from $\{1, 2, \ldots, m^{(t+1)}\}$. Then randomly select k effects out of the $m^{(t+1)}$ effects associated with the new QTL including its main effect and $m^{(t+1)} - 1$ interactions with the existing QTLs. Set the effect indicators corresponding to these k effects to 1. Update $\gamma^{(t)}$ to $\gamma^{(t+1)}$.

- Denote the k effects by $\beta_* = (\beta_{*1}, \ldots, \beta_{*k})$. The effects are sequentially proposed by the normal distributions:

$$N(\mu_{*j}, \sigma^2_{*j}), \ j = 1, \ldots, k,$$

where

$$\mu^+_{*j} = \frac{\mu_\beta/\sigma^2_\beta + \sum_{i=1}^n \delta_{i*j}(y_i - \eta_i^{(t)} - \sum_{l=1}^{j-1} \beta_{*l}\delta_{i*l})/\sigma^{2(t)}}{1/\sigma^2_\beta + \sum_{i=1}^n \delta^2_{i*j}/\sigma^{2(t)}},$$

$$\sigma^2_* = \frac{1}{1/\sigma^2_\beta + \sum_{i=1}^n \delta^2_{i*j}/\sigma^{2(t)}}.$$

When $j = 1$, $\sum_{l=1}^{j-1} = 0$ by convention. Here δ_{i*j} is of the form $\delta_{im^{(t+1)}}$ if the jth proposed effect is the main effect or $\delta_{im^{(t+1)}}\delta_{il}^{(t)}$ for some $1 \le l \le m^{(t)}$ if the jth proposed effect is an interaction.

- Augment $(d_Q^{(t)}, Q^{(t)}, \beta^{(t)})$ with the proposed components and denote the augmented parameters by $(\tilde{d}_Q^{(t)}, \tilde{Q}^{(t)}, \tilde{\beta}^{(t)})$. All the proposed components are accepted or rejected together with acceptance probability $\alpha_a = \min(r_a, 1)$ where

$$r_a = \frac{\pi(m^{(t+1)}, \tilde{d}_Q^{(t)}, \tilde{Q}^{(t)}, \gamma^{(t+1)}, \tilde{\beta}^{(t)}, \sigma^{2(t)}|y)}{\pi(m^{(t)}, d_Q^{(t)}, Q^{(t)}, \gamma^{(t)}, \beta^{(t)}, \sigma^{2(t)}|y)} \frac{1}{q^*}.$$

Here q^* is the proposal probability given by

$$q^* = p(d_{Qm^{(t+1)}})p(Q_*)q_+(\beta^*),$$

where β_* and Q_* are respectively the proposed effects and genotypes of the new QTL,

$$q_+(\beta^*) = q_+(\beta_{*1})q_+(\beta_{*2}|\beta_{*1})\cdots q_+(\beta_{*k}|\beta_{*1}\cdots\beta_{*k-1}),$$

$q_+(\beta_{*j}|\beta_{*1}\cdots\beta_{*j-1})$ being the density of $N(\mu_{*j}^+, \sigma_{*j}^2)$. From the specification of the priors, r_a simplifies to

$$r_a = \frac{\phi(\boldsymbol{y}|\tilde{\boldsymbol{d}}_{\scriptscriptstyle Q}^{(t)}, \tilde{\boldsymbol{Q}}^{(t)}, \boldsymbol{\gamma}^{(t+1)}, \tilde{\boldsymbol{\beta}}^{(t)}, \sigma^{2(t)})}{\phi(\boldsymbol{y}|\boldsymbol{d}_{\scriptscriptstyle Q}^{(t)}, \boldsymbol{Q}^{(t)}, \boldsymbol{\gamma}^{(t)}, \boldsymbol{\beta}^{(t)}, \sigma^{2(t)})} \frac{p(\boldsymbol{\beta}_*)p(\boldsymbol{Q}_*)\lambda}{2m^{(t+1)}q(\boldsymbol{\beta}_*)m^{(t+1)}}.$$

If an existing QTL is to be deleted, then

- Choose the deleted QTL randomly from the existing QTLs. Remove all effects related to this QTL.

- Set the corresponding effect indicators to zero. Set $m^{(t+1)} = m^{(t)} - 1$. Denote the values of the parameters after deleting the components related to the deleted QTL by $(\tilde{\boldsymbol{d}}_{\scriptscriptstyle Q}^{(t)}, \tilde{\boldsymbol{Q}}^{(t)}, \tilde{\boldsymbol{\beta}}^{(t)})$. The move is accepted with probability $\alpha_d = \min(r_d, 1)$ where

$$r_d = \frac{\phi(\boldsymbol{y}|\tilde{\boldsymbol{d}}_{\scriptscriptstyle Q}^{(t)}, \tilde{\boldsymbol{Q}}^{(t)}, \boldsymbol{\gamma}^{(t+1)}, \tilde{\boldsymbol{\beta}}^{(t)}, \sigma^{2(t)})}{\phi(\boldsymbol{y}|\boldsymbol{d}_{\scriptscriptstyle Q}^{(t)}, \boldsymbol{Q}^{(t)}, \boldsymbol{\gamma}^{(t)}, \boldsymbol{\beta}^{(t)}, \sigma^{2(t)})} \frac{2m^{(t)}q_-(\boldsymbol{\beta}_*)m^{(t)}}{p(\boldsymbol{\beta}_*)p(\boldsymbol{Q}_*)\lambda},$$

where $\boldsymbol{\beta}_*$ and $\boldsymbol{Q}_*$ are, respectively, the effects and the genotypes of the removed QTL, $q_-(\boldsymbol{\beta}_*)$ is the density of the normal distribution obtained similarly to $q_+(\boldsymbol{\beta}_*)$ by changing $N(\mu_{*j}^+, \sigma_*^2)$ to $N(\mu_{*j}^-, \sigma_*^2)$ where

$$\mu_{*j}^- = \frac{\mu_\beta/\sigma_\beta^2 + \sum_{i=1}^n \delta_{i*j}(y_i - \eta_i^{(t)} + \sum_{l=1}^{j-1} \beta_{*l}\delta_{i*l})/\sigma^{2(t)}}{1/\sigma_\beta^2 + \sum_{i=1}^n \delta_{i*j}^2/\sigma^{2(t)}}.$$

Updating configuraton with move Type d: With equal probability, determine whether to add two new QTLs or delete two existing QTLs.

If two new QTLs are to be added, then

- Generate two new locations $d_{{\scriptscriptstyle Q}m^{(t)}+1}$ and $d_{{\scriptscriptstyle Q}m^{(t)}+2}$ uniformly from the unoccupied intervals.

- Generate the proposal genotypes of individual i at the new QTLs from $p(\delta_{im^{(t)}+l}|d_{{\scriptscriptstyle Q}m^{(t)}+l}, d_{{\scriptscriptstyle M}(m^{(t)}+l)-}, d_{{\scriptscriptstyle M}(m^{(t)}+l)+}, M_{i(m^{(t)}+l)-}, M_{i(m^{(t)}+l)+})$. $l = 1, 2$.

- Generate the interaction effect β_* from $N(\mu_{a*}, \sigma_*^2)$ by setting $\delta_{i*1} = \delta_{im^{(t)}+1}\delta_{im^{(t)}+2}$

- Augment $(\boldsymbol{d}_{\scriptscriptstyle Q}^{(t)}, \boldsymbol{Q}^{(t)}, \boldsymbol{\beta}^{(t)})$ with the proposed components and denote the augmented parameters by $(\tilde{\boldsymbol{d}}_{\scriptscriptstyle Q}^{(t)}, \tilde{\boldsymbol{Q}}^{(t)}, \tilde{\boldsymbol{\beta}}^{(t)})$. All the proposed components are accepted or rejected together with acceptance probability $\alpha_a = \min(r_a, 1)$ where

$$r_a = \frac{\phi(\boldsymbol{y}|\tilde{\boldsymbol{d}}_{\scriptscriptstyle Q}^{(t)}, \tilde{\boldsymbol{Q}}^{(t)}, \boldsymbol{\gamma}^{(t+1)}, \tilde{\boldsymbol{\beta}}^{(t)}, \sigma^{2(t)})}{\phi(\boldsymbol{y}|\boldsymbol{d}_{\scriptscriptstyle Q}^{(t)}, \boldsymbol{Q}^{(t)}, \boldsymbol{\gamma}^{(t)}, \boldsymbol{\beta}^{(t)}, \sigma^{2(t)})} \frac{p(\boldsymbol{\beta}_*)p(\boldsymbol{Q}_*)\lambda^2}{2^{2m^{(t)}+3}q_+(\boldsymbol{\beta}_*)(m^{(t)}+1)(m^{(t)}+2)},$$

where $\boldsymbol{\beta}_*$ and $\boldsymbol{Q}_*$ are respectively the proposed effects and geno-types of the new QTLs.

If two existing QTLs are to be deleted, then

- Randomly choose two existing QTLs to delete. Remove all the effects related to the two QTLs.

- Set the corresponding effect indicators to zero. Denote the values of the parameters after deleting the components related to the deleted QTLs by $(\tilde{\boldsymbol{d}}_{\scriptscriptstyle Q}^{(t)}, \tilde{\boldsymbol{Q}}^{(t)}, \tilde{\boldsymbol{\beta}}^{(t)})$. The move is accepted with probability $\alpha_d = \min(r_d, 1)$ where

$$r_d = \frac{\phi(\boldsymbol{y}|\tilde{\boldsymbol{d}}_{\scriptscriptstyle Q}^{(t)}, \tilde{\boldsymbol{Q}}^{(t)}, \boldsymbol{\gamma}^{(t+1)}, \tilde{\boldsymbol{\beta}}^{(t)}, \sigma^{2(t)})}{\phi(\boldsymbol{y}|\boldsymbol{d}_{\scriptscriptstyle Q}^{(t)}, \boldsymbol{Q}^{(t)}, \boldsymbol{\gamma}^{(t)}, \boldsymbol{\beta}^{(t)}, \sigma^{2(t)})} \frac{2^{2m^{(t)}-1}q_-(\boldsymbol{\beta}_*)m^{(t)}(m^{(t)}-1)}{p(\boldsymbol{\beta}_*)p(\boldsymbol{Q}_*)\lambda^2},$$

where $\boldsymbol{\beta}_*$, $\boldsymbol{Q}_*$ and k_* are, respectively, the effects, the genotypes and the number of effects of the removed QTLs.

The inference on the true number of QTLs and the QTL effects based on the MCMC sequence $(\boldsymbol{d}_{\scriptscriptstyle Q}^{(t)}, \boldsymbol{Q}^{(t)}, \boldsymbol{\beta}^{(t)})$ is the same as that given in § 7.2.1.2 for models without epistatic effects. The discussion on the inference for epistatic models is therefore omitted.

7.2.3 Example: Mapping QTL Affecting Heading in Two-Row Barley

An experiment is carried out in the North American Barley Genome Mapping Project to characterize QTL that affect seven agronomic traits in a DH population from a two-row barley cross [183]. A population of 150 random DH lines is produced from the F_1 generated by a cross between single plants of two North American two-row barley varieties: Harrington and TR306. More than 200 marker loci are mapped in the Harrington/TR306 DH population. A subset of 127 markers is chosen to provide a base map for the mapping analysis. The 150 DH lines are grown in 30 environmental conditions. Due to suspected errors in data acquisition, 5 DH lines are discarded. The seven agronomic traits are: yield, heading, maturity, height, lodging, kernel weight and test weight.

The Bayesian approach described in § 7.2.2 is applied to the mapping of the heading data in [212]. The heading is measured by the number of days from planting until emergence of 50% of heads on main tillers and 145 DH lines are used in the mapping. For each DH line, the average heading over 29 environmental conditions is taken as the original phenotypic value (y_i). The phenotypic values are standardized by the formula $(y_i - \bar{y})/s$ where $\bar{y}$ and s are respectively the sample mean and sample standard deviation of the 145 phenotypic values.

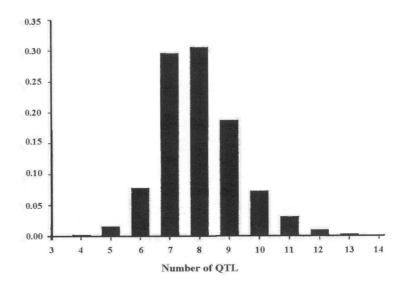

FIGURE 7.4: Posterior probability of QTL number for the barley data (reproduced from Figure 4 in [212]).

The prior for the number of QTL is taken as uniform $(0, 20)$. The priors for all main and interaction effects are taken as $N(0, 1)$. The prior for the overall mean is taken as $N(0, 2)$. The prior for the phenotypic error variance σ^2 is taken as uniform $(0, s^2)$. The priors for the indicator vector γ and QTL positions d_Q are as given in § 7.2.2. The tuning parameter a in the updating for QTL positions is taken as $a = 2$. The MCMC starts with no QTL in the model and 3×10^5 cycles are run. The first 2000 cycles are discarded. In the remaining cycles, one in every 15 cycles are sampled, resulting final sample of size 2×10^4.

From the MCMC sample, the posterior expectation of the number of QTL is estimated as 7.98. The posterior probability density of the number of QTL is depicted in Figure 7.4. The posterior probability that there are at least seven QTL is about 0.92. The empirical QTL intensity function, which is defined in (7.9), is given in Figure 7.5. There are 7 large modes of the intensity function that fall into the 95% high posterior density (HPD) region, which indicate that there are 7 QTL. The estimated 95% HPD regions of the 7 QTL locations and their modes are given in Table 7.2. For more details of the posterior analysis, see [212].

7.2.4 Missing Marker Genotypes and Other Issues

So far, we have implicitly assumed that there are no missing marker genotypes. However, in practical problems, it is almost unavoidable that some

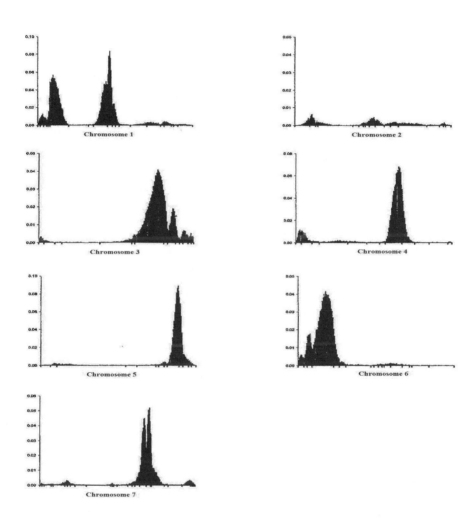

FIGURE 7.5: Posterior QTL intensity for the barley data (reproduced from Figure 5 in [212]).

TABLE 7.2: The estimated 95% HPD regions of QTL locations and the modes of the locations (a location is indicated by its genetic distance (cM) from the left end of the chromosome).

QTL	Chromosome	95% HPD Region	Mode of location
1	1	$6 - 35$	21
2	1	$84 - 107$	98
3	3	$125 - 163$	142
4	4	$114 - 136$	126
5	5	$167 - 187$	177
6	6	$10 - 44$	30
7	7	$175 - 199$	191

individuals will have genotypes at some markers missing by one reason or another. In the Bayesian models, missing marker genotypes can be treated as random variables like the unobservable QTL genotypes. Thus there is an additional component of the unknowns which a prior must be imposed on and which must be updated in the MCMC cycles.

First, let us consider the prior on the missing marker genotypes. Suppose that for individual i, among markers j to $j+l$, the genotypes at markers $j+1$ to $j+l-1$ are missing. Let M_{ij} and M_{ij+l} denote the observed genotypes and $M^*_{ij+1}, \ldots, M^*_{ij+l-1}$ the missing genotypes. The prior on the missing genotypes is specified as

$$p(M^*_{ij+1}|M_{ij}, M_{ij+l})p(M^*_{ij+2}|M^*_{ij+1}, M_{ij+l}) \cdots p(M^*_{ij+l-1}|M^*_{ij+l-2}, M_{i,j+l}),$$

where $p(M^*_{ij+r}|M^*_{ij+r-1}, M_{ij+l})$ is determined the same as

$$p(\delta_{im(t+1)}|d_{Qm(t+1)}, d_{MM(t+1)-}, d_{MM(t+1)+}, M_{im(t+1)-}, M_{im(t+1)+}).$$

In the current notation, the locations of the markers, which are known, are omitted. If the first missing genotype occurs at the end of a chromosome, say the left end, and the lth marker is the first to the right whose genotype is observed, the prior on the first missing genotype is given by $p(M^*_{i1}|M_{il})$ and for the other missing genotypes the priors are still in the same form. For example, in the case of backcross design, if M_{il} is homozygous then

$$p(M^*_{i1}|M_{il}) = \begin{cases} 1 - r, & \text{if } M^*_{i1} \text{ is homozygous;} \\ r, & \text{otherwise,} \end{cases}$$

where r is the recombination fraction between the first and the $l$the marker.

In the MCMC cycles, to update a missing marker genotype, it is generated from its full conditional posterior probability given the genotypes of its two flanking loci, either QTL or marker loci, in the Gibbs sampler, or proposed by its prior probability given the genotypes of the two flanking loci in the

Metropolis-Hastings algorithm and the reversible jump MCMC steps. The updating of the missing marker genotypes is incorporated into the steps for the updating of QTL genotypes.

The MCMC algorithms presented in this chapter are not unique ones. Many variations of these algorithms can be made. For example, in the reversible MCMC procedures, different move types can be considered. In adding or deleting QTL, instead of adding or deleting one QTL at a time, a random number of QTL can be added or deleted. As long as the dimension-matching requirement discussed in § 2.8.3 is met in determining the acceptance probability of the moves, the procedure is a valid one. Although the algorithms in this chapter can be readily used, the readers should treat them only as illustrating examples. The readers can develop their own algorithms following the illustrated principles for their special problems to satisfy their special needs.

7.3 Bayesian QTL Mapping for Outbred Offsprings

The Bayesian QTL mapping for outbred offsprings is essentially the same as that for inbred line crosses. The main difference is that, instead of marker genotypes, the haplotypes of both offsprings and their parents must be considered in the mapping for outbred offsprings. Since the phases of marker genotypes are in general unknown, an additional component — the prior on the marker haplotypes — needs to be incorporated into the the general model (7.1). The posterior density of the Bayesian model for mapping outbred offsprings is then given by

$$\pi(m, \boldsymbol{d}_{\text{\tiny Q}}, \boldsymbol{Q}, \boldsymbol{G}, \boldsymbol{\beta}, \sigma^2 | \boldsymbol{y}) \tag{7.14}$$

$$\propto \quad p(\boldsymbol{\beta}|m)p(\boldsymbol{d}_{\text{\tiny Q}}|m)p(m)p(\sigma^2)p(\boldsymbol{G})p(\boldsymbol{Q}|\boldsymbol{d}_{\text{\tiny Q}}, \boldsymbol{d}_{\text{\tiny M}}, \boldsymbol{G}) \prod_{i=1}^{n} \phi(y_i|m, \boldsymbol{Q}_i, \boldsymbol{\beta}, \sigma^2),$$

where $\boldsymbol{G}$ denotes the set of phase-known genotypes of all the individuals. In Bayesian models for outbred offsprings, except the components $p(\boldsymbol{G})$ and $p(\boldsymbol{Q}|\boldsymbol{d}_{\text{\tiny Q}}, \boldsymbol{d}_{\text{\tiny M}}, \boldsymbol{G})$, all the other components are specified exactly the same as those in the models for inbred line crosses. In § 7.3.1, we consider the specification of these two components. In § 7.3.2, we discuss the adaptation of the MCMC algorithms for inbred line crosses to the case of outbred offsprings.

7.3.1 The Prior on Marker Haplotypes and QTL Genotypes

The prior on marker haplotypes. A method used in the computation of Mendelian likelihood considered in [113] can be adapted for the specification of the prior on $\boldsymbol{G}$. The adaptation is described in this section.

Let $G_j = (H_j^M, H_j^F)$ denote the pair of haplotypes of individual j (an offspring or a parent) where H_i^M and H_i^F are the haplotypes inherited from its female and male parents, respectively. Let $\text{Tran}(G_c|G_m, G_f)$ denote the transmission probability that a mother with haplotype G_m and a father with haplotype G_f produce a child with haplotype G_c. Then the prior on $\boldsymbol{G}$ is given by

$$p(\boldsymbol{G}) = \prod_j p(G_j) \prod_{(i,k,l)} \text{Tran}(G_c|G_m, G_f),$$

where the product on j is taken over all founders and the product on (c, m, f) is taken over all parent-offspring triples. The prior probability $p(G_j)$ for founders is specified under Hardy-Weinberg and linkage equilibrium and is factorized as

$$P(G_j) = \prod_s P(G_{js}) = \prod_s P(G_{js}^M) P(G_{js}^F)$$

where the product is taken over all markers, G_{js} is the phase-known genotype at mark s and G_{js}^M and G_{js}^F are its two alleles. The transmission probability $\text{Tran}(G_c|G_m, G_f)$ can be factorized as the product of gamete transmission probabilities:

$$\text{Tran}(G_c|G_m, G_f) = \text{Tran}(H_c^M|G_m)\text{Tran}(H_c^F|G_f).$$

The gamete transmission probabilities are computed as follows. Obviously, each gamete transmission probability can be factorized as the product of the transmission probabilities over different chromosomes. On a single chromosome, suppose that there are l markers at which the parent's genotypes are heterozygous and they are located at positions $r_1, r_2, \ldots, r_l$, where $1 < r_1 < r_2 < \cdots < r_l < t$ and t is the total number of markers on the chromosome. Then the gamete transmission probability on this chromosome is further factorized as a product of $l + 1$ probabilities as, say,

$$
\begin{aligned}
\text{Tran}(H_c^M|G_m) \;=\; & \text{Tran}[(H_{c,1}^M, \ldots, H_{c,r_1}^M)|(G_{m,1}, \ldots, G_{m,r_1})] \\
& \times \text{Tran}[(H_{c,r_1+1}^M, \ldots, H_{c,r_2}^M)|(G_{m,r_1}, \ldots, G_{m,r_2}), H_{c,r_1}] \\
& \times \text{Tran}[(H_{c,r_2+1}^M, \ldots, H_{c,r_3}^M)|(G_{m,r_1}, \ldots, G_{m,r_3}), H_{c,r_2}] \\
& \times \cdots \\
& \times \text{Tran}[(H_{c,r_{l-1}+1}^M, \ldots, H_{c,r_l}^M)|(G_{m,r_{l-1}}, \ldots, G_{m,r_l}), H_{c,r_{l-1}}] \\
& \times \text{Tran}[(H_{c,r_l+1}^M, \ldots, H_{c,t}^M)|(G_{m,r_l+1}, \ldots, G_{m,t}), H_{c,r_l}],
\end{aligned}
$$

where the first factor equals $\frac{1}{2}$, which is the probability that the allele H_{c,r_1}^M is inherited from the parent at marker r_1, and the last factor is 1, since the genotypes of parent at markers $r_l + 1$ onwards are all homozygous. Let $\theta_{r_j, r_{j+1}}$ denote the recombination fraction between marker r_j and marker r_{j+1}. Then

$$
\begin{aligned}
& \text{Tran}[(H_{c,r_j+1}^M, \ldots, H_{c,r_{j+1}}^M)|(G_{m,r_1}, \ldots, G_{m,r_3}), H_{c,r_2}] \\
& = \begin{cases} \theta_{r_j, r_{j+1}}, & \text{if } H_{c,r_j} \text{ and } H_{c,r_{j+1}} \text{ are a recombination,} \\ 1 - \theta_{r_j, r_{j+1}}, & \text{otherwise,} \end{cases} \\
& j = 1, \ldots, l - 1.
\end{aligned}
$$

The method described above is applicable for any type of outbred offsprings. To illustrate, we consider an example adapted from [164] in the following. The example concerns with an F_2 full-sib family with three offsprings. The haplotypes across four markers, denoted as M1, M2, M3 and M4, of the family are shown in Figure 7.6. At markers M1 and M4, both parents have homozygous genotypes. Let θ_{23} be the recombination fraction between M2 and M3. The transmission probability of the paternal haplotype of offspring 1 is computed as follows:

$$
\begin{aligned}
&\text{Tran}[(A, A, D, E)|(A, A, C, E)\text{-}(A, B, D, E)] \\
=\ &\text{Tran}[(A, A)|(A, A)\text{-}(A, B)]\text{Tran}[D|(A, C)\text{-}(B, D), A]\text{Tran}[E|(E\text{-}E), D] \\
=\ &\frac{1}{2}\theta_{23}.
\end{aligned}
$$

Similar computation yields

Offspring 1, maternal:
$$
\text{Tran}[(E, A, G, G)|(E, E, F, G)\text{-}(E, A, G, G)] = \frac{1}{2}(1 - \theta_{23}),
$$

Offspring 2, paternal:
$$
\text{Tran}[(A, A, C, E)|(A, A, C, E)\text{-}(A, B, D, E)] = \frac{1}{2}(1 - \theta_{23}),
$$

Offspring 2, maternal:
$$
\text{Tran}[(E, E, F, G)|(E, E, F, G)\text{-}(E, A, G, G)] = \frac{1}{2}(1 - \theta_{23}),
$$

Offspring 3, paternal:
$$
\text{Tran}[(A, B, D, E)|(A, A, C, E)\text{-}(A, B, D, E)] = \frac{1}{2}(1 - \theta_{23}),
$$

Offspring 3, maternal:
$$
\text{Tran}[(E, E, G, G)|(E, E, F, G)\text{-}(E, A, G, G)] = \frac{1}{2}\theta_{23}.
$$

The prior on QTL genotypes. The prior on QTL genotypes is determined, in principle, in the same way as that for inbred offsprings. Given the marker haplotypes, the probability of the genotype of a QTL depends only on its two flanking markers and its location between the two markers. The number and type of the QTL genotypes are determined by the outbred cross. For individual offspring i, denote by G_{iq} its genotype with known allele origins at the QTL and H_{il} and H_{ir} its genotypes with known allele origins at the left and right flanking markers respectively. Let θ_{lq}, θ_{qr} and θ_{lr} be the recombination fractions between the left marker and the QTL, between the QTL and the right marker, and between the left and right markers respectively. The prior

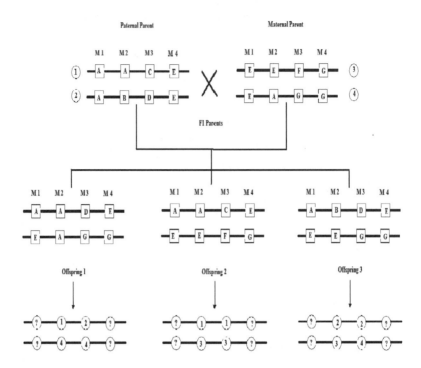

FIGURE 7.6: Genotypes of four marker positions in an F_2 full-sib family with three offsprings (where the numbers indicate the origin of the alleles).

on G_{iq} is given by

$$
\begin{aligned}
&p(G_{iq}|H_{il}, H_{ir}, \theta_{lq}, \theta_{qr}, \theta_{lr}) \\
&= \frac{p(G_{iq}|H_{il}, \theta_{lq})p(H_{ir}|G_{iq}, \theta_{qr})}{p(H_{ir}|H_{il}, \theta_{lr})} \\
&= \frac{p(G_{iq}^{M}|H_{il}^{M}, \theta_{lq})p(H_{ir}^{M}|G_{iq}^{M}, \theta_{qr})}{p(H_{ir}^{M}|H_{il}^{M}, \theta_{lr})} \times \frac{p(G_{iq}^{F}|H_{il}^{F}, \theta_{lq})p(H_{ir}^{F}|G_{iq}^{F}, \theta_{qr})}{p(H_{ir}^{F}|H_{il}^{F}, \theta_{lr})}.
\end{aligned}
$$

For example, in the example of F_2 full-sib family, as shown in Figure 7.6, there are four QTL genotypes: 13, 14, 23 and 24. The probabilities of these four genotypes occurring in offspring 1 are given below:

$$13: \quad \frac{(1-\theta_{lq})\theta_{qr}}{\theta_{lr}} \frac{\theta_{lq}\theta_{qr}}{1-\theta_{lr}},$$

$$14: \quad \frac{(1-\theta_{lq})\theta_{qr}}{\theta_{lr}} \frac{(1-\theta_{lq})(1-\theta_{qr})}{1-\theta_{lr}},$$

$$23: \quad \frac{\theta_{lq}(1-\theta_{qr})}{\theta_{lr}} \frac{\theta_{lq}\theta_{qr}}{1-\theta_{lr}},$$

$$24: \quad \frac{\theta_{lq}(1-\theta_{qr})}{\theta_{lr}} \frac{(1-\theta_{lq})(1-\theta_{qr})}{1-\theta_{lr}}.$$

7.3.2 The MCMC Procedures for Mapping with Outbred Offsprings

The MCMC algorithms for various Bayesian models described in § 7.2 can be adapted to the corresponding Bayesian models for mapping with outbred offsprings.

The MCMC procedures for outbred crosses have the same steps as those for the corresponding inbred line crosses. When a MCMC procedure for an inbred line cross is adapted to the one for an outbred cross, the main modification is on the update of $(\boldsymbol{d}_{\mathrm{Q}}^{(t)}, \boldsymbol{Q}^{(t)}, \boldsymbol{G}^{(t)}, \boldsymbol{\beta}^{(t)}, \sigma^{2(t)})$ while the number of QTLs remains the same. We elaborate the details of its update as follows. Note that the QTL effect vector $\boldsymbol{\beta}$ now has a different content from that in the case of inbred line crosses. Let n_{QTL} be the number of QTL genotypes. Let δ_{jl} be the dummy variable defined for QTL j as follows:

$$\delta_{jl} = \begin{cases} 1, \text{if the } l\text{th genotype}, \\ 0, \text{otherwise}; \end{cases} \quad l = 2, \ldots, n_{\mathrm{QTL}}.$$

The QTL effect on the trait of offspring i is then given by

$$\eta_i = \beta_0 + \sum_{j=1}^{m} \sum_{l=2}^{n_{\mathrm{QTL}}} \beta_{jl} \delta_{ijl}.$$

The components of $\boldsymbol{\beta}$ consist of β_0 and the β_{jl}'s. The update of $(\boldsymbol{d}_{\mathrm{Q}}^{(t)}, \boldsymbol{Q}^{(t)}, \boldsymbol{G}^{(t)}, \boldsymbol{\beta}^{(t)}, \sigma^{2(t)})$ is broken up into the following steps:

Step 1: Update $(\boldsymbol{d}_{\mathrm{Q}}^{(t)}, \boldsymbol{Q}^{(t)})$ to $(\boldsymbol{d}_{\mathrm{Q}}^{(t+1)}, \tilde{\boldsymbol{Q}}^{(t)})$ one QTL at a time. For QTL j, propose the position d_j^* in the same way as in MCMC algorithm I, propose the genotype Q_{ij}^* for the ith offspring by the prior distribution given in § 7.3.1. Accept the proposal with probability $\alpha = \min\{r, 1\}$ where

$$r = \frac{\pi(d_j^* | \boldsymbol{d}_{\mathrm{Q}-j}^{(t)}, \tilde{\boldsymbol{Q}}^{(t,j)}, \boldsymbol{\beta}^{(t)}, \sigma^{2(t)}) u_j(d_j^*, d_{\mathrm{Q}j}^{(t)})}{\pi(d_{\mathrm{Q}j}^{(t)} | \boldsymbol{d}_{\mathrm{Q}-j}^{(t)}, \tilde{\boldsymbol{Q}}^{(t,j-1)}, \boldsymbol{\beta}^{(t)}, \sigma^{2(t)}) u_j(d_{\mathrm{Q}j}^{(t)}, d_j^*)}.$$

Here $\tilde{\boldsymbol{Q}}^{(t,j-1)}$ is the most updated QTL genotype after the $(j-1)$th QTL is considered, $\tilde{\boldsymbol{Q}}^{(t,j)}$ is obtained by replacing the components of $\tilde{\boldsymbol{Q}}^{(t,j-1)}$ corresponding to the jth QTL with the proposed genotypes, and the other factors are the same as those in the MCMC algorithm I. If the proposal is accepted, update $d_{qj}^{(t)}$ to $d_{qj}^{(t+1)} = d_j^*$ and $\tilde{\boldsymbol{Q}}^{(t,j-1)}$ to $\tilde{\boldsymbol{Q}}^{(t,j)}$; otherwise, skip QTL j and consider the next QTL.

Step 2: After the QTL positions are updated, update $\tilde{\boldsymbol{Q}}^{(t)}, \boldsymbol{G}^{(t)}$ to $\boldsymbol{Q}^{(t+1)}, \boldsymbol{G}^{(t+1)}$ as follows. Propose $\boldsymbol{G}^*$ by first generating the haplotypes of the founders using the prior probability $p(G_j)$ as given in § 7.3.1 and then generating the haplotypes of the offsprings by the transmission probabilities $\mathrm{Tran}(G_c|G_m, G_f)$. Propose $\boldsymbol{Q}^{(t+1)}$ by their prior distributions (the conditional distributions given $\boldsymbol{G}^{(t+1)}$). The proposal is accepted with probability $\alpha = \min\{r, 1\}$ where

$$r = \frac{\pi(m^{(t)}, \boldsymbol{d}_q^{(t+1)}, \boldsymbol{Q}^*, \boldsymbol{G}^*, \boldsymbol{\beta}^{(t)}, \sigma^{2(t)}|\boldsymbol{y})}{\pi(m^{(t)}, \boldsymbol{d}_q^{(t+1)}, \tilde{\boldsymbol{Q}}^t, \boldsymbol{G}^{(t)}, \boldsymbol{\beta}^{(t)}, \sigma^{2(t)}|\boldsymbol{y})}.$$

Step 3: Update $(\boldsymbol{\beta}^{(t)}, \sigma^{2(t)})$ to $\boldsymbol{\beta}^{(t+1)}, \sigma^{2(t+1)}$ by their full conditional posterior distributions as in MCMC algorithm I.

7.4 Bayesian QTL Mapping for Ordinal Polytomous Traits

In Chapter 5, we treated the mapping of general polytomous traits (including binary traits) using generalized linear models with multinomial distributions in the multiple interval mapping method. The framework of the Bayesian approach discussed in § 7.1 still applies to polytomous traits when the conditional distribution of the traits in (7.1) is given by the density of the multinomial distributions. A main difference between the models for polytomous traits and that for Gaussian traits is that the full conditional posterior distributions of the QTL effect parameters no longer have a standard form even if normal priors are imposed on the effect parameters. Thus the Gibbs sampler cannot be applied to update the states of these parameters in the MCMC simulation. But this does not cause any difficulty. We can simply replace the Gibbs samplers with the Metropolis-Hastings algorithms for updating the effect parameters. Thus the MCMC procedures developed for Gaussian traits in the previous sections can be directly applied to the Bayesian models for polytomous traits with some straightforward modifications.

In this section, we consider an alternative version of the Bayesian models for ordinal polytomous traits. We introduced in § 5.7.1 a liability threshold mechanism to derive the link function of the generalized linear models for ordinal polytomous traits. In the liability threshold mechanism, the categories of an ordinal polytomous trait are determined by certain threshold values of a latent continuous variable. Apart from being used as the mechanism to derive the link function, the latent variable does not play any further role in the generalized linear models. However, the latent continuous variable can be explicitly incorporated into the Bayesian models as intermediate nuisance parameters, which gives rise to the version of Bayesian models for ordinal polytomous traits described in this section.

Let z denote the polytomous trait and y the latent continuous variable. For convenience, we call the latent continuous variable y the latent trait. Suppose that z takes values $0, 1, \ldots, K$, which represent $K + 1$ ordered categories. Under the liability threshold mechanism, the categories of z are determined as follows:

$$z = r, \quad \text{if and only if} \quad \theta_{r-1} < y \le \theta_r, \ r = 0, \ldots, K,$$

where the θ_r's are unknown threshold values satisfying $-\infty = \theta_{-1} < \theta_0 < \theta_1 < \cdots < \theta_K = +\infty$. We denote by $\boldsymbol{\theta}$ the vector $(\theta_0, \ldots, \theta_{K-1})$. In the generalized linear model approach of § 5.7.1, a K-dimensional multinomial vector is introduced to represent z. In the Bayesian models to be developed, this multinomial vector is no longer needed.

For an individual i, the values of the polytomous trait and the latent trait are denoted by z_i and y_i respectively. We assume that the latent traits y_i have the same normal distributions as those assumed for Gaussian traits in the previous sections such as (7.2) and (7.12) under the backcross designs. Without loss of generality, we can assume that the variance of y_i is 1, since in general we can consider y_i/σ instead of y_i as the latent trait. Let η_i be a generic notation for the QTL effects determined by the QTL genotypes of individual i. The content of η_i is determined by the particular model under consideration.

If we take the latent trait as if it is the observed trait, then the posterior density of the unknown parameters of interest is still given by (7.1). In the current context, it becomes

$$\pi(m, \boldsymbol{d}_{\text{Q}}, \boldsymbol{Q}, \boldsymbol{\beta}|\boldsymbol{y}) \tag{7.15}$$

$$\propto \ p(\boldsymbol{\beta}|m)p(\boldsymbol{d}_{\text{Q}}|m)p(\boldsymbol{\gamma}|m)p(m)p(\boldsymbol{Q}|\boldsymbol{d}_{\text{Q}}, \boldsymbol{d}_{\text{M}}, \boldsymbol{M}) \prod_{i=1}^{n} \phi(y_i|m, \boldsymbol{Q}_i, \boldsymbol{\beta}, \boldsymbol{\gamma}),$$

where the effect indication vector $\boldsymbol{\gamma}$ is taken as a constant vector when the epistatic effects are not considered in the model and $p(\boldsymbol{\gamma}|m)$ is set to 1, and

$$\phi(y_i|m, \boldsymbol{Q}_i, \boldsymbol{\beta}) = \frac{1}{\sqrt{2\pi}} \exp\{-\frac{1}{2}(y_i - \eta_i)^2\}.$$

However, the latent traits y_i are not observed, a further layer, i.e., the conditional density of the z_i given y_i is needed to complete the posterior density. Since the z_i's are completely determined by the y_i's and the threshold vector $\boldsymbol{\theta}$ and are independent from each other, the conditional density of $\boldsymbol{z} = (z_1, \ldots, z_n)$ is of the form

$$p(\boldsymbol{z}|\boldsymbol{y}, \boldsymbol{\theta}) = \prod_{i=1}^{n} p(z_i|y_i, \boldsymbol{\theta}).$$

Thus the full posterior density given the observed traits is as follows:

$$\pi(m, \boldsymbol{d}_{\scriptscriptstyle Q}, \boldsymbol{Q}, \boldsymbol{\beta}, \boldsymbol{\gamma}, \boldsymbol{y}, \boldsymbol{\theta}|\boldsymbol{z}) \tag{7.16}$$

$$\propto \quad p(\boldsymbol{\beta}|m)p(\boldsymbol{d}_{\scriptscriptstyle Q}|m)p(\boldsymbol{\gamma}|m)p(m)p(\boldsymbol{Q}|\boldsymbol{d}_{\scriptscriptstyle Q}, \boldsymbol{d}_{\scriptscriptstyle M}, \boldsymbol{M}) \prod_{i=1}^{n} \phi(y_i|m, \boldsymbol{Q}_i, \boldsymbol{\beta}, \boldsymbol{\gamma})$$

$$\times p(\boldsymbol{\theta}) \prod_{i=1}^{n} p(z_i|y_i, \boldsymbol{\theta}).$$

The components of the full posterior density are specified in the following. The prior $p(\boldsymbol{\theta})$ of the threshold vector is the density of the order statistics of a sample of size K from a uniform distribution over a specified interval, say $[L, \ U]$. The conditional density of z_i given y_i and $\boldsymbol{\theta}$ takes the form:

$$p(z_i|y_i, \boldsymbol{\theta}) = \sum_{r=0}^{K} I\{\theta_{r-1} < y_i \leq \theta_r\}I\{z_i = r\}.$$

The other components are specified the same as in § 7.2.1 and § 7.2.2, respectively, for models without and with epistatic effects.

The Bayesian model described above covers the models with or without epistatic effects and under inbred line cross designs and outbred cross designs. It includes as a special case the model for ordered binary traits which is considered in [211]. The components $p(\boldsymbol{\beta}|m)$, $p(\boldsymbol{d}_{\scriptscriptstyle Q}|m)$, $p(m)$, $p(\boldsymbol{\theta})$ and $\prod_{i=1}^{n} p(z_i|y_i, \boldsymbol{\theta})$ are common for all particular models. Only components $p(\boldsymbol{Q}|\boldsymbol{d}_{\scriptscriptstyle Q}, \boldsymbol{d}_{\scriptscriptstyle M}, \boldsymbol{M})$ and $\prod_{i=1}^{n} \phi(y_i|m, \boldsymbol{Q}_i, \boldsymbol{\beta}, \boldsymbol{\gamma})$ change as the particular model varies. The probabilities $p(\boldsymbol{Q}|\boldsymbol{d}_{\scriptscriptstyle Q}, \boldsymbol{d}_{\scriptscriptstyle M}, \boldsymbol{M})$ are determined by the designs. In inbred line crosses, the formulae of these probabilities are given in Table 3.4, Table 3.5 and Table 3.6 for backcross, intercross and RIL designs respectively. For outbred crosses, these probabilities are determined by the pedigree structures. The linear form of QTL effects η_i in $\prod_{i=1}^{n} \phi(y_i|m, \boldsymbol{Q}_i, \boldsymbol{\beta}, \boldsymbol{\gamma})$ depends on whether or not epistatic effects are considered. For example, under backcross designs, η_i takes the linear form in (7.2) when epistatic effects are not considered, it takes the linear form in (7.12) when epistatic effects are considered.

The MCMC procedure for sampling from (7.16) can be simply obtained by adding an additional step for the updating of $\boldsymbol{y}$ and $\boldsymbol{\theta}$ to the MCMC procedures developed in the previous sections. The updating of $\boldsymbol{y}$ and $\boldsymbol{\theta}$ can be done simultaneously using the Metropolis-Hastings algorithm. The new values of the components of $\boldsymbol{\theta}$ can be proposed in the same way as that for QTL locations described in § 7.2.1.1. The new values of y_i's can be proposed from their full conditional posterior distribution which are truncated normal distributions with densities given below:

$$p(y_i|\boldsymbol{\theta}, z_i = r) = \frac{\phi(y_i - \eta_i)}{\Phi(\theta_r - \eta_i) - \Phi(\theta_{r-1} - \eta_i)} I\{\theta_{r-1} < y_i \leq \theta_r\},$$
$$r = 0, 1, \ldots, K,$$

where $\phi(\cdot)$ and $\Phi(\cdot)$ are, respectively, the probability density function and the cumulative distribution function of the standard normal distribution.

The inference procedures on the QTL locations and effects are the same as in the previous sections and hence are omitted here.

Chapter 8

Multi-trait QTL Mapping and eQTL Mapping

In many QTL mapping studies, observations on several traits are recorded together with the genotypes of markers. Very often, such traits are correlated and there are common chromosome regions that affect multiple traits. Although the statistical methods described in previous chapters can be applied to each trait one-by-one, such approach does not take into account the correlation among multiple traits and loses power in detecting QTL with pleiotropic effects. A QTL has pleiotropic effect if it affects several traits simultaneously. Statistical methods especially for mapping multiple traits are demanded. Many methods for multi-trait QTL mapping have been developed in the literature, ranging from simple extensions of single-trait approaches to sophisticated multi-trait approaches designed specifically for multi-trait QTL mapping. In this chapter, we discuss some of these methods. In § 8.1, we deal with methods based on single-QTL-models, including one-trait-at-a-time approaches, meta-trait methods and multivariate composite interval mapping. In § 8.2, we consider more sophisticated approaches which are based on multi-QTL models such as multivariate sparse partial least square regression, multivariate sequential procedures, and multivariate Bayesian approaches.

In biological and medical studies, it is crucial to understand the genetic architecture behind the variation of a quantitative trait or the susceptibility to a complex disease. Variation in gene expression is a more important underlying mechanism. The gene expression level is directly modified by polymorphism in regulatory loci in the genome. The accurate identification of the loci regulating gene expression levels plays an important role in revealing the whole picture of the genetic architecture. Gene expression levels can be considered as quantitative traits which can be mapped to the regulating loci. These loci are referred to as expression QTL (eQTL). In principle, methods for multi-trait QTL mapping can be used for eQTL mapping. However, in a typical eQTL mapping study, there are usually thousands of transcripts. Because of the the sheer number of transcripts, the methods for multi-trait QTL mapping cannot be directly applied. They need to be either modified or combined with other techniques to tackle the challenge raised by eQTL mapping. Section 8.3 is devoted to the special issues and methods for eQTL mapping.

In multi-trait QTL mapping or eQTL mapping, the measurements of multiple traits (or gene expression levels) and the genotypes of markers over the genome for each subject (experiment unit) are obtained. Throughout this chapter, the following notation will be used. The data is denoted by $\{(y_{ti}, x_{i1}, \ldots, x_{iM}) : t = 1, \ldots, T; i = 1, \ldots, n\}$ where i is the index of individual subject, y_{ti} is the measurement on the t-th trait (or transcript) of individual i, x_{ij} is the covariate representing the genotype of individual i at marker j, T is the number of traits and n is the sample size. Let $\boldsymbol{y}_t$ denote the vector of the measurements on the t-th trait. Let $\boldsymbol{x}_j$ denote the vector of the jth covariate values.

8.1 Multi-trait QTL Mapping I: Single-QTL Model Approaches

In this section, we consider relatively straightforward approaches for multi-trait QTL mapping. Either single-trait methods are applied directly to multi-trait QTL mapping, § 8.1.1, or the multi-trait QTL mapping problem is reduced to a single-trait problem, § 8.1.2, or single-trait methods are extended to the case of multi-trait QTL mapping, § 8.1.3.

8.1.1 One-trait-at-a-time Method

A naive approach to multi-trait QTL mapping is to apply single-trait methods to each trait one-by-one with or without adjustment for multi-trait effect. This approach has been taken by researchers who are mainly genetics-oriented, see, e.g., [4], [11], [12], [156]. Any single-trait method discussed in previous chapters can be used in this way. Since most commonly used method in this context is multiple tests based on single-trait-by-single-marker models, the approach is presented in terms of multiple tests below.

For the sake of convenience, suppose we have a backcross design. For a fixed j, the genotypes at marker j partition the experiment population into two groups, one with the heterozygous genotype and the other with the homozygous genotype, the corresponding x_{ij} taking values 0 and 1 respectively. Suppose that for trait t, y_{ti} has a p.d.f. $f(y_{ti}, \mu_{t0}, \sigma)$ if i belongs to the group with heterozygous genotype, and a p.d.f. $f(y_{ti}, \mu_{t1}, \sigma)$ if i belongs to the group with homozygous genotype.

For each pair (t, j), a test statistic is computed for testing $H_0 : \mu_{t0} = \mu_{t1}$ against $H_1 : \mu_{t0} \neq \mu_{t1}$. The test statistic commonly used in the genetics literature is the LOD score given below:

$$\log_{10} \left(\frac{\prod_{i:x_{ij}=0} f(y_{ti}, \hat{\mu}_{t0}, \hat{\sigma}_1) \prod_{i:x_{ij}=1} f(y_{ti}, \hat{\mu}_{t1}, \hat{\sigma}_1)}{\prod_{i=1}^{n} f(y_{ti}, \hat{\mu}, \hat{\sigma}_0)} \right),$$

where $\hat{\mu}$, and $\hat{\sigma}_0$ are the MLE of $\mu = \mu_{t0} = \mu_{t1}$ and σ under H_0, and $\hat{\mu}_{t0}, \hat{\mu}_{t1}$ and $\hat{\sigma}_1$ are the MLE of μ_{t0}, μ_{t1} and σ under H_1.

The LOD score is equivalent to the classical likelihood ratio test statistic and the t-statistic for the comparison of two groups. The LOD score is compared with an overall critical value c_α common for all pairs (t, j). The critical value c_α is determined such that overall trait-wise and marker-wise type I error is controlled below α. The critical value can be obtained either by a permutation approach, see [40], or by Bonferroni adjustment.

The one-trait-at-a-time approach is simple and straightforward. But in general the power of the approach is low, especially, when the critical value is adjusted to control trait-wise and marker-wise type I error. The approach is more useful as a preliminary screening step for more sophisticated methods, see § 8.2.

8.1.2 Meta-trait Method

Often multiple traits are affected by common pleiotropic loci. Combining multiple traits into meta-traits that account for the total variation in the traits more effectively will allow a better chance to map to pleiotropic loci. A meta-trait is a linear combination of the original traits derived by certain statistical principles. In a meta-trait method, the meta-traits are derived first, then mapping methods applicable to ordinary traits are applied to meta-traits.

Different kinds of meta-traits have been considered in the literature. The most widely used meta-traits are obtained by using principal components and independent components. Principal components and independent components have been used to process microarray expression data, see, e.g., [2] [210] [114] [64] [146]. The application of principal components and independent components in multi-trait QTL mapping can be found in, e.g., [197] [13] [195] [8] [72] [110]. The principal components and independent components are described briefly as follows.

Principal component analysis is a classical statistical method [3]. The principal components of a random vector are linear forms of the random vector which are mutually independent. The sum of their variances is the same as the sum of the variances of the original elements of the random vector. Though the exact principal components are unknown, they can be estimated from the data. Let Y be the $n \times T$ matrix of the observed values of the traits. Suppose Y is appropriately standardized such that its columns have sample mean 0 and variance 1. The estimated principal components are obtained from the eigenvectors of the matrix $Y^\tau Y$. Denote by $\boldsymbol{q}_1, \ldots, \boldsymbol{q}_T$ the eigenvectors corresponding to their eigenvalues ordered from the largest to the smallest. Then the jth estimated principal component is given by $z_j = \boldsymbol{q}_j^\tau \boldsymbol{y}$ where $\boldsymbol{y}$ is the vector of traits. The first principal component is the linear form that has the largest variance among all liner forms, the second principal component is the linear form that has the second largest variance, and so on.

Independent component analysis is developed mainly by computer scientists for feature extraction and signal separation, etc.; see [93]. The independent components are also linear forms of $\boldsymbol{y}$ but derived from a different principle. In the independent component analysis, it is assumed that the observed traits are linear mixtures of certain latent independent non-Gaussian random variables, i.e.,

$$y_t = \sum_{j=1}^{p} a_{tj} s_j, \ t = 1, \ldots, T,$$

where the s_j's are unobservable, mutually independent and non-Gaussian. The independent component analysis attempts to extract the latent variables s_j's in the form of linear combinations $\boldsymbol{b}_j^\tau \boldsymbol{y}$ by minimizing their Gaussianity. The Gaussianity is measured by some contrast functions such as kurtosis and mutual information; for details, see [93]. The independent components obtained depend on the contrast function used.

The principal components and the independent components capture different features of the original variables. In a meta-trait method, the meta-traits can be taken either as the principal components or the independent components. However, in a particular problem, it is beneficial to use both principal components and independent components as meta-traits. There are situations where some QTL can be detected with principal components but not with independent components, or vice versa; see [13].

In general, the meta-traits capture the variation in the data more effectively than the original traits. For example, in most of cases, the first few principal components explain a major proportion of the total variation in the data. As a result, with the meta-traits explaining the larger portion of the total variation, a pleiotropic QTL can be detected with more power. Though there is no difference in the analysis of the original traits and of the meta-traits, there is an important difference between the original trait analysis and meta-trait analysis. With the original traits, it is necessary to consider all the traits one by one. However, in meta-trait analysis, one can focus on only a few meta-traits that explain most of the variation in the data.

The limitation of a meta-trait method is that all the QTL are treated as pleiotropic QTL. For a detected QTL, it cannot tell whether it is a pleiotropic one or a non-pleiotropic one, and it cannot tell which particular traits are associated with the QTL either.

8.1.3 Multivariate Composite Interval Mapping

The multivariate composite interval mapping is a straightforward extension of the single-trait composite interval mapping described in § 4.3. The basic idea of composite interval mapping, as discussed in § 4.3, is to adjust background effects caused by other genetic and/or environmental factors while carrying out interval mapping. Usually, in a particular study, the environmental factors are given and the genetic factors are selected from the available markers. The first task in composite interval mapping is to determine the adjusting markers. In the context of multi-trait mapping, we can take the markers identified as QTL by the methods discussed in the preceding sections as adjusting factors. In the following description, we assume that the adjusting factors are given and denote by x_{ij}^*, $i = 1, \ldots, n; j = 1, \ldots, K$ the values of the adjusting variables.

Consider a putative QTL on a particular interval flanked by two markers. Let z_i be the indicator representing the genotype of subject i at the putative QTL. In the case of backcross and RIL, z_i is a scaler. In the case of intercross, z_i is a vector of two indicators. The model at the putative QTL for multivariate composite interval mapping is given as follows:

$$y_{ti} = \beta_{0t} + \beta_{1t}^\tau z_i + \sum_{j=1}^K b_{jt} x_{ij} + \epsilon_{ti}, \qquad (8.1)$$

$$t = 1, \ldots, T; \quad i = 1, \ldots, n.$$

where β_{0t} and b_{jt} are scaler parameters, β_{1t} is a scaler in the case of backcross and RIL, a 2-vector in the case of intercross. In matrix notation, the model is expressed as

$$\boldsymbol{y}_i^\tau = \boldsymbol{\beta}_0^\tau + z_i \boldsymbol{\beta}_1^\tau + \boldsymbol{x}_i^{*\tau} B + \boldsymbol{\epsilon}_i^\tau, \quad i = 1, \ldots, n,$$

where $\boldsymbol{y}_i^\tau = (y_{1i}, \ldots, y_{Ti})$, $\boldsymbol{\beta}_0^\tau = (\beta_{01}, \ldots, \beta_{0T})$, $\boldsymbol{\beta}_1^\tau = (\beta_{11}, \ldots, \beta_{1T})$, $\boldsymbol{x}_i^* = (x_{i1}^*, \ldots, x_{iK}^*)^\tau$, $\boldsymbol{\epsilon}_i = (\epsilon_{i1}, \ldots, \epsilon_{iT})^\tau$ and $B = (b_{jt})|_{j=1,\ldots,K;t=1,\ldots,T}$ is a $K \times T$ parameter matrix. It is assumed that $\boldsymbol{\epsilon}_i$'s are i.i.d. multivariate normal vectors with mean vector zero and variance-covariance matrix

$$\Sigma = \begin{pmatrix} \sigma_{11} & \cdots & \sigma_{1T} \\ \cdots & \cdots & \cdots \\ \sigma_{T1} & \cdots & \sigma_{TT} \end{pmatrix}.$$

In the context of QTL mapping, the parameters $\boldsymbol{\beta}_0$, B and Σ are nuisance parameters. Whether or not the putative QTL is really a QTL is decided by testing $H_0 : \boldsymbol{\beta}_1 = 0$ versus $H_1 : \boldsymbol{\beta}_1 \neq 0$.

In the procedure of the multi-trait composite interval mapping, like the single-trait case, the whole genome is partitioned by the available markers into consecutive intervals, each interval is divided by equally spaced positions, and at each position (considered as a putative QTL), the hypothesis as described above is tested. A common critical value (or an adjusted significance level) is used to control the overall Type I error of the tests. Those positions at which the null hypothesis is rejected are claimed as detected QTLs.

In the model described above, the z_i's (the genotypes of the putative QTL) are not observed. But they can be inferred from the two flanking markers of the interval. Given the genotypes of the flanking markers, the probabilities of the genotypes of the putative QTL are given in Tables 3.4, 3.5 and 3.6, respectively, for backcross, intercross and recombination inbred lines. There are two versions of multivariate composite interval mapping: a likelihood approach based on mixture models, which is considered in [97], and a linear regression approach, which is considered in [24]. The two versions differ in their treatment of z_i's. The likelihood approach treats the z_i's as latent random variables, and the regression approach replaces z_i's with their conditional expectations conditioning on the flanking markers. In what follows, we describe the specific details of the two approaches. For the sake of convenience, we assume the backcross design in the description. The extension to the case of intercross and RIL designs are straightforward.

Likelihood approach

In the case of backcross, z_i is defined as follows:

$$z_i = \begin{cases} 1, & \text{if the genotype of subject } i \text{ is homozygous;} \\ 0, & \text{if the genotype of subject } i \text{ is heterozygous.} \end{cases}$$

Let x_{li} and x_{ri} be the genotypes of subject i at the left and right flanking markers respectively. They are defined in the same way as z_i. Denote by p_i the conditional probability that $z_i = 1$ given x_{li} and x_{ri}. In addition to x_{ri} and x_{li}, p_i depends on the position of the putative QTL and the genetic distance between the two flanking markers. Let r and γ denote respectively the recombination fraction between the left flanking marker and the putative QTL and that between the two flanking markers. In fact, p_i is obtained by a common function $p(x_{li}, x_{ri}, r, \gamma)$, see Table 3.4. The z_i follows a Bernoulli distribution $\mathcal{B}(1, p_i)$. Given z_i and the adjusting factors (which are considered non-random values), the vector $\boldsymbol{y}_i$ follows a multivariate normal distribution with mean $\boldsymbol{\beta}_0 + \boldsymbol{\beta}_1 z_i + B^\tau \boldsymbol{x}_i^*$ and variance-covariance matrix Σ. Thus, the joint p.d.f. of z_i and $\boldsymbol{y}_i$ are given by

$$p_i^{z_i}(1 - p_i)^{1-z_i} \phi(\boldsymbol{\beta}_0 + \boldsymbol{\beta}_1 z_i + B^\tau \boldsymbol{x}_i^*, \Sigma, \boldsymbol{y}_i),$$

where $\phi(\boldsymbol{\mu}, \Sigma, \boldsymbol{y})$ denotes the p.d.f. of a multivariate normal distribution with mean vector $\boldsymbol{\mu}$ and variance-covariance matrix Σ. From the joint p.d.f., we obtain the marginal p.d.f. of $\boldsymbol{y}_i$ as

$$f_i(\boldsymbol{y}_i | \boldsymbol{\beta}_0, \boldsymbol{\beta}_1, B, \Sigma) = p_i \phi(\boldsymbol{\beta}_0 + \boldsymbol{\beta}_1 + B^\tau \boldsymbol{x}_i^*, \Sigma, \boldsymbol{y}_i) + (1 - p_i)\phi(\boldsymbol{\beta}_0 + B^\tau \boldsymbol{x}_i^*, \Sigma, \boldsymbol{y}_i).$$

Under the null hypothesis $H_0 : \boldsymbol{\beta}_1 = 0$, the p.d.f. of $\boldsymbol{y}_i$ reduces to

$$f_{i0}(\boldsymbol{y}_i | \boldsymbol{\beta}_0, B, \Sigma) = \phi(\boldsymbol{\beta}_0 + B^\tau \boldsymbol{x}_i^*, \Sigma, \boldsymbol{y}_i).$$

Thus, under $H_1 : \boldsymbol{\beta}_1 \neq 0$, the log likelihood function is given by

$$l_1(\boldsymbol{\beta}_0, \boldsymbol{\beta}_1, B, \Sigma) = \sum_{i=1}^{n} \ln[p_i \phi(\boldsymbol{\beta}_0 + \boldsymbol{\beta}_1 + B^\tau \boldsymbol{x}_i^*, \Sigma, \boldsymbol{y}_i) + (1 - p_i)\phi(\boldsymbol{\beta}_0 + B^\tau \boldsymbol{x}_i^*, \Sigma, \boldsymbol{y}_i)],$$

and, under $H_0 : \boldsymbol{\beta} = 0$, the log likelihood function is given by

$$l_0(\boldsymbol{\beta}_0, B, \Sigma) = \sum_{i=1}^{n} \ln[\phi(\boldsymbol{\beta}_0 + B^\tau \boldsymbol{x}_i^*, \Sigma, \boldsymbol{y}_i)].$$

At the given putative QTL, the null hypothesis is tested by the likelihood ratio statistic

$$\chi = 2[l_1(\hat{\boldsymbol{\beta}}_0^{(1)}, \hat{\boldsymbol{\beta}}_1^{(1)}, \hat{B}^{(1)}, \hat{\Sigma}^{(1)}) - l_0(\hat{\boldsymbol{\beta}}_0^{(0)}, \hat{B}^{(0)}, \hat{\Sigma}^{(0)})],$$

where $(\hat{\boldsymbol{\beta}}_0^{(k)}, \hat{\boldsymbol{\beta}}_1^{(k)}, \hat{B}^{(k)}, \hat{\Sigma}^{(k)})$ are obtained by maximizing l_k with respect to $(\boldsymbol{\beta}_0, \boldsymbol{\beta}_1, B, \Sigma)$, $k = 0, 1$. Under the null hypothesis, the statistic χ has an asymptotic χ^2- distribution with degrees of freedom T. The critical value of χ is adjusted by the Bonferroni correction. Taking into account that the tests on the same interval are highly correlated, it is sugested in [97] that the critical value is taken as $\chi^2_{\alpha/M,T}$ to control the overall Type I error at level α, where M is the number of intervals.

We now turn to the computation issue. Let $X = (\boldsymbol{x}_1^*, \cdots, \boldsymbol{x}_n^*)^\tau$ and $Y = (\boldsymbol{y}_1, \cdots, \boldsymbol{y}_n)^\tau$. Augment X to $\tilde{X} = (\mathbf{1}, X)$. The MLEs under the null hypothesis have the following explicit forms:

$$\begin{pmatrix} \hat{\boldsymbol{\beta}}_0^{(0)} \\ \hat{B}^{(0)} \end{pmatrix} = (\tilde{X}^\tau \tilde{X})^{-1} \tilde{X}^\tau Y,$$

$$\hat{\Sigma}^{(0)} = \frac{1}{n}[Y - \mathbf{1}\hat{\boldsymbol{\beta}}_0^{(0)} - X\hat{B}^{(0)}]^\tau [Y - \mathbf{1}\hat{\boldsymbol{\beta}}_0^{(0)} - X\hat{B}^{(0)}].$$

The MLEs under the alternative hypothesis is computed by an ECM algorithm [134] that is a special version of general EM algorithms, which is described as follows. In the framework of EM algorithms, the unobserved z_i's are treated as missing data. The complete data consists of the missing data and the observed data. The complete data (if none is missing) provides the log likelihood function:

$$l_c(\boldsymbol{\beta}_0, \boldsymbol{\beta}_1, B, \Sigma) = \sum_{i=1}^{n} \{z_i \ln(p_i) + (1 - z_i) \ln(1 - p_i) + \ln[\phi(\boldsymbol{\beta}_0 + \boldsymbol{\beta}_1 z_i + B^\tau \boldsymbol{x}_i^*, \Sigma, \boldsymbol{y}_i)]\}.$$

Let $\boldsymbol{v}_i = \boldsymbol{\beta}_0 + B^\tau \boldsymbol{x}_i^* - \boldsymbol{y}_i$. The above likelihood is then expressed as

$$\begin{aligned} l_c(\boldsymbol{\beta}_0, \boldsymbol{\beta}_1, B, \Sigma) &= \sum_{i=1}^{n} z_i \ln(p_i) + (1 - z_i) \ln(1 - p_i) \\ &\quad - \frac{n}{2} \ln|\Sigma| - \frac{1}{2} \sum_{i=1}^{n} [z_i^2 \boldsymbol{\beta}_1^\tau \Sigma^{-1} \boldsymbol{\beta}_1 + 2 z_i \boldsymbol{\beta}_1^\tau \Sigma^{-1} \boldsymbol{v}_i + \boldsymbol{v}_i^\tau \Sigma^{-1} \boldsymbol{v}_i]. \end{aligned}$$

Note that the $\boldsymbol{v}_i$'s do not involve any missing data and that $z_i^2 = z_i$ since z_i's are indicator functions.

Like the general EM algorithm, the ECM algorithm alternates between an *E step* and an *M step* iteratively until convergence occurs. At the E step, the conditional expectation of the complete log likelihood conditioning on the observed data and given parameter values is computed. From the expression of the complete log likelihood give above, the E step reduces to the computation of $\tilde{z}_i = E(z_i|\boldsymbol{y}_i, \boldsymbol{\beta}_0^{(k)}, \boldsymbol{\beta}_1^{(k)}, B^{(k)}, \Sigma^{(k)})$ at the $(k+1)$th cycle of the iterations. By the Bayes formula, $\tilde{z}_i$ is given by

$$\tilde{z}_i = \frac{p_i\phi(\boldsymbol{\beta}_0^{(k)} + \boldsymbol{\beta}_1^{(k)} + B^{(k)\tau}\boldsymbol{x}_i^*, \Sigma^{(k)}, \boldsymbol{y}_i)}{p_i\phi(\boldsymbol{\beta}_0^{(k)} + \boldsymbol{\beta}_1^{(k)} + B^{(k)\tau}\boldsymbol{x}_i^*, \Sigma^{(k)}, \boldsymbol{y}_i) + (1-p_i)\phi(\boldsymbol{\beta}_0^{(k)} + B^{(k)\tau}\boldsymbol{x}_i^*, \Sigma^{(k)}, \boldsymbol{y}_i)}.$$

Let $Z = (\mathbf{1}, \tilde{\boldsymbol{z}})$, where $\tilde{\boldsymbol{z}} = (\tilde{z}_1, \cdots, \tilde{z}_n)^\tau$. Let $\mathcal{B} = (\boldsymbol{\beta}_0, \boldsymbol{\beta}_1)^\tau$. At the M step, the parameter matrices $\mathcal{B}$, B and Σ are updated in the following order:

$$\begin{aligned}
\mathcal{B}^{(k+1)} &= (Z^\tau Z)^{-1} Z^\tau (Y - XB^{(k)}), \\
B^{(k+1)} &= (X^\tau X)^{-1} X^\tau (Y - Z\mathcal{B}^{(k+1)}), \\
\Sigma^{(k+1)} &= \frac{1}{n} [Y - Z\mathcal{B}^{(k+1)} - XB^{(k+1)}]^\tau [Y - Z\mathcal{B}^{(k+1)} - XB^{(k+1)}].
\end{aligned}$$

In the calculation of the sub matrix $\tilde{\boldsymbol{z}}\tilde{\boldsymbol{z}}^\tau$ of ZZ^τ, the diagonal element $\tilde{z}_i^2$ is replaced by $\tilde{z}_i$.

Regression approach

In the regression approach, the unobserved putative QTL genotypes z_i's are replaced by their conditional expectations given the genotypes of the flanking markers; that is, model (8.1) is replaced by the following model

$$y_{ti} = \beta_{t0} + \boldsymbol{\beta}_{t1}^\tau p_i + \sum_{j=1}^{K} b_{tj} x_{ij} + \epsilon_{ti}, \tag{8.2}$$

$$t = 1, \ldots, m; \quad i = 1, \ldots, n,$$

where $p_i = E(z_i|x_{li}, x_{ri}, r, \gamma)$ is the conditional probability that the genotype is homozygous at the putative QTL. In matrix form, (8.2) is expressed as

$$Y = \mathbf{1}\boldsymbol{\beta}_0^\tau + \boldsymbol{p}\boldsymbol{\beta}_1^\tau + XB + \Xi,$$

where $\boldsymbol{p} = (p_1, \cdots, p_n)^\tau$.

Let $V = (\mathbf{1}, \boldsymbol{p}, X)$ and $V_0 = (\mathbf{1}, X)$. Define

$$\begin{aligned}
E &= Y^\tau [I - V(V^\tau V)^{-1} V^\tau] Y, \\
H &= Y^\tau [V(V^\tau V)^{-1} V^\tau - V_0(V_0^\tau V_0)^{-1} V_0^\tau] Y.
\end{aligned}$$

The null hypothesis $H_0 : \boldsymbol{\beta}_1 = 0$ is tested by using either Roy, Pillai or Lawley-Hotelling statistics that are functions of the eigenvalues of $E^{-1}H$; see page 379 of [149]. For example, the Lawley-Hotelling statistic is $\text{tr}E^{-1}H$. To control the overall Type I error at level α, the Bonferroni corrected level for each individual test is taken as α/M, where M is the number of intervals. For the critical values of the Lawley-Hotelling statistic, see pages 185, 186 and 379 of [149].

In the multivariate composite interval mapping described above, it does not distinguish between a pleiotropic QTL and a non-pleiotropic QTL in the mapping procedure. Whether or not a detected QTL is a non-pleiotropic one can be determined by testing whether or not any of the components of the corresponding β_1 is zero. The testing can be done by likelihood ratio test in the likelihood approach, and by a general linear hypothesis test in the regression approach.

The composite interval mapping model considers a single putative QTL at a time. The mapping procedure is essentially a one-dimensional scan over the genome. It can be extended to consider more than one putative QTLs at a time, as was done in [24]. However, the mapping procedure is no longer a one-dimensional scan. It loses the appeal of the composite interval mapping as a one-dimensional scan approach. For the consideration of more than one putative QTL simultaneously, the approaches in the next section are more appropriate and more efficient.

Some problem would occur if the interval under consideration is flanked by at least one of the adjusting markers. In this situation, a common practice is to exclude the marker from the set of adjusting markers.

If the design is RIL, the description of the approaches is exactly the same except the z_i's are defined differently: z_i equals 1 if the genotype is one homozygote, and 0 the other homozygote. Instead of Table 3.4, Table 3.6 is used to compute the p_i's. If the design is intercross, z_i is replaced by a vector (z_{i1}, z_{i2}). Let QQ, Qq and qq be the generic notation for the genotypes. Then z_{i1} takes 1 if the genotype is QQ, and 0 otherwise; z_{i2} takes 1 if the genotype is Qq, and 0 otherwise. The conditional expectations $E(z_{i1})$ and $E(z_{i2})$ are, respectively, conditional probabilities of genotype QQ and Qq which are given in Table 3.5. The vector p is then replaced by a two-column matrix P.

8.2 Multi-trait QTL Mapping II: Multi-QTL Model Approaches

The methods discussed in § 8.1 are based on single-QTL models, i.e., the putative QTLs are considered one at a time. However, quantitative traits are usually affected by multiple QTLs. Methods with single-QTL models have a potential loss of power in the detection of QTL and a potential inflation of false discovery rate. Methods based on multiple-QTL models are more appropriate and have advantages over single-QTL models. In this section, we discuss methods that treat all the traits and all the markers simultaneously. In § 8.2.1, a partial least squares regression approach adapted for the multi-trait QTL mapping is presented. In § 8.2.2, two sequential methods — multi response sparse regression (MRSR) and simultaneous variable selection (SVS) — are described. In § 8.2.3, we discuss two Bayesian approaches: heirachical Bayesian linear models and Bayesian seemingly unrelated regression. The two non-Bayesian methods are applicable only when the covariates (the genotypes of markers in the context of QTL mapping) are numerical variables. These methods are originally developed as part of the tools for eQTL mapping. In the context of eQTL mapping, the commonly used markers are SNPs. The genotypes are represented by 0, 1, and 2 accordingly as the genotypes contain 0, 1 and 2 non-wild alleles. If the genetic modes of all the SNPs are assumed additive, which is a good approximation even when the modes of some SNPs are not additive, the genotypes such represented can be treated as numerical variables.

8.2.1   Sparse Partial Least Squares Regression

We first give a brief introduction to partial least squares regression (PLS). PLS is an alternative approach to ordinary least squares regression in ill-conditioned linear models and was introduced in [200]. The multivariate version of PLS is described as follows. Let Y be the $n \times T$ trait matrix and X be the $n \times M$ marker genotype matrix. Suppose the columns of Y and X are centered such that each of them sum up to zero. The PLS approach first finds q linear combinations of the columns of X such that they are most correlated with the columns of Y. The linear combinations of the covariates are taken as the estimates of q latent variables satisfying $Y = U\Omega + \Xi$ where Ω is a $T \times q$ matrix of regression coefficients. Denote these linear combinations by XW where W is the $M \times q$ matrix of combination coefficients. Once the matrix W is obtained, Y is regressed on $U = XW$ to obtain $\hat{\Omega} = (W^\tau X^\tau XW)^{-1}W^\tau X^\tau Y$. Then the PLS estimate of the regression coefficients of Y regressed on X is given by $\hat{B} = W\hat{\Omega}$. The columns of W are determined sequentially. For $1 \leq k \leq q$, the columns of W are derived as

$$
\begin{aligned}
\boldsymbol{w}_k \;=\; & \operatorname{argmax}_{\boldsymbol{w}} \boldsymbol{w}^\tau X^\tau Y^\tau Y X \boldsymbol{w}, \\
& \text{subject to } \boldsymbol{w}^\tau \boldsymbol{w} = 1, \; \boldsymbol{w}^\tau S_{XX}\boldsymbol{w}_j = 0, \\
& j = 1, \ldots, k-1,
\end{aligned}
$$

where S_{XX} is the sample covariance matrix of X. The theoretical properties of PLS under various conditions have been investigated in [39] [85] [138] [172].

The PLS regression reduces the dimension of the regressors to q, the number of linear combinations of the original covariates. But it does not select variables, it keeps all the original covariates. To adapt PLS regression for variable selection in multivariate multiple regression models, the sparse PLS regression (SPLS) method was proposed in [39]. A procedure using SPLS regression for eQTL mapping was developed in [38]. The SPLS regression is an iterated PLS regression procedure that achieves dimensionality reduction and variable selection simultaneously.

A variable selection step is incorporated into each iteration of the SPLS procedure. The variable selection is done through a mechanism of penalized minimization. Let $\tilde{Y}$ be the current matrix of residuals. At the initial step, $\tilde{Y} = Y$. Let $V = X^\tau \tilde{Y}^\tau \tilde{Y} X$. The penalized minimization is

$$\min_{w,\tilde{w}} \{ -\kappa wVw + (1-\kappa)(\tilde{w}-w)^\tau V(\tilde{w}-w) + \lambda_1 \|\tilde{w}\|_1 + \lambda_2 \|\tilde{w}\|_2^2 \},$$
$$\text{subject to} \quad w^\tau w = 1, \tag{8.3}$$

where κ is a positive number less than $1/2$, $\|\cdot\|_j$ is the L_j-norm, $j = 1, 2$. With certain fixed values of λ_1 and λ_2, the minimization of (8.3) results in that the components of $\tilde{w}$ corresponding to small components of w become zero. The idea of the above penalized minimization is to find the most correlated linear combination and to chop off its non-significant components simultaneously. Problem (8.3) is solved by alternatively iterating between solving for w with fixed $\tilde{w}$ and solving for $\tilde{w}$ with fixed w. With fixed $\tilde{w}$ and $0 < \kappa < 1/2$, the problem for solving w becomes

$$\min_{w} \{ (Z^\tau w - \kappa' Z^\tau \tilde{w})^\tau (Z^\tau w - \kappa' Z^\tau \tilde{w}) \} \quad \text{subject to } w^\tau w = 1, \tag{8.4}$$

where $Z = X^\tau \tilde{Y}$ and $\kappa' = (1-\kappa)/(1-2\kappa)$. With fixed w, the problem for solving $\tilde{w}$ becomes

$$\min_{\tilde{w}} \{ (Z^\tau \tilde{w} - Z^\tau w)^\tau (Z^\tau \tilde{w} - Z^\tau w) + \lambda_1 \|\tilde{w}\|_1 + \lambda_2 \|\tilde{w}\}_2^2 \}. \tag{8.5}$$

The constrained least squares problem (8.4) can be solved by the method of Lagrange multipliers. In the penalized problem (8.5), a large value of λ_2 is needed to tackle the low rank of the $M \times M$ matrix $Z^\tau Z$ whose rank is less than or equal to T. An approximation, as considered in [38], to the solution of (8.5) is a soft thresholded estimator given by

$$\tilde{w}_i = (|w_i| - \eta \max_{1 \leq j \leq p} |w_j|) I\{ |w_i| \geq \eta \max_{1 \leq j \leq p} |w_j| \} \text{sign}(w_i), i = 1, \ldots, p, \tag{8.6}$$

where η, $0 \leq \eta \leq 1$, is intrinsically related to λ_1.

We now describe the method of SPLS regression. The covariates involved in the SPL regression of an iteration are referred to as the active variables. Let $\mathcal{A}$ be the index set of the active variables. Let $X_{\mathcal{A}}$ be the submatrix of X consisting of the columns corresponding to the active variables. Let $\hat{\mathcal{B}}$ denote the PLS estimate of the matrix of regression coefficients of Y regressed on $X_{\mathcal{A}}$. The procedure of SPLS regression is as follows.

Initial step: Set $\hat{\mathcal{B}} = 0$, $\mathcal{A} = \phi$, the empty set, $\tilde{Y} = Y$ and $k = 1$.

Interative steps: While $k \leq K$,

    (a) Find $\tilde{w}$ by solving (8.3) with $V = X^{\tau}\tilde{Y}^{\tau}\tilde{Y}X$.

    (b) Update $\mathcal{A}$ as $\mathcal{A} = \{i : \tilde{w}_i \neq 0\} \cup \{i : \hat{\beta}_i \neq 0\}$ where $\hat{\beta}_i$ is the ith row of $\hat{\mathcal{B}}$.

    (c) Fit the PLS regression model with response matrix $\tilde{Y}$, covariate matrix $X_{\mathcal{A}}$ and $q = k$. Obtain $\hat{\mathcal{B}}_{\mathcal{A}}$, the estimated matrix of regression coefficients of $\tilde{Y}$ regressed on $X_{\mathcal{A}}$.

    (d) Update $\hat{\mathcal{B}} \leftarrow \hat{\mathcal{B}}_{\mathcal{A}}$, $\tilde{Y} \leftarrow Y - X_{\mathcal{A}}\hat{\mathcal{B}}_{\mathcal{A}}$, and $k \leftarrow k + 1$.

The same thresholding parameter η of the thresholded estimator (8.6) is used throughout the above procedure. Thus there are two tuning parameters, η and K, for the SPLS regression procedure. These tuning parameters are chosen by cross-validation (CV).

For multi-trait QTL mapping, the application of the SPLS regression procedure to the trait matrix Y and marker matrix X yields K selected markers and an $T \times K$ matrix of estimated regression coefficients. Each of the K markers is deemed associated with one or more traits. The significance of the association is reflected in the magnitude of the estimated regression coefficients. To identify statistically significant association between traits and markers, simultaneous confidence intervals are constructed for the regression coefficients of Y on the K markers. A bootstrap procedure is used for the construction of these confidence intervals. Re-samples of size n are drawn at random from the original n records of individuals with replacement. For each re-sample, a PLS regression model is fitted to the re-sampled trait matrix and genotype matrix of the K selected markers to obtain a set of bootstrapped PLS estimates of the regression coefficients. The simultaneous confidence intervals are constructed based on the empirical distribution of the bootstrapped PLS estimates.

The computation algorithm for the implementation of the SPLS regression procedure is provided by Chun and Keles [39] in an R package at `http://cran.r-project.org/web/packages/spls`.

8.2.2   Multivariate Sequential Procedures

A sequential procedure selects variables one-at-a-time sequentially. There are many sequential procedures in the literature for variable selection with univariate response models. These include forward selection (FS) (or orthogonal matching pursuit, OMP, by another terminology) [143] [184] [187] [196], Lasso solution path [181] and least angle regression (LAR) [49], to name but a few. These methods have been extended to multivariate response models in a variety of ways. The multivariate versions of these methods can be used for multi-trait QTL mapping.

A sequential procedure produces a sequence of models

$$(\mathcal{A}_1, \hat{B}_1), \ldots, (\mathcal{A}_k, \hat{B}_k), \ldots, (\mathcal{A}_K, \hat{B}_K),$$

where $\mathcal{A}_k$, which is referred to as the active set at step k, is a subset of $\{1, \ldots, p\}$ and $\hat{B}_k$ is the estimated matrix of regression coefficient in the model

$$Y = X_{\mathcal{A}_k} B_k + \mathcal{E}.$$

Model selection procedures can be used to select among this sequence of models. The markers with indices in the active set of the selected model are then identified as detected QTLs.

In this subsection, we discuss two multi-response sequential procedures: multi-response sparse regression (MRSR) [166] [167], which is an extension of the LAR algorithm [49], and simultaneous variable selection (SVS) [168] [188] [189], which is an extension of Lasso [181]. The subsection is concluded with a discussion on model selection procedures.

Multi Response Sparse Regression

The multi-response sparse regression (MRSR) is an extension of the LAR algorithm proposed in [166] and [167]. The idea of the LAR algorithm is as follows. Starting with a current estimate $\hat{\mu}$, $\hat{\mu} = 0$ at the initial step, the variables having the maximum absolute correlation with the current residual $\boldsymbol{y} - \hat{\mu}$ form an active set. Then the current estimate moves along a direction $\boldsymbol{u}$, which has equal angles with the variables in the active set, in the form $\hat{\mu} + \gamma \boldsymbol{u}$. The absolute correlation of the variables in the active set with $\boldsymbol{y} - (\hat{\mu} + \gamma \boldsymbol{u})$ decreases by the same amount as γ increases. There is a minimum value of γ, say $\tilde{\gamma}$, such that there are new variables that have the same absolute correlation with $\boldsymbol{y} - (\hat{\mu} + \tilde{\gamma} \boldsymbol{u})$ as those in the active set. The current estimate is then updated as $\hat{\mu} + \tilde{\gamma} \boldsymbol{u}$ and the active set is updated by adding the new variables. The process continues this way until a certain step determined by some model selection criterion. The MRSR follows the same idea but is more elaborate. We describe the details of MRSR in the following.

Let $\| \cdot \|_p$, $p = 1, 2, \infty$, denote the L_p-norm of a vector. We use the same notation as in the previous section. Denote by $\mathcal{A}_k$ the active set, $X_{\mathcal{A}_k}$ the corresponding matrix and $\hat{Y}_k$ the estimate of EY at the kth step of the MRSR procedure. When $k = 1$, $\mathcal{A}_1 = \phi$ and $\hat{Y}_1 = 0$. The active set and the estimate of EY are updated at the kth step as follows:

(a) For $j \notin \mathcal{A}_k$, compute $c_{k,j} = \|(Y - \hat{Y}_k)^\tau x_j\|_p$ and $c_k = \max_{j \notin \mathcal{A}_k} c_{k,j}$. Update $\mathcal{A}_k$ to $\mathcal{A}_{k+1}$ as

$$\mathcal{A}_{k+1} = \mathcal{A}_k \cup \{j : c_{k,j} = c_k\}.$$

(b) Obtain the ordinary least squares estimate

$$\tilde{Y}_{k+1} = X_{\mathcal{A}_{k+1}} (X_{\mathcal{A}_{k+1}}^\tau X_{\mathcal{A}_{k+1}})^{-1} X_{\mathcal{A}_{k+1}}^\tau Y.$$

Determine a step length γ_k, $0 < \gamma_k < 1$, and update the estimate $\hat{Y}_k$ to $\hat{Y}_{k+1}$ as

$$\hat{Y}_{k+1} = (1 - \gamma_k)\hat{Y}_k + \gamma_k \tilde{Y}_{k+1}.$$

We now consider the determination of the step length γ_k. Let

$$c_{k+1,j}(\gamma) = \|[Y - (1-\gamma)\hat{Y}_k - \gamma\tilde{Y}_{k+1}]^\tau x_j\|_p.$$

Since $X_{\mathcal{A}_{k+1}}^\tau Y = X_{\mathcal{A}_{k+1}}^\tau \tilde{Y}_{k+1}$, it is straightforward to verify that, for $j \in \mathcal{A}_{k+1}$,

$$c_{k+1,j} = (1-\gamma)c_k. \tag{8.7}$$

For $j \notin \mathcal{A}_{k+1}$,

$$c_{k+1,j} = \|\boldsymbol{u}_{kj} - \gamma\boldsymbol{v}_{kj}\|_p, \tag{8.8}$$

where $\boldsymbol{u}_{kj} = (Y - \hat{Y}_k)^\tau x_j$ and $\boldsymbol{v}_{kj} = (\tilde{Y}_{k+1} - \hat{Y}_k)^\tau x_j$. For each $j \notin \mathcal{A}_{k+1}$ there is a unique $\gamma_{kj} \in (0,1)$ such that (8.7) and (8.8) are equal (see Theorem 1 in [167]). Then γ_k is taken as

$$\gamma_k = \min_{j \notin \mathcal{A}_{k+1}} \gamma_{kj}.$$

The computation of γ_{kj} with the three different norms are presented next.

(i) L_1-*norm*: The γ_{kj} is computed as

$$
\begin{aligned}
\gamma_{kj} &= \max_\gamma\{\gamma : \|\boldsymbol{u}_{kj} - \gamma\boldsymbol{v}_{kj}\|_1 \leq (1-\gamma)c_k\} \\
&= \max_\gamma\{\gamma : \sum_{l=1}^m s_l(u_{kjl} - \gamma v_{kjl}) \leq (1-\gamma)c_k\} \\
&= \min{}^+\left\{\frac{c_k - \sum_{l=1}^m s_l u_{kjl}}{c_k - \sum_{l=1}^m s_l v_{kjl}}\right\},
\end{aligned}
$$

where $s_l = \pm 1$ and $\min^+$ indicates that the minimum is taken over all positive terms among the 2^m possible entries as s_l's vary.

(ii) L_2-*norm*: The γ_{kj} is computed as

$$
\begin{aligned}
\gamma_{kj} &= \min_\gamma\{\gamma : \|\boldsymbol{u}_{kj} - \gamma\boldsymbol{v}_{kj}\|_2^2 = (1-\gamma)^2 c_k^2\} \\
&= \min{}^+\left\{\frac{b \pm \sqrt{b^2 - ac}}{a} : \begin{array}{l} a = c_k^2 - \|\boldsymbol{v}_{kj}\|_2^2, \\ b = c_k^2 - \boldsymbol{u}_{kj}^\tau\boldsymbol{v}_{kj}, \\ c = c_k^2 - \|\boldsymbol{u}_{kj}\|_2^2. \end{array}\right\}.
\end{aligned}
$$

(iii) L_∞-*norm*: The γ_{kj} is computed as

$$
\begin{aligned}
\gamma_{kj} &= \max_\gamma\{\gamma : \|\boldsymbol{u}_{kj} - \gamma\boldsymbol{v}_{kj}\|_\infty \leq (1-\gamma)c_k\} \\
&= \max_\gamma\{\gamma : \pm(u_{kjl} - \gamma v_{kjl}) \leq (1-\gamma)c_k, 1 \leq l \leq m\} \\
&= \min_{1 \leq l \leq m}{}^+\left\{\frac{c_k + u_{kjl}}{c_k + v_{kjl}}, \frac{c_k - u_{kjl}}{c_k - v_{kjl}}\right\}.
\end{aligned}
$$

The MRSR procedure with L_1-norm is not computationally feasible if T is large, since the computation of γ_{kj} scales as $O(2^T)$. The problem does not appear with L_2 and L_∞ norms; with these two norms, the computation of γ_{kj} scales as $O(T)$ only.

There are two special cases of the MRSR procedure. If $\gamma_k = 1$ for all k, the MRSR becomes a multi-response version of OMP considered in [186]. If $T = 1$, the procedure reduces to the single response LAR algorithm. Although the updating of the current estimate in MRSR is described in a different manner, it is equivalent to that of the LAR algorithm. The equivalence is easily seen from (8.7) which implies that the current estimate indeed moves along the direction having equal angles with the variables in the active set.

Simultaneous variable selection

The regularized regression methods for single-response models are extended to multi-response models independently by many authors in the signal processing community for sparse signal reconstruction, see, e.g., [42] [129] [185]. The extended versions are referred to as simultaneous variable selection (SVS) methods, since the methods force a common sparse covariate structure for all the response variables. Path-tracing algorithms for the regularized regression with L_∞ and L_2 penalties have been developed in [188] [189] and [168] respectively. These path-tracing algorithms allow the regularized multi-response regression to be implemented sequentially and hence can readily be used for multi-trait QTL mapping. For the sake of convenience, the method of regularized multi-response regression with L_q, $q = \infty, 2$, penalty is referred to as L_q-SVS method in the following.

The $n \times T$ response matrix Y, the $n \times M$ matrix of marker genotypes X and the $M \times T$ matrix of regression coefficients B are expressed, in terms of columns and rows, as

$$Y = (\boldsymbol{y}_1, \ldots, \boldsymbol{y}_T) = (\boldsymbol{y}_{(1)}, \ldots, \boldsymbol{y}_{(n)})^\tau,$$
$$X = (\boldsymbol{x}_1, \ldots, \boldsymbol{x}_M) = (\boldsymbol{x}_{(1)}, \ldots, \boldsymbol{x}_{(n)})^\tau,$$
$$B = (\boldsymbol{b}_1, \ldots, \boldsymbol{b}_T) = (\boldsymbol{b}_{(1)}, \ldots, \boldsymbol{b}_{(M)})^\tau,$$

where a bold face letter with a plain index denotes a column vector, and a bold face letter with a bracketed index denotes the transpose of a row vector. The L_q-SVS method is formulated as the minimization of the following penalized sum of squares:

$$\frac{1}{2}\|Y - XB\|_F^2 + \lambda \sum_{j=1}^{M} \|\boldsymbol{b}_{(j)}\|_q, \quad q = \infty \text{ or } 2, \tag{8.9}$$

where $\|\cdot\|_F$ is the Frobenius matrix norm defined as the sum of the squared entries of the matrix. When $T = 1$, problem (8.9) reduces to Lasso. A particular value of λ implicitly determines a threshold for $\|\boldsymbol{b}_{(j)}\|_q$. When (8.9) is minimized with the particular value of λ, the coefficients with $\|\boldsymbol{b}_{(j)}\|_q$ below the threshold will be forced to zero, thus achieving the goal of variable selection. If λ is large enough then all the coefficient vectors will be forced to zero. As λ decreases, the number of non-zero coefficient vectors increases. A path-tracing algorithm solves the solution of (8.9) as a function of λ. As λ varies, it produces the whole solution path. The penalized sum of squares (8.9) can be reformulated as a penalized sum of squares with a single response vector. The operation $\text{vec}(\cdot)$ on a matrix produces a vector by stacking the columns of the matrix. For example, $\text{vec}(Y) = (\boldsymbol{y}_1^\tau, \ldots, \boldsymbol{y}_T^\tau)^\tau$. Using this operation, (8.9) can be expressed as

$$\frac{1}{2}\|\text{vec}(Y - XB)\|_2^2 + \lambda \sum_{j=1}^M \|\boldsymbol{b}_{(j)}\|_q$$

$$= \|\text{vec}(Y) - \text{diag}(X, \cdots, X)\text{vec}(B)\|_2^2 + \lambda \sum_{j=1}^M \|\boldsymbol{b}_{(j)}\|_q, \qquad (8.10)$$

where $\text{diag}(X, \cdots, X)$ is a diagonal block matrix with T identical blocks X. The penalty in (8.10) is not put on individual components of $\text{vec}(B)$ but on groups of the components, each group corresponding to one $\boldsymbol{x}_j$. The single response problem of this type is referred to as group variable selection; see, e.g., [6] [216]. Path-tracing algorithms have been developed for various group variable selection problems; see, e.g., [89] [105] [119] [133] [144]. The techniques of the path-tracing algorithms for group variable selection can be adapted for solving problem (8.10).

In the following, we describe the path-tracing algorithm developed in [168] for L_2-SVS. The algorithm is of the nature of general group variable selection but, by taking into account the special structure of (8.10), is more efficient. Let

$$f(B) = \frac{1}{2}\|Y - XB\|_F^2.$$

Problem (8.9) is equivalent to

$$\text{minimize}_B f(B) \text{ subject to } \sum_{j=1}^M \|\boldsymbol{b}_{(j)}\|_2 \leq r. \qquad (8.11)$$

A small r in (8.11) corresponds to a large λ in (8.9). In turn, (8.11) is equivalent to the following problem

$$\text{minimize}_{B,\boldsymbol{s}} f(B) \text{ subject to } \sum_{j=1}^M s_j \leq r, \ \|\boldsymbol{b}_{(j)}\|_2 \leq s_j, \ j = 1, \ldots, M. \quad (8.12)$$

The solution to problem (8.12) with a fixed r can be approximated by solving a logarithmic barrier reformulation. For a positive number ν, define

$$F_\nu(B, \boldsymbol{s}) = f(B) - \frac{\nu}{M} \sum_{j=1}^M \ln(s_j^2 - \|\boldsymbol{b}_{(j)}\|_2^2) - \nu \ln(r - \sum_{j=1}^M s_j).$$

Instead of solving problem (8.12), the logarithmic barrier reformulation is to find B and s to minimize $F_\nu(B, s)$.

It has been shown [17] that the function $f(B)$ evaluated at the solution of the logarithmic barrier reformulation is at most 3ν apart from that evaluated at the solution of (8.12). Let $u_\nu(B, s)$ denote the vector of the first derivatives and $A_\nu(B, s)$ the matrix of the second derivatives of $F_\nu(B, s)$. The algorithm for solving the logarithmic barrier reformulation presented in [168] is as follows.

Algorithm 1 (Barrier method).

Initialize

Specify starting point (B, s), tolerance $\epsilon_{\text{MIN}} > 0$, initial value $\nu = \nu_{\text{MAX}} > 0$, final value $\nu_{\text{MIN}} \in (0, \nu_{\text{MAX}})$, and a fixed fraction $\alpha \in (0, 1)$.

Repeat

$\text{vec}([\hat{B}, \hat{s}]) = -A_\nu(B, s)^{-1} u_\nu(B, s)$,

$\epsilon = u_\nu(B, s)^\tau \text{vec}([\hat{B}, \hat{s}])$,

quit if $\epsilon < \epsilon_{\text{MIN}}$ and $\nu = \nu_{\text{MIN}}$.

$\tilde{\gamma} = \max\{\gamma \leq 1 : \sum_{j=1}^M (s_j + \gamma \hat{s}_j) < r, s_j + \gamma \hat{s}_j > \|b_{(j)} + \gamma \hat{b}_{(j)}\|_2, \forall j\}$

$B \leftarrow B + \tilde{\gamma}\hat{B}, \quad s \leftarrow s + \tilde{\gamma}\hat{s}, \quad \nu \leftarrow \max\{\alpha\nu, \nu_{\text{MIN}}\}$.

The idea of the path-tracing algorithm developed in [168] is as follows. First, a sequence of dense enough r values is pre-determined, say, $r^{[1]} < r^{[2]} < \cdots < r^{[K]}$. Then, starting with $k = 1$, an active set $\mathcal{A}_k$ is determined and Algorithm 1 is used to solve the logarithmic barrier reformulation problem with $r = r^{[k]}$ and covariate matrix $X_{\mathcal{A}_k}$ sequentially. The solution to the problem with $r = r^{[k]}$, after some scaling, is used as the starting point in Algorithm 1 for solving the problem with $r = r^{[k+1]}$. The matrix $X_{\mathcal{A}_k}$ is formed by the columns of X with indices in $\mathcal{A}_k$.

Let $\lambda^{[0]} = \max_{1 \leq j \leq p}\{\|Y^\tau x_j\|_2\}$. Without loss of generality, suppose $x_{\tilde{j}}$ is the unique covariate that achieves the maximum. Define

$$\lambda(r) = \lambda^{[0]} - r\|x_{\tilde{j}}\|_2^2 \quad \text{and} \quad b_{(j)}(r) = \begin{cases} (r/\lambda^{[0]})Y^\tau x_{\tilde{j}}, & j = \tilde{j}, \\ 0, & j \neq \tilde{j}. \end{cases} \quad (8.13)$$

The path-tracing algorithm in [168]) is presented below.

Algorithm 2 (Path-tracing method).

Initialize

Specify sequence $r^{[1]} < r^{[2]} < \cdots < r^{[K]}$ and parameters ϵ_{MIN}, , ν_{MIN}, α as needed in Algorithm 1.

Compute $\lambda^{[1]} = \lambda(r^{[1]})$ and $b_{(j)}^{[1]} = b_{(j)}(r^{[1]})$ according to (8.13). Let $B^{[1]} = (b_{(1)}^{[1]}, \ldots, b_{(M)}^{[1]})^\tau$.

Compute $s_j^{[1]} = \|\boldsymbol{b}_{(j)}^{[1]}\|_2 + 10^{-4}$ for $j = 1, \ldots, M$.

Define $\mathcal{A}_1 = \{j : \|(Y - XB^{[1]})^\tau \boldsymbol{x}_j\|_2 = \lambda^{[1]}\}$.

Set $k = 1$.

while $k < K$

$$\tilde{\zeta} = \frac{2s_j^{[k]}}{p[s_j^{[k]2} - \|\boldsymbol{b}_{(j)}^{[k]}\|_2^2]}, \quad \zeta_j = \frac{\tilde{\zeta}_j r^{[k+1]}}{1 + \sum_{j \in \mathcal{A}_k} \tilde{\zeta}_j s_j^{[k]}}, \quad j \in \mathcal{A}_k,$$

$$\nu_{\text{MAX}} = \lambda^{[k]} [r^{[k+1]} - \sum_{j \in \mathcal{A}_k} \zeta_j s_j^{[k]}].$$

Run Algorithm 1 with $r = r^{[k+1]}$ and covariance matrix $X_{\mathcal{A}_k}$ using starting values $\boldsymbol{b}_{(j)} = \zeta_j \boldsymbol{b}_{(j)}^{[k]}$, $s_j = \zeta_j s_j^{[k]}$, $j \in \mathcal{A}_k$ to produce solution $B_{\mathcal{A}_k}^{[k+1]}, \boldsymbol{s}_{\mathcal{A}_k}^{[k+1]}$.

Compute

$$\lambda^{[k+1]} = \max_{1 \le j \le M} \|(Y - X_{\mathcal{A}_k} B_{\mathcal{A}_k}^{[k+1]})^\tau \boldsymbol{x}_j\|_2,$$

$$\mathcal{A}_{k+1} = \{j : \|(Y - X_{\mathcal{A}_k} B_{\mathcal{A}_k}^{[k+1]})^\tau \boldsymbol{x}_j\|_2 > \lambda^{[k+1]} - 0.1\lambda^{[1]}\}.$$

$k \leftarrow k + 1$.

The first r value, i.e., $r^{[1]}$, in the above algorithm is determined such that there is at least one more index j in $\mathcal{A}_1$ in addition to $\tilde{j}$. This value is determined as follows. For $j = 1, \ldots, M$, let

$$a_j = (\boldsymbol{x}_{\tilde{j}}^\tau \boldsymbol{x}_{\tilde{j}})^2 - (\boldsymbol{x}_{\tilde{j}}^\tau \boldsymbol{x}_j)^2,$$

$$b_j = \lambda^{[0]} \boldsymbol{x}_{\tilde{j}}^\tau \boldsymbol{x}_{\tilde{j}} - \frac{1}{\lambda^{[0]}} (\boldsymbol{x}_{\tilde{j}}^\tau \boldsymbol{x}_j) \boldsymbol{x}_{\tilde{j}}^\tau Y Y^\tau \boldsymbol{x}_j,$$

$$c_j = \lambda^{[0]2} - \|Y^\tau \boldsymbol{x}_j\|_2^2.$$

Then, $r^{[1]}$ is determined as

$$r^{[1]} = \min_{j \ne \tilde{j}} r_j, \quad \text{where} \quad r_j = \overset{+}{\min}(b_j \pm \sqrt{b_j^2 - a_j c_j})/a_j.$$

Model selection methods for the sequential procedures

A final step of the the sequential procedures discussed above is to select a model from the resultant sequence of models:

$$(\mathcal{A}_1, \hat{B}_1), \ldots, (\mathcal{A}_k, \hat{B}_k), \ldots, (\mathcal{A}_K, \hat{B}_K),$$

There are two standards for the selection: cross-validation (CV) and BIC-type criteria. The CV aims to select a model that has the smallest prediction error. A BIC-type criterion emphasizes the contribution of the covariates in the model. In general, the model selected by CV contains more covariates than that selected by a BIC-type criterion. The model selected by a BIC-type criterion is more accurate in terms of whether or not the covariates in the model are causal ones. If the puspose of a QTL mapping study is to find a model for prediction or classification, CV is preferred. If the purpose is to detect the true QTL, a BIC-type criteria is more appropriate. In the following, we briefly discuss these two approaches for model selection in the sequential procedures.

(i) Cross Validation. In order for the computation to be feasible, a d-fold CV is usually used instead of the original CV for some integer d, see § 2.10. The d-fold CV procedure is as follows. The data is partitioned at random into d equal size portions. Each portion is, in turn, used for validation and the remaining portions are used to build up models. Let $Y^{(-j)}$ and $X^{(-j)}$ denote the matrices obtained from Y and X respectively by omitting the rows with indices in the jth portion. Let

$$(\mathcal{A}_1^{(-j)}, \hat{B}_1^{(-j)}), \ldots, (\mathcal{A}_k^{(-j)}, \hat{B}_k^{(-j)}), \ldots, (\mathcal{A}_K^{(-j)}, \hat{B}_K^{(-j)})$$

be the sequence of models produced by the sequential procedure using the data with the jth portion left out. Then the CV scores are given by

$$\mathrm{CV}(k) = \sum_{j=1}^{d} \|Y^{(-j)} - X_{\mathcal{A}_k^{(-j)}}^{(-j)} \hat{B}_k^{(-j)}\|_F^2, \quad k = 1, \ldots, K.$$

Let $k^* = \mathrm{argmin}\{\mathrm{CV}(k)\}$. The chosen model is then $(\mathcal{A}_{k^*}, \hat{B}_{k^*})$.

(ii) BIC-type criterion. In the approach of CV, the coefficients of the models are generated by the sequential procedure. They are shrunken estimates. The shrunken coefficients play a role in determining the total prediction power of the covariates. However, the shrunken estimates, in general, do not fully count the effects of the covariates on the response variables. The full effects of the covariates are more properly reflected by the ordinary least squares estimates (which are the same as the maximum likelihood estimates when the errors are assumed normally distributed). When a BIC-type criterion is used to select the final model, the models produced by the sequential procedures should be re-fitted by ordinary least squares methods. Then the model selection criterion is calculated for each of those models. The model with the smallest value of the model selection criterion is chosen as the final model. When M is large, an appropriate BIC-type criterion is the EBIC discussed in § 2.10.

8.2.3 Bayesian Approaches

In principle, the Bayesian approaches for mapping single traits discussed in Chapter 7 can be extended to the case of multi-trait QTL mapping. The extension is straightforward, though it could be tedious. Bayesian approaches have also been developed especially for multi-trait QTL mapping; see [7] [96] [122] [136] [194] [208]. In this section, we present two such Bayesian approaches: hierarchical Bayesian linear models [96] and Bayesian seemingly unrelated regression models [7]. These approaches take markers as surrogates of QTLs as in the case of dense markers considered in Chapter 6. A linear model is formulated to describe the relationship between the traits and marker genotypes, prior distributions are imposed on the parameters of the linear model, and the inference on QTL is then based on the posterior distribution of the parameters. The two approaches differ in the formulation of the linear model and the specification of the prior distributions.

Hierarchical Bayesian linear model

In a hierarchical Bayesian linear model, the relationship between the traits and marker genotypes is assumed as

$$y_{ti} = \alpha_t + \sum_{j=1}^{p} \gamma_{tj} x_{ij} + \epsilon_{ti}, \ t = 1, \ldots, T, \ j = 1, \ldots, M, \ i = 1, \ldots, n,$$

where ϵ_{ti} is i.i.d. with normal distribution $N(0, \sigma^2)$. After centering the trait and genotype values such that $\sum_{i=1}^{n} y_{ti} = 0$ and $\sum_{i=1}^{n} x_{ij} = 0$, the above model can be expressed in matrix form as

$$\boldsymbol{y}_t = \sum_{j=1}^{M} \gamma_{tj} \boldsymbol{x}_j + \boldsymbol{\epsilon}_t, \tag{8.14}$$

where $\boldsymbol{y}_t = (y_{t1}, \ldots, y_{tn})^\tau$, $\boldsymbol{x}_j = (x_{1j}, \ldots, x_{nj})^\tau$ and $\boldsymbol{\epsilon}_t = (\epsilon_{t1}, \ldots, \epsilon_{tn})^\tau$. By an abuse of notation, the centered and non-centered quantities are denoted by the same notation. In (8.14), $\boldsymbol{\epsilon} \sim N(\boldsymbol{0}, R\sigma^2)$ where R is a positive definite matrix with diagonal entries $1 - \frac{1}{n}$ and off-diagonal entries $\frac{1}{n}$. Thus, given all the γ_{tj} variables, $\boldsymbol{y}_t$'s are independent with p.d.f.

$$p(\boldsymbol{y}_t | \boldsymbol{\gamma}_t, \sigma^2) = \phi(\boldsymbol{y}_t; \sum_{j=1}^{M} \gamma_{tj} \boldsymbol{x}_j, R\sigma^2), \tag{8.15}$$

where $\boldsymbol{\gamma}_t = (\gamma_{t1}, \ldots, \gamma_{tp})^\tau$ and $\phi(\boldsymbol{y}, \mu_y, \Sigma_y)$ denotes the p.d.f. of a multivariate normal distribution with mean vector μ_y and variance-covariance matrix Σ_y. The prior distributions of the parameters $\boldsymbol{\gamma}_t$ and σ^2 in the above model are assigned in hierarchical layers, hence the name hierarchical Bayesian linear models.

First layer of priors:

(i1) Prior for γ_{tj}'s:

$$p(\gamma_{tj} | \eta_{tj}, \sigma_j^2) = (1 - \eta_{ij}) \phi(\gamma_{tj}; 0, \delta) + \eta_{tj} \phi(\gamma_{tj}; 0, \sigma_j^2),$$

where $\phi(x; \mu, \tau)$ denotes the p.d.f. of the normal distribution with mean μ and variance τ, and $\delta = 10^{-4}$.

(i2) Prior for σ^2: A vague prior is assigned to σ^2 as

$$p(\sigma^2) \propto \frac{1}{\sigma^2}.$$

Second layer of priors:

(ii1) Prior for η_{tj}'s:

$$p(\eta_{tj}|\rho_j) = \rho_j^{\eta_{tj}}(1 - \rho_j)^{1-\eta_{tj}}, \text{i.e., } \eta_{tj} \sim \text{Bernoulli}(1, \rho_j).$$

(ii2) Prior for σ_j^2:

$$p(\sigma_j^2) = \text{Inv-}\chi^2(\sigma_j^2; d_0, w_0), \text{ i.e., } \sigma_j^2 \sim \text{Inv-}\chi^2(d_0, w_0),$$

where d_0 and w_0 are fixed degrees of freedom and scale parameter respectively. They are specified as $d_0 = 5, w_0 = 50$ in [96]. A random variable X follows an Inv-$\chi^2(d, w)$ distribution (inverse χ^2 distribution with degrees of freedom d and scale parameter w) iff $\frac{1}{X}$ follows a $\chi^2(d, w)$-distribution, see [10].

Third layer of priors:

(iii) Prior for ρ_j's:
$$p(\rho_j) = \text{Beta}(\rho_j; 1, 1),$$

where $\text{Beta}(x; 1, 1)$ is the p.d.f. of the Beta distribution with parameters $\alpha = \beta = 1$.

With the priors assigned above and (8.15), the posterior distribution of the parameters are derived as

$$\frac{\prod_{t=1}^{T} \left\{ p(\boldsymbol{y}_t|\boldsymbol{\gamma}_t, \sigma^2) \prod_{j=1}^{M} \left[p(\gamma_{tj}|\eta_{tj}, \sigma_j^2) p(\eta_{tj}|\rho_j) p(\sigma_j^2) p(\rho_j) \right] \right\} p(\sigma^2)}{m(Y)}, \quad (8.16)$$

where $m(Y)$ is the marginal joint distribution of $\boldsymbol{y}_t$'s. The Bayesian inference is based on the posterior distribution. In the context of QTL mapping, the most important quantities of the posterior distribution are the posterior means of η_{tj}'s, ρ_j's and γ_{tj}'s. The posterior mean of η_{tj} is the posterior probability that trait t is associated with marker j. The posterior mean of ρ_j is the proportion of traits that are associated with marker j. The posterior mean of γ_{tj} reflects the magnitude of the effect of marker j on trait t.

An MCMC algorithm based on Gibbs sampler is developed in [96] for the generation of pseudo-random samples from the posterior distribution. Starting with some chosen initial values of the unknown parameters, the MCMC algorithm iterates and one iteration generates one observation of the parameters.

Within each iteration, each parameter is generated in turn by its conditional distribution given the data and the most updated values of the other parameters. The observations generated by a certain number of iterations from the beginning are discarded. In the remainder of the iterations, one observation in every 10 iterations is saved to form the pseudo-random sample until a desired sample size is reached. The following are the steps in each iteration of the MCMC algorithm:

1. Generate η_{tj} from Bernoulli$(1, \pi_{tj})$ with

$$\pi_{tj} = \frac{\rho_j \phi(\gamma_{tj}; 0, \sigma_j^2)}{\rho_j \phi(\gamma_{tj}; 0, \sigma_j^2) + (1 - \rho_j) \phi(\gamma_{tj}; 0, \delta)}.$$

2. Generate γ_{tj} from normal distribution $N(\mu_\gamma, \sigma_\gamma^2)$ with

$$\mu_\gamma = \left[\boldsymbol{x}_j^T R^{-1} \boldsymbol{x}_j + \frac{\sigma^2}{\eta_{tj}\sigma_j^2 + (1 - \eta_{tj})\delta} \right]^{-1} \boldsymbol{x}_j^T R^{-1} \Delta \boldsymbol{y}_t,$$

$$\sigma_\gamma^2 = \left[\boldsymbol{x}_j^T R^{-1} \boldsymbol{x}_j + \frac{\sigma^2}{\eta_{tj}\sigma_j^2 + (1 - \eta_{tj})\delta} \right]^{-1} \sigma^2,$$

where $\Delta \boldsymbol{y}_t = \boldsymbol{y}_t - \sum_{l \neq j}^{M} \boldsymbol{x}_l \gamma_{tl}$.

3. Generate from σ_j^2 from the inverse χ^2 distribution

$$\text{Inv-}\chi^2 \left(\sum_{t=1}^{T} \eta_{tj} + d_0, \sum_{t=1}^{T} \eta_{tj} \gamma_{tj}^2 + w_0 \right).$$

4. Generate σ^2 from the inverse χ^2 distribution

$$\text{Inv-}\chi^2 \left(mn, \sum_{t=1}^{T} (\boldsymbol{y}_t - \sum_{j=1}^{M} \boldsymbol{x}_j \gamma_{tj})^T R^{-1} (\boldsymbol{y}_t - \sum_{j=1}^{M} \boldsymbol{x}_j \gamma_{tj}) \right).$$

5. Generate ρ_j from the Beta distribution

$$\text{Beta} \left(\sum_{t=1}^{T} \eta_{tj} + 1, m - \sum_{t=1}^{T} \eta_{tj} + 1 \right).$$

Suppose the MCMC algorithm generates a posterior sample of size L. Then the posterior means of η_{tj}, γ_{tj} and ρ_j are approximated, respectively, by

$$\bar{\eta}_{tj} = \frac{1}{L} \sum_{l=1}^{L} \eta_{tj}^{(l)}, \tag{8.17}$$

$$\bar{\gamma}_{tj} = \frac{1}{L} \sum_{l=1}^{L} \gamma_{tj}^{(l)}, \tag{8.18}$$

and $$\bar{\rho}_j = \frac{1}{L} \sum_{l=1}^{L} \rho_j^{(l)}. \tag{8.19}$$

The existence of association between a trait and a marker (i.e, the marker is a QTL of the trait) is determined as follows. Let q_0 be a properly specified threshold. If $\bar{\eta}_{tj} \geq q_0$ then an association between trait t and marker j is claimed. This inference is similar to the differential expression analysis for microarray expression data [141]. The threshold can be specified in a way such that the false discovery rate (FDR) is controlled at a pre-specified level α. Such a threshold is obtained as

$$q_0 = \min\{q : \frac{\sum_{\bar{\eta}_{tj} \geq q} \bar{\eta}_{tj}}{\sum_{\bar{\eta}_{tj} \geq q} I\{\bar{\eta}_{tj} \geq q\}} > 1 - \alpha\}; \tag{8.20}$$

that is, q_0 is the smallest threshold such that the average of the $\bar{\eta}_{tj}$'s that are larger than or equal to the threshold is bigger than $1 - \alpha$; see [48] [141].

Bayesian seemingly unrelated regression

The Bayesian seemingly unrelated regression (BSUR) approach proposed in [7] is another Bayesian approach for multi-trait QTL mapping. The BSUR approach is developed in [7] originally for experimental crosses. The set of observed markers is augmented by including additional loci that are equally spaced over the whole genome. These additional loci are referred to as pseudo-markers. The genotypes of the pseudo-markers are unobserved and treated as unknown parameters. The conditional probabilities of the pseudo-marker genotypes given the observed markers are used as the priors for these unknowns in the Bayesain framework. The conditional probabilities can be computed by using Table 3.4, 3.5 or 3.6 according to whether the experiment is backcross, intercross or recombinant inbred line. However, the BSUR approach can also be applied to non-experimental cross data with or without the inclusion of the pseudo-markers. For the sake of convenience, we assume that each marker has two different alleles in the population and the genetic mode of each marker is additive. Thus the genotype of each marker can be expressed by a single variable.

The BSUR formulates a different linear model for the traits from that of the Bayesian hierarchical linear model. It assumes that the number of QTL for each trait is at most L, L being taken as the upper bound of the number of detectable QTL with high probability for a given data set. The value of L is determined on the basis of initial analysis on each trait. The positions of the L loci for trait t are denoted by $\boldsymbol{\delta}_t = (\delta_{t1}, \ldots, \delta_{tL})^\tau$. Denote the marker genotypes of individual i at these L loci by $\boldsymbol{x}_{t(i)}$. The relationship between trait t and the L markers is described by the linear model:

$$y_{ti} = \mu_t + \boldsymbol{x}_{t(i)}^\tau \boldsymbol{\beta}_t \circ \boldsymbol{\eta}_t + \epsilon_{ti}, \ t = 1, \ldots, m, \ i = 1, \ldots, n, \qquad (8.21)$$

where $\boldsymbol{x}_{t(i)} = (x_{t1i}, \ldots, x_{tLi})^\tau$, $\boldsymbol{\beta}_t = (\beta_{t1}, \ldots, \beta_{tL})^\tau$, $\boldsymbol{\eta}_t = (\eta_{t1}, \ldots, \eta_{tL})^\tau$, $\eta_{tj} = 1$ or $0, j = 1, \ldots, L$, and $\boldsymbol{\beta}_t \circ \boldsymbol{\eta}_t$ is a pointwise product, i.e., $\boldsymbol{\beta}_t \circ \boldsymbol{\eta}_t = (\beta_{t1}\eta_{t1}, \ldots, \beta_{tL}\eta_{tL})^\tau$. In matrix form, (8.21) can be expressed as

$$\boldsymbol{y}_{(i)} = \boldsymbol{\mu} + X_i \boldsymbol{\beta} \circ \boldsymbol{\eta} + \boldsymbol{\epsilon}_i, \ i = 1, \ldots, n, \qquad (8.22)$$

where $\boldsymbol{y}_{(i)} = (y_{1i}, \ldots, y_{Ti})^\tau$, $X_i = \text{diag}(\boldsymbol{x}_{1(i)}^\tau, \ldots, \boldsymbol{x}_{T(i)}^\tau)$, $\boldsymbol{\mu} = (\mu_1, \ldots, \mu_T)^\tau$, $\boldsymbol{\beta} = (\boldsymbol{\beta}_1^\tau, \ldots, \boldsymbol{\beta}_T^\tau)^\tau$, $\boldsymbol{\eta} = (\boldsymbol{\eta}_1^\tau, \ldots, \boldsymbol{\eta}_T^\tau)^\tau$ and $\boldsymbol{\epsilon}_i = (\epsilon_{1i}, \ldots, \epsilon_{Ti})^\tau$. It is assumed that $\boldsymbol{\epsilon}_i$'s are i.i.d. normal random vectors with mean $\boldsymbol{0}$ and variance-covariance matrix Σ. Thus

$$p(\boldsymbol{y}_i | \boldsymbol{\mu}, \boldsymbol{\beta}, \Sigma, \boldsymbol{\delta}, \boldsymbol{\eta}) \sim N(\boldsymbol{\mu} + X_i \boldsymbol{\beta} \circ \boldsymbol{\eta}, \ \Sigma), \ i = 1, \ldots, n. \qquad (8.23)$$

In the linear model above, the traits have different sets of markers and are seemingly unrelated, but are indeed related because of the dependence among the components of $\boldsymbol{\epsilon}_i$, hence the seemingly unrelated regression name [217].

The priors on the unknown parameters are described in [7]. The prior on Σ is assigned through a non-informative prior on Σ^{-1}: $p(\Sigma^{-1}) \sim |\Sigma^{-1}|^{-(1+T)/2}$. The priors on trait associated parameters are independent across traits. The priors on $(\boldsymbol{\delta}_t, \mu_t, \boldsymbol{\beta}_t, \boldsymbol{\eta}_t)$ are as follows.

Prior on $\boldsymbol{\delta}_t$: The prior is a uniform distribution over the space of all L-combinations of the positions of the markers in the marker set (including pseudo-markers if the marker set is augmented); that is, if the total number of markers is M, the probability on each L-combination is $1/\binom{M}{L}$.

Prior on $\boldsymbol{\beta}_t$ and $\boldsymbol{\eta}_t$: The joint prior of $\boldsymbol{\beta}_t$ and $\boldsymbol{\eta}_t$ is assigned as

$$p(\boldsymbol{\beta}_t, \boldsymbol{\eta}_t) = \prod_{j=1}^{L} \{w_t^{\eta_{tj}} (1 - w_t)^{1 - \eta_{tj}} [(1 - \eta_{tj}) I_0 + \eta_{tj} \phi(\beta_{tj}, 0, \sigma_t^2)]\}, \qquad (8.24)$$

where I_0 is a point mass at 0, w_t is a fixed number between 0 and 1, and σ_t^2 is a random variable having a inverse χ^2 distribution Inv-$\chi^2(\nu_t, s_t^2)$. It is recommended in [7] that $\nu_t = 6$ and $s_t^2 = (\nu_t - 2)hV_t/(\nu_t V_{tj})$, where h is a small constant between 0.05 to 0.2, V_t is the sample variance of $\{y_{ti}, i = 1, \ldots, n\}$, and V_{tj} is the variation of the y_{ti}'s explained by the jth marker for trait t. It is also recommended to set $w_t = l_t/L$ where l_t is the prior expected number of detectable QTL for trait t which can be set to the number of QTL detected by a single-trait mapping method applied to trait t. For details and more general cases on the markers, see [214].

Prior on μ_t: It is assigned as the normal distribution with mean and variance being the sample mean and sample variance of the tth trait.

Note that the prior (8.24) is similar to the prior assigned on $\boldsymbol{\beta}_t$ in the hierarchical Bayesian linear model, but it differs in the probability that $\eta_{tj} = 1$. In the hierarchical Bayesian linear model, this probability depends on the jth marker, not on the trait, and measures the proportion of traits that are associated with the jth marker. In the above, this probability depends on the tth trait, not on the marker, and measures the proportion of markers associated with the tth trait.

With the priors assigned above, the joint posterior distribution of the unknown parameters is obtained as

$$p(\boldsymbol{\mu}, \boldsymbol{\beta}, \boldsymbol{\sigma}, \Sigma^{-1}, \boldsymbol{\delta}, \boldsymbol{\eta} | \boldsymbol{y})$$
$$\propto \prod_{i=1}^{n} p(\boldsymbol{y}_i | \boldsymbol{\mu}, \boldsymbol{\beta}, \Sigma^{-1}, \boldsymbol{\delta}, \boldsymbol{\eta}) p(\boldsymbol{\beta}_t, \boldsymbol{\eta}_t) \phi(\mu_t, \bar{\boldsymbol{y}}_t, S_t^2) |\Sigma^{-1}|^{-(1+T)/2}. \quad (8.25)$$

Similar to the algorithm for the Bayesian hierarchical linear model, the random parameters can be generated one at a time by a MCMC algorithm and each parameter is generated according to its conditional distribution given the data and the most updated values of all the other parameters. The conditional distributions of the parameters are described in detail in [7].

The inference on QTL is based on the posterior samples generated by the MCMC algorithm. The most important posterior samples are those of $\boldsymbol{\delta}_t$ and $\boldsymbol{\eta}_t$ which determine the number of QTL and their positions. As proposed in [7], among all the positions identified in the posterior sample, the posterior probability that a locus is a QTL is estimated as its frequency in the posterior sample, i.e, the proportion that it is sampled with the corresponding indicator equal to 1. The evidence for a locus to be a QTL is measured by the Bayes factor (BF) which is the ratio of the posterior odds and prior odds of being a QTL. For example, suppose that the estimated posterior probability of locus j being a QTL of trait t is $\hat{w}_{tj}$, then the Bayes factor for this locus is $\frac{\hat{w}_{tj}/(1-\hat{w}_{tj})}{w_t/(1-w_t)}$, where w_t is the prior probability that a locus is a QTL of trait t. A locus is claimed as a QTL if $2\ln \mathrm{BF}$ of the locus exceeds a threshold.

Determination of the threshold. In the simulation studies reported in [7], the threshold is determined by simulation under the genetic structure of the simulation studies with the setting of no QTLs. Each simulation replicate produces a maximum $2\ln \mathrm{BF}$. The 95th percentile of the empirical distribution of the simulated maximum $2\ln \mathrm{BF}$ is taken as the threshold. For real QTL mapping data, no guidance is given in [7]. However, the permutation approach [40] can be used in a similar way to that for the simulation studies in [7]. Instead of simulating the data in each replication, let each trait be permuted and reassigned to the individuals and the BSUR procedure is carried out for the permuted traits, which produces a maximum $2\ln BF$. Then the empirical distribution of maximum $2\ln BF$ produced from a large enough number of permutations can be taken as the reference distribution.

The BSUR procedure can be implemented by using the R package R/qtlbim [207].

8.3   eQTL Mapping Approaches

QTL mapping detects DNA variants that are associated with quantitative traits. However, DNA variants do not affect quantitative traits directly. Instead, they affect quantitative traits through complex biological networks with intermediate molecular endophenotypes such as those of gene expression, protein and metabolism. Without an understanding of the connection among these intermediate endophenotypes, the association between DNA variants and quantitative traits is devoid of biological contents.

There is a great number of genes which are differentially expressed among individuals. The variation in gene expression is a more important mechanism underlying phenotypic variation in morphological, physiological and behavioral traits as well as disease susceptibility. A connection between DNA variants and gene expression levels not only provides more understanding of the biological network but also enhances the mapping of the above mentioned quantitative traits. Gene expression levels themselves can be treated as quantitative traits. The DNA variants that affect gene expression levels are referred to as expression quantitative trait loci (eQTL). The most commonly used DNA variants in eQTL mapping are SNPs. Other DNA variants include copy number variation (CNV), insertions and deletions, short tandem repeats and single amino acid repeats, etc. eQTL mapping aims to identify DNA variates that affect gene expression levels. For a general review on eQTL mapping, the reader is referred to [35] [41] [45] [128] [137].

In principle, any method for QTL mapping can be applied for eQTL mapping. However, there are specific features in eQTL mapping that entail particular treatments. In § 8.3.1, we give a brief discussion on the specific features of eQTL mapping. In the subsequent sub sections, we consider statistical methods specific for eQTL mapping. In § 8.3.2, we present eQTL mapping methods based on single-QTL models, i.e., models that assume the existence of at most one eQTL per transcript. Multi-QTL-model methods are discussed in § 8.3.3. The multi-QTL-model methods require the dimensionality reduction of the transcripts in the original eQTL mapping data. Statistical methods for clustering transcripts are discussed in § 8.3.4.

8.3.1 Specific Features of eQTL Mapping

An eQTL mapping study involves measurement of expression levels of thousands of genes on each individual as well as the individual's genotypes at DNA variants which are usually also huge in number. There are special features in eQTL mapping which do not present in ordinary QTL mapping. The understanding of these special features is helpful. Statistical methods for the analysis of eQTL mapping data can be made more efficient while these special features are taken into account. In the following, we briefly discuss these special features.

cis-acting and trans-acting eQTL. Since the location of an expressed target gene is known, the physical distances between the target gene and the SNPs under study are known. There are two types of loci that affect the expression of a target gene: *cis*-acting eQTL and *trans*-acting eQTL. There is no rigorous definition for these two types of eQTL. Roughly, a *cis*-acting eQTL is near the target gene. Typically, it is considered to be located on the same chromosome as and to be within 100 kb upstream and downsteam of the target gene; see [71] [41] and the references therein. A *trans*-acting eQTL is more distant from the target gene. It can occur anywhere in the genome. Usually it is located on a different chromosome from that of the target gene. The terminology of *cis*- and *trans*-acting does not have any functional significance. It is probably more appropriate to refer to *cis*- and *trans*-acting eQTL as *local* and *distal* eQTL respectively, as suggested in [150]. However, we still use the terminology because it is more commonly used in the literature.

In general, a *cis*-acting eQTL has a larger effect than a *trans*-acting eQTL. Therefore, the sample size required for detecting a *cis*-acting eQTL is less than that for detecting a *trans*-acting eQTL. As a consequence, in many small sample eQTL mapping studies, more *cis*-acting eQTL are detected than *trans*-acting eQTL; see [51] [75] [155]. However, this does not necessarily imply that there are more *cis*-acting eQTL than *trans*-acting eQTL.

A *trans*-acting eQTL usually affects multiple target genes. Hence the effects of *trans*-acting eQTL are pleiotropic. Furthermore, *trans*-acting eQTL usually occur in the so-called *hot spots*. A *hot spot* is a region in the genome that contains significantly more co-regulators than can occur at random.

When the information on the location of the target genes and the DNA variants is incorporated into statistical analysis, the power of detecting eQTL can be enhanced.

Large number of responses, large number of covariates and small sample size. In eQTL mapping studies, in addition to the large number of DNA variants, the number of differentially expressed genes is also very large, unlike multi-trait QTL mapping where the number of traits are usually only a few. Because of this, the statistical methods for multi-trait QTL mapping cannot be directly applied for eQTL mapping. The reduction of the dimensionality of the differentially expressed genes becomes compelling. In addition, the sample size is usually small, especially in human eQTL mapping studies, because of the difficulty to obtain tissues for the studies. This adds a further layer of complexity to eQTL mapping.

Correlation among gene expression levels. Expression levels of many genes are highly correlated. There are two causes for the correlation. Either the genes are located close to each other or they share similar functions. Thus, the correlations among the expression levels have a cluster nature; that is, the expression levels are clustered such that the expression levels within a cluster is highly correlated due to the reasons mentioned above. The information on the correlation of the expression levels is helpful in several aspects. The fact that genes sharing similar functions are usually co-regulated by *trans*-acting eQTLs in *hot spots* helps in identifying pleiotropic eQTL and distinguishing causal associations from spurious associations. The increased power of detecting QTL by multi-trait mapping methods is expected only when the traits are correlated. Because of the cluster nature of the correlations, it is not necessary to map all the traits together. It is more appropriate to map only the expression levels within the same cluster together. Mapping expression levels cluster by cluster has two obvious advantages: it reduces the dimensionality of the problem and it achieves a higher power in detecting eQTL.

8.3.2 Approaches Based on Single-QTL Models

In the early studies of eQTL mapping, two major approaches were employed: transcript-based approach and marker-based approach. The transcript-based approach stems from the ordinary QTL mapping. In fact, it is the one-trait-at-a-time approach discussed in § 8.1.1. The marker-based approach stems from microarray data analysis on differentially expressed genes. At each marker, the individuals are classified into different groups according to their genotypes at the marker, the marker-based approach identifies, at each marker, the transcripts which are differentially expressed among the groups. The transcript-based approach and marker-based approach are also combined to inference on eQTLs simultaneously over all transcripts and all markers. These approaches are simple and can be easily implemented. Beyond their historical interest, they can be used for preliminary analysis before more sophisticated approaches are applied. There is a nature common in these approaches; that is, they are based on models that assume at most one eQTL for each transcript. In this sub section, we consider in detail some single-QTL-model approaches. For other single-QTL-model approaches, see the review in [101].

The following notation will be used throughout. For individual i, the log-transformed expression level of transcript t is denoted by y_{ti} and the genotype at marker j is denoted by x_{ij}, $i = 1, \ldots, n$, $j = 1, \ldots, M$, and $t = 1, \ldots, T$. Without loss of generality, let x_{ij} take values $0, 1, 2$ representing three genotypes of a SNP. Let $\boldsymbol{y}_t = (y_{t1}, \ldots, y_{tn})^\tau$. Let $\boldsymbol{y}_t^{(0)}, \boldsymbol{y}_t^{(1)}$ and $\boldsymbol{y}_t^{(2)}$ be generic notations for the subvectors of $\boldsymbol{y}_t$ corresponding to the individuals with genotype 0, 1 and 2 respectively at a particular marker. Let μ_{tk} be the expectation of y_{ti} if individual i has genotype k at the particular marker, $k = 0, 1, 2$.

EBarrays. This approach is originally developed in [102] for microarray data analysis. It can be readily used for eQTL mapping. The version for eQTL mapping is described in [101]. We first describe the method in the case that only a single marker is under consideration. EBarrays assumes a Bayessian model with the following assumptions:

(a) Given μ_{tk}, y_{ti}'s are independently distributed with a probability density function $f(y|\mu_{tk}, \theta)$ where θ represents other parameters of the distribution.

(b) The μ_{tk}'s are s i.i.d. random variables with a prior probability density function $\pi(\mu)$. Thus, if transcript t is equivalently expressed (EE), the distribution of $\boldsymbol{y}_t$ is given by

$$f_0(\boldsymbol{y}_t) = \int \left[\prod_i f(y_{ti}|\mu) \right] \pi(\mu)d\mu.$$

If it is differentially expressed (DE), the distribution of $\boldsymbol{y}_t$ is given by

$$f_1(\boldsymbol{y}_t) = f_0(\boldsymbol{y}_t^{(0)})f_0(\boldsymbol{y}_t^{(1)})f_0(\boldsymbol{y}_t^{(2)}).$$

(c) Let z_t be the expression status of transcript t: $z_t = 1$, if transcript t is DE, 0, otherwise. The prior distribution of z_t is given by

$$p(z_t) = p^{z_t}(1 - p)^{1 - z_t}.$$

The EBarrays approach is an empirical Bayes procedure. First, it uses an EM algorithm to obtain the estimate $\hat{p}$ of the hyperparameter p. Second, for each t, calculate the posterior probability that $z_t = 1$ by the Bayes rule as

$$\hat{p}_t = \frac{\hat{p}f_1(\boldsymbol{y}_t)}{(1 - \hat{p})f_0(\boldsymbol{y}_t) + \hat{p}f_1(\boldsymbol{y}_t)}.$$

The transcript t is claimed DE with respect to the particular marker if $1 - \hat{p}_t$ is smaller than a threshold.

In the multi-marker case, the EBarrays method is implemented as follows. First, the above procedure is applied to each marker. For marker j, let the computed posterior probability be denoted by $\hat{p}_{tj}$. Then, for each transcript, find the maximum

$$\hat{p}_{t_{\mathrm{MAX}}} = \max\{\hat{p}_{tj} : j = 1, \ldots, M\}.$$

When $\hat{p}_{t_{\text{MAX}}}$ exceeds a threshold value, the transcript t is claimed to map to at least one of the markers. The threshold is determined as $1 - k_\alpha$ where

$$k_\alpha = \max\left\{k : \frac{\sum_{t=1}^{T}(1 - \hat{p}_{t_{\text{MAX}}})I\{1 - \hat{p}_{t_{\text{MAX}}} \le k\}}{\sum_{t=1}^{T} I\{1 - \hat{p}_{t_{\text{MAX}}} \le k\}} \le \alpha\right\}. \tag{8.26}$$

This choice of k_α controls the posterior false discovery rate (FDR) below α.

The EBarrays is a special case of the MOM model discussed next. More details of the conditional distribution of y_{ti}, the prior $\pi(\mu)$, the EM algorithm will be discussed when we consider the MOM model.

Mixture over markers (MOM) model. The EBarrays, as well as other transcript-based and marker-based approaches, has an obvious drawback: when it examines whether a transcript maps to a particular marker it ignores the possibility that the transcript could map to other markers. It is more plausible to examine all the markers simultaneously. The MOM model, which is developed in [101], is motivated by this idea. The MOM model makes essentially the same assumptions as EBarrays except that, for any transcript t, its expression status is described by a vector $\boldsymbol{z}_t = (z_{t0}, \ldots, z_{tM})$ where $z_{tj} = 1$, if transcript t maps to marker j, 0, otherwise, and $\sum_{j=0}^{M} z_{tj} = 1$. The assumption (c) is replaced by the following prior on $\boldsymbol{z}_t$:

$$p(\boldsymbol{z}_t) = p_0^{z_{t0}} p_1^{z_{t1}} \cdots p_M^{z_{tM}}, \quad \sum_{j=0}^{M} p_j = 1.$$

When $M = 1$, the MOM model reduces to EBarrays. The MOM model approach proceeds in the same way as EBarrays. It uses an EM algorithm to compute $\hat{p}_j, j = 0, 1, \ldots, M$, the estimates of the hyperparameters. Then compute the posterior probabilities

$$\hat{p}_{tj} = \frac{\hat{p}_j f_1(\boldsymbol{y}_t)}{\sum_{l=0}^{M} \hat{p}_l f_l(\boldsymbol{y}_t)}, \quad j = 0, 1, \ldots, M.$$

Transcript t is claimed as DE if $\hat{p}_{t0}$ is smaller than a threshold.

We now turn to some details of the procedure of the MOM model approach.

1. *Distribution assumptions.* It is reasonable to assume that the log-transformed expression measurement y_{ti} follows a normal distribution given the genotype of the individual i. Thus, at marker j, it is assumed that

$$y_{ti} \sim N(\mu_{tx_{ij}}, \sigma^2).$$

The conjugate prior is assumed for μ_{tk}, i.e.,

$$\mu_{tk} \sim N(\mu_0, \tau_0^2),$$

where μ_0, τ_0^2 are hyperparameters. With the above specification, the density function $f_0(\boldsymbol{y}_t)$ has an explicit form:

$$f_0(\boldsymbol{y}_t) = \int \left[\prod_i f(y_{ti}|\mu) \right] \pi(\mu) d\mu$$

$$= (2\pi\sigma^2)^{-\frac{n-1}{2}} \left[\frac{\tau_0^2}{n\tau_0^2 + \sigma^2} \right]^{1/2}$$

$$\times \exp\left\{ -\frac{1}{2\sigma^2} \left[\sum_i (y_{ti} - \bar{Y}_t)^2 - \frac{(n\tau_0^2 \bar{Y}_t + \sigma^2 \mu_0)^2}{\tau_0^2 (n\tau_0^2 + \sigma^2)} \right] \right\}.$$

For marker j,

$$f_j(\boldsymbol{y}_t) = \prod_k f_0(\boldsymbol{y}_t^{(k)}).$$

For the explicit expression of $f_0(\boldsymbol{y}_t^{(k)})$, the n in the expression of $f_0(\boldsymbol{y}_t)$ is replaced by n_k, the number of individuals with genotype k at marker j.

2. *EM algorithm.* If z_t's were observed then the joint density of $\boldsymbol{y}_t$ and z_t is given by

$$\prod_{j=0}^{M} [p_j f_j(\boldsymbol{y}_t)]^{z_{tj}}$$

and hence the complete log likelihood function given $\{(\boldsymbol{y}_t, z_t)\}$ is

$$\sum_{t=1}^{T} \sum_{j=0}^{M} z_{tj} \ln p_j + \sum_{t=1}^{T} \sum_{j=0}^{M} z_{tj} \ln f_j(\boldsymbol{y}_t)$$

$$\equiv L_1(\boldsymbol{p}; \boldsymbol{Z}) + L_2(\sigma^2, \mu_0, \tau_0^2; \boldsymbol{Z}), \text{ say.}$$

Since L_1 and L_2 involve separate parameters, the ECM [134], a variant of EM, is used to estimate the parameters. The ECM algorithm iterates the following steps until convergence:

(i) Compute the conditional expectation of the z_{tj}'s as

$$\tilde{z}_{tj} = \frac{p_j^{(0)} f_j^{(0)}(\boldsymbol{y}_t)}{\sum_{l=0}^{M} p_j^{(0)} f_j^{(0)}(\boldsymbol{y}_t)},$$

where the superscript $^{(0)}$ indicates that the quantity is calculated at the most updated values of the parameters.

(ii) Maximize $L_1(\boldsymbol{p}, \tilde{\boldsymbol{Z}})$ to update $\boldsymbol{p}$.

(iii) Repeat (i) with the updated values of $\boldsymbol{p}$.

(iv) Maximize $L_2(\sigma^2, \mu_0, \tau_0^2; \tilde{\boldsymbol{Z}})$ to update $\sigma_2, \mu_0, \tau_0^2$.

In the end of the ECM algorithm, not only are the estimated parameters $\hat{p}, \hat{\sigma}^2, \hat{\mu}_0, \hat{\tau}_0^2$ obtained but also the (empirical) posterior probabilities $\hat{p}_{tj} = p(z_{tj} = 1)$ which are given by the final version of the $\tilde{z}_{tj}$'s.

3. *Determination of the threshold.* Similar to EBarrays, the threshold for $\hat{p}_{t0}$ is determined by finding the largest value of k_α such that

$$\frac{\sum_{t=1}^{T} \hat{p}_{t0} I\{\hat{p}_{t0} \le k_\alpha\}}{\sum_{t=1}^{T} I\{\hat{p}_{t0} \le k_\alpha\}} \le \alpha. \qquad (8.27)$$

In the above description, the variance σ^2 is assumed the same for all transcripts. If there is evidence that transcripts are clustered and within each cluster the transcripts are more homogeneous, it then can be assumed that the variance differs from cluster to cluster but remain the same within clusters, as suggested in [101].

MOM proximity model. As discussed in § 8.3.1, eQTL data contains the information on the locations of the markers and the expressed genes. In general, a locus closer to a target gene is more likely to be a regulating locus. To incorporate this information into the mapping, the MOM model is modified in [68] to the so-called MOM proximity model. The major modification is on the assumption of the prior probabilities p_j's. In the setting of MOM model, these prior probabilities are assumed the same for all transcripts. In the MOM proximity model, the relative positions of the transcript and the markers are taken into account and these prior probabilities become transcript dependent.

The MOM proximity model differs from the original MOM model in the following two points.

(i) Instead of assuming the same p_j's for all transcripts, the MOM proximity model assumes, for a particular transcript t, that the prior probabilities have the following structure:

$$p_{tj}(\boldsymbol{\beta}, \gamma) = \frac{\exp(\beta_j + \gamma I\{j \text{ is closest marker to } t\})}{\sum_{j'=0}^{M} \exp(\beta_{j'} + \gamma I\{j' \text{ is closest marker to } t\})},$$

where, for $j = 0$, $I\{j \text{ is closest marker to } t\} = 1$, if none of the markers are close to t. Whether or not a marker can be considered close to t is a subjective matter. As a rule of thumb, a marker can be considered close to t if the marker is less than 100 kb away from t. When $\gamma = 0$, the assumption reduces to that of the MOM model.

(ii) Instead of assuming constant σ^2 (though it might differ from transcript to transcript), the MOM proximity model considers σ^2 as a random variable and imposes a prior $\pi_{\sigma^2}(\sigma^2)$. To distinguish the density functions of the traits from those of the MOM model, let $\tilde{f}_j(\boldsymbol{y}_t)$ denote the density function of trait t at marker j. Because of the additional prior on σ^2, the $\tilde{f}_j$'s have the following form:

$$\tilde{f}_j(\boldsymbol{y}_t) = \int f_j(\boldsymbol{y}_t) \pi_{\sigma^2}(\sigma^2) d\sigma^2, \; j = 0, 1, \ldots, M,$$

where $f_j(\boldsymbol{y}_t)$ is the corresponding density function in the MOM model. In [68], the prior π_{σ^2} is specified as a discrete distribution such that σ has equal mass over a given number of points uniformly chosen from an interval $[0, \sigma^*]$, where σ^* is determined by estimating the variances of a subset of genes considered to be equivalently expressed.

The other aspects of the MOM proximity model are the same as the MOM model, except that the ECM algorithm needs to be modified for the estimation of the hyperparameters $\boldsymbol{\beta}$ and γ. In the above modeling, if a marker, say j, is not associated with any transcripts, then the MLE of β_j tends to $-\infty$, which causes a convergence problem. To circumvent this problem, it is suggested in [68] that in the ECM algorithm, the first component of the complete likelihood $L_1(\boldsymbol{p}(\boldsymbol{\beta}, \gamma), \tilde{\boldsymbol{Z}})$ is replaced by a penalized version $L_1(\boldsymbol{p}(\boldsymbol{\beta}, \gamma), \tilde{\boldsymbol{Z}}) - \lambda \|\boldsymbol{\beta}\|_2^2$ with a λ in the range 10^{-4} to 10^{-2}.

For any particular transcript t, the MOM approaches described above do not identify definitely its eQTL when $\hat{p}_{t0} \leq k_\alpha$, i.e., when the transcript is claimed differentially expressed. However, the whole set of posterior probabilities $\{\hat{p}_{tj} : t = 1, \ldots, T; j = 1, \ldots, M\}$ provides evidence for eQTL hot spots. An ad hoc method is used in [101] for the identification of eQTL hot spots. The method goes as follows. At each marker j, the posterior probabilities are averaged over all differentially expressed transcripts; that is

$$\hat{p}_j = \frac{\sum_{t=1}^{T} \hat{p}_{tj} I\{\hat{p}_{t0} \leq k_\alpha\}}{\sum_{t=1}^{T} I\{\hat{p}_{t0} \leq k_\alpha\}}$$

is computed. Then the $\hat{p}_j$'s are plotted against the markers in the genome. The peaks of the plot are identified as eQTL hot spots. See Figure 3 in [101].

8.3.3   Multi-QTL Model Methods

The single-QTL model approaches described in § 8.3.2 are relatively simple and easy to implement. However, there are two obvious drawbacks of those approaches. First, the assumption that there is at most one eQTL for each transcript is far from realistic. Second, the approaches analyze the transcripts essentially one at a time without taking into account the rich correlation relations among the expression levels. Approaches based on multi-QTL models that accommodate multi transcripts are more efficient. In this section, we provide a general methodology of multi-trait multi-QTL model methods for eQTL mapping.

In principle, the multi-trait QTL mapping methods presented in § 8.2 can all be used for eQTL mapping. In fact, those methods are originally developed as parts of the methodologies for eQTL mapping rather than ordinary QTL mapping. However, those methods cannot be directly applied for the raw eQTL mapping data due to the large number of transcripts. Some preliminary treatments for reducing the dimensionality of the data are needed in order that those methods can be applied.

A multi-trait mapping approach is more efficient than a single trait mapping approach only when the traits are related. It is not necessary to use a multi-trait approach when the traits under study are not related at all or only have very weak relations. This points the way to reduce the dimensionality of the eQTL mapping data. Not all the transcripts are related unless they are regulated by certain common loci or located close enough in the genome. It becomes natural to group together those transcripts which are strongly related to each other and then to apply the multi-trait mapping approaches to each group of the transcripts.

In § 8.3.4, we will discuss various statistical clustering methods for grouping the transcripts. A multi-trait multi-QTL model method for eQTL mapping is a combination of one of the clustering methods in § 8.3.4 and one of the multi-trait mapping methods discussed in § 8.2. For example, in [38], either the k-means or a hierarchical clustering approach is used for clustering the transcripts, and then the sparse partial least squares regression approach in § 8.2.1 is used for each cluster. Each combination provides us with a method, thus we have a rich class of multi-trai multi-QTL model methods for eQTL mappping.

8.3.4 Clustering of Transcripts

Besides its own interest, a statistical clustering analysis of transcripts serves two purposes in eQTL mapping. First, it effectively reduces the dimensionality of all the transcripts under study to that of the clusters. Second, it groups the transcripts into clusters so that they can be more efficiently mapped.

Statistical clustering methods can be broadly classified into two categories: hierarchical and non-hierarchical. In this section, we present some of the statistical clustering methods commonly used in genetic studies. For a comprehensive coverage of clustering analysis, the reader is referred to [82].

A key element in clustering analysis is a measure on the dissimilarity (or distance) between any individuals. In the context of eQTL mapping, an individual stands for a transcript. If individuals are represented by vectors in $\mathcal{R}^n$, the measures on the dissimilarity between two individuals include the L_1 and L_2 *distances* or any other *metrics*. Let $\boldsymbol{x}, \boldsymbol{y}$ be the vectors representing two individuals and the L_1 and L_2 distances are defined respectively as

$$\|\boldsymbol{x} - \boldsymbol{y}\|_1 = \sum_{j=1}^{n} |x_j - y_j| \text{ and } \|\boldsymbol{x} - \boldsymbol{y}\|_2 = \sqrt{\sum_{j=1}^{n} (x_j - y_j)^2}.$$

But the dissimilarity measures are not limited to metrics. The dissimilarity measure does not necessarily satisfy the triangular inequality required of a metric. For example, any monotone function of $1 - |r|$, where r is the Pearson correlation between two points, can be taken as a measure of dissimilarity.

Without loss of generality, we assume the individuals are represented by vectors in $\mathcal{R}^n$ throughout our discussion. The essential goal of clustering analysis is to group individuals into clusters such that the individuals within clusters are more similar or related (with less dissimilarity) than the individuals between clusters.

Hierarchical clustering. There are two strategies for hierarchical clustering: agglomerative and divisive. The agglomerative strategy starts with the state that every individual forms a cluster of singletons and then recursively merges the pair of clusters with the least dissimilarity into a new cluster until eventually all the individuals are merged into a single cluster. The divisive strategy starts with the single cluster containing all the individuals and then recursively splits one of the clusters into two new clusters until eventually every cluster becomes a singleton. The agglomerative strategy is more commonly used in the genetics literature than the divisive strategy. In the following, we describe the agglomerative strategy in more detail.

In an agglomerative clustering algorithm, an inter-cluster dissimilarity measure must be defined. Let A and B denote two clusters and $d(x, y)$ denote the dissimilarity measures between individuals x and y. The inter-cluster dissimilarity measure between A and B are most commonly defined by taking the minimum, maximum or average dissimilarity between individuals in A and individuals in B; that is, the inter-cluster dissimilarity measure is defined as $d_{\min}(A, B) = \min\{d(x, y) : x \in A, y \in B\}$, $d_{\max}(A, B) = \max\{d(x, y) : x \in A, y \in B\}$ or $d_{\text{ave}}(A, B) = \text{Average}\{d(x, y) : x \in A, y \in B\}$. At each step of an agglomerative clustering algorithm, the inter-cluster dissimilarity measure is computed for each pair of the existing clusters and the pair of clusters with the least dissimilarity are merged into a single cluster. This produces one less cluster at the next step. When $d_{\min}$ is used as the dissimilarity measure, the algorithm is called a *single linkage agglomerative clustering*. When $d_{\max}$ is used, the algorithm is called a *complete linkage agglomerative clustering*. When d_{ave} is used, the algorithm is called a *average linkage agglomerative clustering*.

For any of the above inter-cluster dissimilarity measures, the agglomerative clustering algorithm is as follows.

Agglomerative Clustering Algorithm:

1. Form clustering C_0 that consists of T clusters with one individual each. Compute the inter-cluster dissimilarity measure for each pair of the clusters.

2. Given clustering $C_j (j \geq 0)$ with the inter-cluster dissimilarity measures, merge the pair of clusters with the smallest dissimilarity measure together to form a single cluster. This updates C_j to clustering C_{j+1} that has one less cluster than C_j. Compute the inter-cluster dissimilarity measure between the newly merged cluster and each of the other clusters. The dissimilarity measure of other pairs remains unchanged.

3. Repeat the process until all clusters are merged into a single cluster.

Define the diameter of a cluster as the largest dissimilarity among its members. Single linkage agglomerative clustering algorithms tend to produce clusters with very large diameters. They tend to suffer the drawback that the produced clusters are not necessarily compact; that is, the individuals within a cluster might not necessarily be similar to each other. In contrast, complete linkage agglomerative clustering algorithms tend to produce compact clusters with small diameters. But they tend to suffer the drawback that the individuals within clusters might not necessarily be more similar than between clusters. Average linkage agglomerative clustering algorithms offer a compromise between the single linkage and complete linkage algorithms. But they are not scale-invariant; that is, they depend on the numerical scale on which the dissimilarity is measured, while the single and complete linkage algorithms do not depend on the numerical scale. When choosing among these algorithms for a particular problem, the above mentioned features must be taken into account. In the clustering analysis of expression levels, the scale of measurement is of less concern and the average linkage agglomerative algorithms are more popularly used; see [44] [65] [70] [215] [219].

The output of a hierarchical clustering algorithm can be presented in a dendrogram like that illustrated in Figure 8.1, which is the clustering of USA states according to the number of arrests for four major crimes. The dendrogram helps to understand the systemic dissimilarity structure. But, it does not provide a definite clustering. To obtain a definite clustering, the dendrogram must be cut at certain level. The level to cut the dendrogram can be determined by several methods depending on the purpose of the clustering. It can be determined by a pre-specified number of clusters. The dendrogram is then cut at the level where the number of clusters equal the pre-specified number. If the number of clusters cannot be pre-specified, it can be determined by a certain statistic. The dendrogram is cut at the level where the statistic computed at that level exceeds a statistically determined threshold. The gap statistic based on within-cluster dissimilarity proposed in [182] can be used for this purpose. Suppose in a clustering there are k clusters $A_j, j = 1, \ldots, k$. The within-cluster dissimilarity for the clustering is defined as

$$W(k) = \sum_{j=1}^{k} \sum_{i,i' \in A_j} d(\boldsymbol{y}_i, \boldsymbol{y}_{i'}).$$

The gap statistic is given by $|\ln W(k) - \ln EW(k)|$ where $EW(k)$ is the expectation of $W(k)$ obtained by the same algorithm when the x_i's are uniformly distributed over a rectangle containing the data. $EW(k)$ can be approximated by simulation. The dendrogram is then cut at the level where the gap statistic achieves its maximum.

k-means Clustering. The k-means algorithm is the most popular non-hierarchical clustering methods. A generic k-means algorithm is as follows.

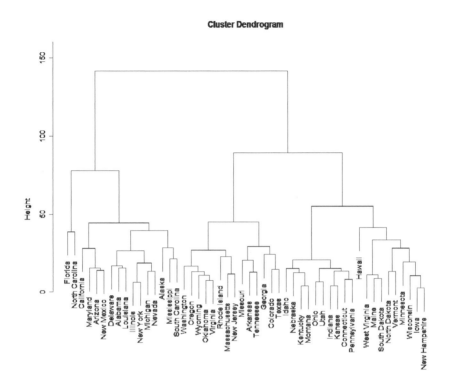

FIGURE 8.1: The cluster dendrogram of US arrests.

k-means Clustering Algorithm

1. Determine an initial clustering with k clusters (if there are no clear clues on how the initial clusters can be determined, the clusters can be obtained by a random partition). Calculate the centroid (to be defined below) of each cluster.

2. Move each individual to the cluster whose centroid is least dissimilar to the individual according to the dissimilarity measure. Re-calculate the centroids of the clusters after all individuals are reassigned.

3. Repeat step 2 until no individual is reassigned to a new cluster.

If the L_2 distance is used as the dissimilarity measure, the centroid of a cluster is defined as the mean vector of the members in the cluster. Thus the name *k-means clustering*. For a general dissimilarity measure, the centroid of a cluster is defined as the member of the cluster that has the least total dissimilarity with all the other members in the cluster. For an individual i in cluster A, the total dissimilarity of i with all the other members is defined as

$$\sum_{j:j\in A,j\neq i} d(\boldsymbol{y}_i, \boldsymbol{y}_j).$$

The centroid of cluster A is then

$$\boldsymbol{y}^* = \operatorname{argmin}_{i\in A} \sum_{j:j\in A,j\neq i} d(\boldsymbol{y}_i, \boldsymbol{y}_j).$$

With the centroid defined above, the k-means algorithm is referred to as a k-medoids algorithm in [82].

In the k-means algorithm, the number k is a fixed pre-specified number. In certain practical problems, k is given by the context of the problem. However, in general, k has to be determined by a data-driven technique. The gap statistic described earlier can be used in a similar way to determine k. Consider a sequence of k values: $\{k_0, k_0 + 1, k_0 + 2, \ldots, K\}$. For each k, run the k-means algorithm to obtain the clustering with k clusters, and calculate the within-cluster dissimilarity $W(k)$ for the clustering. By simulation, obtain the approximation to $EW(k)$. Then the optimal number of clusters is determined as

$$k^* = \operatorname{argmax}_{k=k_0,\ldots,K} |\ln W(k) - \ln EW(k)|.$$

If k^* is near k_0 or K, then consider more values of k smaller than k_0 or more values greater than K.

Modularity Clustering. Modularity clustering is a clustering approach recently developed in the study of networked systems; see [73] [139] [140] [151] [173]. Networked systems include neural networks, worldwide web networks, social networks, biological networks and so on. A network can be represented by a graph of vertices and edges. Four networks are illustrated in Figure 8.2. In the graphs, vertices represent the units in the network, and edges represent relationships between the units. Two vertices have a direct relationship if and only if they are connected by an edge. If the edges connecting vertices do not appear in a random pattern then there is a community structure of the network. In other words, a community structure corresponds to a statistically surprising arrangement of edges. In a true community structure, vertices are arranged in groups, the number of edges falling within groups is more than what can be expected if the edges are arranged at random. The community structure can be quantified by a measure called *modularity* [140]. Modularity clustering is essentially a combinatorial clustering method. It considers all possible community structures, i.e., partitions of the vertices, and aims to maximize the modularity over all possible community structures.

A graph can be represented by a matrix A called the *affinity matrix*. The vertices index the rows and columns of A. The entries a_{ij}'s of A indicate the connectedness of the vertices, $a_{ij} = 1$ or 0, 1 indicating that there is an edge connecting vertices i and j, 0 indicating no connection. With such an affinity

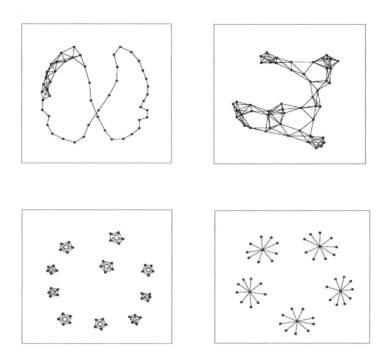

FIGURE 8.2: Graphs of four networks.

matrix, the graph is called an unweighted graph. In general, the entries of the affinity matrix can take values between 0 and 1, i.e., $0 \leq a_{ij} \leq 1$. Then the graph is called a weighted graph. The magnitude of a_{ij} reflects the strength of relationship between i and j. The measure of modularity is originally defined for unweighted graphs; see [73] [139] [140]. It is extended to weighted graphs in [173]. Let the vertex set of a graph be partitioned as $P = \{V_1, \ldots, V_k\}$. Given the affinity matrix A of the graph, the modularity of the partition is defined as

$$Q(P) = \sum_{c=1}^{k} \left[\frac{\mathbf{1}^{\tau} A[V_c, V_c] \mathbf{1}}{\mathbf{1}^{\tau} A \mathbf{1}} - \left(\frac{\mathbf{1}^{\tau} A[V_c, V] \mathbf{1}}{\mathbf{1}^{\tau} A \mathbf{1}} \right)^2 \right],$$

where $A[X, Y]$ denotes the sub matrix of A whose rows and columns are indexed by the vertices in X and Y respectively. For an unweighted graph, the modularity of P is simply the difference between the total number of edges connecting vertices within V_c's and what would be expected in the same partition with edges placed at random up to a factor of constant. In principle, the modularity clustering maximizes $Q(P)$ over all possible partitions P. The al-

gorithm for the implementation of the modularity clustering will be presented
later.

A modified version of modularity clustering called *modulated modularity
clustering* is considered in [173], which is particularly useful for the clustering
of transcripts. Let r_{ij} be the Pearson correlation coefficient between $\boldsymbol{y}_i$ and
$\boldsymbol{y}_j$. Define $d_{ij} = \sqrt{2(1 - |r_{ij}|)}$. In the modulated modularity clustering, the
entry of the affinity matrix is defined as

$$a_{ij} = \exp\left\{-\frac{d_{ij}^2}{2\sigma^2}\right\}.$$

The Gaussian transformation of d_{ij} has the effect to strengthen the connectedness of those highly correlated pairs and weaken the connectedness of those
weakly correlated pairs so that the true community structure can emerge.
The parameter σ is introduced to modulate the extent of this effect. Thus
the modularity measure depends on the partition P as well as the modulating
parameter σ and is denoted by $Q(P, \sigma)$. The modulated modularity clustering
then maximizes $Q(P, \sigma)$ with respect to both P and σ.

In the following, we present the procedure for the implementation of the
modulated modularity clustering described in [139] and [173]. First, we consider the case when σ is fixed. The procedure with fixed σ is the same as that
for the implementation of the original modularity clustering. Because of the
huge number of possible partitions even when the number of vertices is small,
the complete maximization of $Q(P, \sigma)$ over all P is impossible. A divisive spectral approach is developed in [139] to provide an approximate solution. The
approach starts with the partition of V into two clusters by maximizing the
modularity. Once an optimal partition is obtained, each of the two clusters
is further partitioned into two clusters by the same process. The procedure
continues recursively until the partition of any clusters does not increase the
modularity further.

A matrix called modularity matrix is defined as

$$B = A - \frac{A\mathbf{1}\mathbf{1}^{\tau}A}{\mathbf{1}^{\tau}A\mathbf{1}}.$$

For an arbitrary partition $P = \{V_1, V_2\}$ of V. Let $s_i = 1$ if the ith vertex is
assigned to V_1, and $s_i = -1$ if the ith vertex is assigned to V_2. It is shown in
[139] that

$$Q(P, \sigma) = \frac{1}{4m}\boldsymbol{s}^{\tau}B\boldsymbol{s},$$

where $\boldsymbol{s} = (s_1, \ldots, s_T)^{\tau}$ and $2m = \mathbf{1}^{\tau}A\mathbf{1}$ which is fixed. Thus maximizing
$Q(P, \sigma)$ with respect to P is equivalent to maximizing $\boldsymbol{s}^{\tau}B\boldsymbol{s}$ under the constraint that the components of $\boldsymbol{s}$ take values 1 or -1. If this constraint is
relaxed, the solution is simply the eigenvector, say $\boldsymbol{y}$, of B corresponding to
its largest eigenvalue. If zero is the largest eigenvalue, V will not be divided

and V is not divisible. Otherwise, an approximate solution is obtained by assigning the positive entries of y to V_1 and the remaining entries to V_2. The approximation is refined by the so-called Kernighan-Lin algorithm [103] as follows. Each of the vertices is moved from its own cluster to the other cluster one at a time to form a new partition. After every vertex is moved once and only once, there are altogether T partitions (including the original one). Among these T partitions, the one with the largest modularity is identified. If the partition with the largest modularity is different from the original one, the process is repeated until no further increment of modularity can be made. By an abuse of notation, the optimal partition is denoted by $P = \{V_1, V_2\}$ as well. This completes the first step of the divisive spectral procedure.

In the next step, let $\tilde{B}_j = B[V_j, V_j], j = 1, 2$ and define

$$B_j = \tilde{B}_j - \mathrm{Diag}(\tilde{B}_j \mathbf{1}), \ \ j = 1, 2,$$

where $\mathrm{Diag}(a)$ denotes the diagonal matrix with the components of a as the diagonal entries. The process for partitioning B is repeated for $B_j, j = 1, 2$. If B_j is un-divisible, the corresponding V_j remains un-partitioned in the subsequent steps. Otherwise, B_j is divided further to produce two daughter clusters. The procedure continues until no daughter clusters are divisible.

The above procedure is summarized in the following algorithm.

Modularity Clustering Algorithm:

1. Let $B^{(1)} = B$. Calculate the maximum eigenvalue and the corresponding eigenvector of $B^{(1)}$. If the maximum eigenvalue is zero, stop. Otherwise, assign the positive entries of the eigenvector to $V_1^{(1)}$ and the rest to $V_2^{(1)}$. Refine the partition $P = \{V_1^{(1)}, V_2^{(1)}\}$ by the Kernighan-Lin algorithm. Form the modularity matrices

$$B_j^{(2)} = B^{(1)}[V_j, V_j] - \mathrm{Diag}(B^{(1)}[V_j, V_j]\mathbf{1}), \ \ j = 1, 2.$$

2. For $k \geq 2$, if, for all $j = 1, \ldots, j_k$, $B_j^{(k)}$ is not divisible, where j_k is the number of clusters derived at step $k - 1$, stop. Otherwise, for each divisible $B_j^{(k)}$, repeat the process in step 1 with $B^{(1)}$ replaced by $B_j^{(k)}$ to obtain a partition $P_j = \{V_{j1}, V_{j2}\}$. Form two daughter modularity matrices

$$B_{jl}^{(k+1)} = B_j^{(k)}[V_{jl}, V_{jl}] - \mathrm{Diag}(B_j^{(k)}[V_{jl}, V_{jl}]\mathbf{1}), \ \ l = 1, 2.$$

Re-arrange all the un-divisible matrices $B_j^{(k)}$ and all the daughter matrices $B_{jl}^{(k+1)}$ as

$$B_j^{(k+1)}, j = 1, \ldots, j_{k+1}.$$

In the above algorithm, the modularity $G(P, \sigma)$ is maximized with respect to P while σ is fixed. In the modulated modularity clustering, the modularity $G(P, \sigma)$ is maximized with respect to both P and σ. The following strategy is adopted in [173].

Modulated Modularity Clustering Algorithm:

1. Specify a range of σ and a fine grid over the range (a range between 0.05 and 0.50 by steps of 0.001 is suggested in [173]). For each value of σ, apply the modularity clustering algorithm but skip the Kernighan-Lin refinement at each bi-partition. Let $\hat{\sigma}$ be the value that achieves the maximum modularity. If $\hat{\sigma}$ is close to one of the ends of the specified range, extend the range at that end and search for more σ values. Otherwise, take $\hat{\sigma}$ as the optimal value.

2. Fully run the modularity clustering algorithm (without skipping the Kernighan-Lin refinements) with $\sigma = \hat{\sigma}$.

In [173], a comparison of the modulated modularity clustering is made with average linkage agglomerative hierarchical clustering and three spectral clustering methods by a simulation study. The distance $d_{ij} = \sqrt{2(1 - |r_{ij}|)}$ is used in the average linkage agglomerative hierarchical clustering. Let A be the affinity matrix with $\sigma = \hat{\sigma}$ and $D = \mathrm{Diag}(A\mathbf{1})$. Let $L = D - A$. Define $B_1 = L$, $B_2 = D^{-1}L$ and $B_3 = D^{-1/2}LD^{-1/2}$. The three spectral clustering methods are based on one of the matrices B_j's each. The spectral clustering based on a given B_j is as follows. For a specified number of clusters k, the eigenvectors corresponding to the k largest eigenvalues of B_j are extracted into a $p \times k$ matrix. The rows of the matrix are treated as points in $\mathcal{R}^k$ and clustered by the k-means method. The simulation study demonstrated that the modulated modularity clustering outperforms the other clustering methods.

8.3.5   General eQTL Mapping Strategies

In this chapter, we discussed various methods for multi-trait QTL mapping and eQTL mapping. Roughly, the methods can be classified as single-QTL-model based and multi-QTL-model based approaches. In general, single-QTL-model based approaches are simple, easy to implement and less restricted by the dimensionality. But they lack power and efficiency. On the other hand, multi-QTL-model based approaches are more sophisticated, require more computation and, usually, are subject to the constraint of dimensionality. But they gain in power and efficiency. Among the various methods, none of them is more superior than the others. One method might be better under a certain underlying data structure, but another might be better under another different underlying data structure. Since the underlying data structure is unknown in practical problems, what is the best strategy for a practical mapping problem? We offer a general guideline on strategies for eQTL mapping to end this chapter. The guideline also applies to ordinary multi-trait QTL mapping.

In an eQTL mapping problem, we are faced with the difficulty of high-dimensionality: a huge number of transcripts (usually, in thousands) and a huge number of markers (sometimes, it is even much bigger than the number of transcripts). As the first step, we need to reduce the dimensionality of the problem so that the data can be more efficiently and meaningfully analyzed. The single-QTL-model based approaches are useful for this purpose. Suppose both T, the number of transcripts, and M, the number of markers, are huge. The following strategy can be helpful in the dimensionality reduction. (i) *Transcript-wise screening on markers.* For each transcript t, use the single-trait-single-marker model approach considered in § 8.1.1 to screen the markers. For instance, compute the LOD at each marker, order the markers according to their LOD scores from the largest to the smallest and keep only the first M_0 markers. The number M_0, which is much smaller than M, can be taken the same for all transcripts or different from transcript to transcript. It can be pre-specified (e.g., a number which can be tackled in the subsequent more sophisticated methods) or determined by some statistical procedure (e.g., the number of LOD scores exceeding an individual critical value at a given level). This screening procedure is equivalent to the sure independence screening (SIS) advocated in [56]. (ii) *Screening on transcripts.* Use the MOM model (or MOM proximity model) described in § 8.3.2 to screen transcripts. Order the transcripts according to the posterior probabilities $\hat{p}_{t0}$ from the smallest to the largest and keep the first $T_0(<< T)$ transcripts. As with the screening of markers, the number T_0 can be pre-specified or determined by setting the false discovery rate at a certain level.

The next step is to cluster the reduced T_0 transcripts. Any of the methods in § 8.3.4 can be used at this step.

As the last step, apply any of the multi-QTL-model approaches in § 8.2 to each cluster of the transcripts.

It should be noted that different clustering methods produce different clusters and that the final results obtained by using different multi-QTL model approaches also vary. The result obtained by using any particular method is not necessarily better than those by using other methods. It is profitable to apply all possible methods to the same data and draw a conclusion by considering all the results together.

8.3.6 Case Study: eQTL Mapping with GEO Mouse Data GSE3330

In an experiment reported in [111], an F2-ob/ob mouse sample is generated from two founder strains: C57BL/6J (B6) and BTBR. The two strains differ in diabetes susceptibility when made obese. B6-ob/ob mice are diabetes resistant but BTBR-ob/ob mice are not. The F2 mice are genotyped for 194 markers distributed over 19 chromosomes. A subset of 60 mice is chosen and, for each mouse, 45,037 mRNA abundance measurements are obtained from the liver using the Affymetrix Gene Chips MOE430. The probe level data are normalized and background-corrected by the robust multi-array average (RMA) method developed in [94]. After removing low abundance transcripts, defined as transcripts with average expression level below the tenth percentile, 40,738 expression levels are left. The data described above are publicly available at www.ncbi.nlm.nih.gov/projects/geo/query/acc.cgi?acc=GSE3330.

The data used for eQTL mapping consists of the 40,738 expression levels for each of the 60 mice together with the genotypes of the mice at 145 markers (a subset of the 194 markers). These data are analyzed by using the following methods: EBarrays and MOM [101] as described in § 8.3.2, Hierarchical Bayesian linear model [96] as described in § 8.2.3, and SPLS regression [38] as described in § 8.2.1. The analyses using these methods are presented in the following.

Analysis using EBarrays. In the method of EBarrays, for each of 40,738 transcripts, the empirical posterior probability of DE is computed at each marker of the 145 markers, and the maximum of the posterior probabilities, i.e., $\hat{p}_t = \max_{j=1}^{145} \hat{p}_{tj}$, is obtained. The threshold value $k_{0.05}$ is determined as

$$k_{0.05} = \max\left\{ k : \frac{\sum_{t=1}^{T}(1 - \hat{p}_t)I\{1 - \hat{p}_t \le k\}}{\sum_{t=1}^{T} I\{1 - \hat{p}_t \le k\}} \le 0.05 \right\},$$

where $T = 40,738$. The transcripts with $\hat{p}_t \ge 1 - k_{0.05}$ are identified as mapping to at least one of the markers. The method of EBarrays identifies 4083 transcripts.

Analysis using MOM. In the method of MOM, the posterior probabilities of EE for all 40,738 transcripts, i.e., the $\hat{p}_{t0}$'s, are simultaneously computed using the EM algorithm discussed in § 8.3.2. The threshold value $k_{0.05}$ is determined such that

$$k_{0.05} = \max\left\{ k : \frac{\sum_{t=1}^{T} \hat{p}_{t0}I\{\hat{p}_{t0} \le k\}}{\sum_{t=1}^{T} I\{\hat{p}_{t0} \le k\}} \le 0.05 \right\}.$$

The transcripts with $\hat{p}_{t0} \le k_{0.05}$ are identified as mapping to at least one of the markers. The method of MOM identifies 3039 transcripts and 84% of the 3039 transcripts are also identified by the method of EBarrays.

Analysis using hierarchical Bayesian linear model. In this analysis, a preliminary step is taken before the hierarchical Bayesian linear model approach is applied. In the preliminary step, the variances across individuals of the transcripts are computed and those transcripts with variance less than 0.12 are excluded from the analysis. This step leaves 1,576 transcripts for the further analysis.

The priors for the hierarchical linear model are specified the same as those in § 8.2.3. A total of 3,000 MCMC cycles are run. The first 1,000 cycles are taken as the burn-in period and are discarded. In the remaining 2,000 cycles, every 10th cycle is sampled and results in a MCMC sample of size 200.

From the MCMC sample, the quantities $\bar{\eta}_{tj}$ and $\bar{\rho}_j$ as defined, respectively, in (8.17) and (8.19) are computed. The threshold value q_0 for $\bar{\eta}_{tj}$'s is determined by formula (8.20) given in § 8.2.3. When $\bar{\eta}_{tj} \geq q_0$, marker j is declared as an eQTL for transcript t. Out of the 1,576 transcripts, 843 are linked to at least one marker. Of the 145 markers, 129 are claimed to control the expression of the transcripts.

The proportion of transcripts associated with each marker, i.e., the $\bar{\rho}_j$'s, is plotted against the markers in Figure 8.3. The five markers with the highest proportions, which suggest hot spots are indicated by triangles.

For more details of the Bayesian analysis, the reader is referred to [96].

Analysis using SPLS regression. The analysis using SPLS regression given in [38] takes three steps. The first step is the same as the preliminary step in the analysis using hierarchical Bayesian linear model. The transcripts are screened according to their variances and the same 1,576 transcripts are left for further analysis.

The second step is a clustering analysis. First, the affinity matrix $A = (a_{ij})$

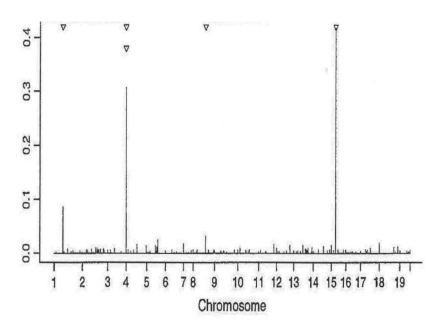

FIGURE 8.3: Hot spot regions for the mouse genome (reproduced from Figure 8 in [96]).

(for the definition of the affinity matrix, see the paragraph on modularity clustering in § 8.3.4) for an undirected and unweighted gene network based on the expression levels of the 1,576 transcripts is constructed using the approach of Gaussian graphical models [157]. There are 95 transcripts which are not connected with any other transcripts, i.e., the entries of the rows in the affinity matrix corresponding to these transcripts are all 0. The remaining 1,481 transcripts are clustered by the average linkage agglomerative clustering method in § 8.3.4 using a dissimilarity measure defined as follows. The dissimilarity measure between transcripts $\boldsymbol{y}_i$ and $\boldsymbol{y}_j$ is taken as

$$d(\boldsymbol{y}_i, \boldsymbol{y}_j) = 1 - \text{TOM}(\boldsymbol{y}_i, \boldsymbol{y}_j),$$

where

$$\text{TOM}(\boldsymbol{y}_i, \boldsymbol{y}_j) = \frac{a_{ij} + \text{number of } k \text{ such that } a_{ik} = 1 \text{ and } a_{jk} = 1}{\min\{\sum_k a_{ik}, \sum_k a_{jk}\}}.$$

The $\text{TOM}(\boldsymbol{y}_i, \boldsymbol{y}_j)$ above is the topological overlap measure given in [145]. The clustering analysis yields 47 clusters.

In the third step, the SPLS regression approach is applied for the mapping. The univariate SPLS regression is applied to each of the 95 transcripts which are unconnected with other transcripts. The multi-variate SPLS is applied to each of the 47 clusters. The results for one of the clusters are reported in [38]. The cluster consists of 83 transcripts. Among the 83 transcripts, 21 transcripts are mapped to at least one marker. A total of 27 markers are associated with the transcripts. Except one marker, all the other markers are associated with at least 11 transcripts; 13 of them are associated with all the 21 transcripts. The markers identified from this cluster are given in Table 8.1. For more details of the analysis, the reader is referred to [38].

In the case study above, we only demonstrated how different eQTL mapping methods are applied to the GEO mouse data without the comparison of the results obtained from the various methods. For the comparison, we refer the reader to the original papers [101], [96], [38].

TABLE 8.1: Markers identified in the mouse
genome for the cluster containing 83 transcripts.

Marker	Map (cM)	No. of mapping transcripts
D2Mit274	69.6	1
D2Mit17	73.9	16
D2Mit106	77.9	19
D2Mit194	85.4	18
D2Mit263	98.7	21
D2Mit51	101.7	20
D2Mit49	104.2	21
D2Mit229	110.5	21
D2Mit148	121.6	21
D5Mit348	6.3	17
D5Mit75	11.8	21
D5Mit267	17.1	21
D5Mit259	43	21
D5Mit9	46	21
D5Mit240	49.1	21
D5Mit136	54.9	21
D8Mit249	58.1	18
D8Mit211	72	21
D8Mit113	77.6	21
D9Mit21	43.8	12
D9Mit207	45.3	17
D9Mit8	58.6	18
D9Mit15	93.6	21
D9Mit18	110.7	17
D15Mit174	0	11
D15Mit136	11.5	17
D15Mit63	21	13

Bibliography

[1] H. Akaike. Information theory and an extension of the maximum likelihood principle. In *Second International Symposium on Information Theory*, Ed. B.N. Petrox and F. Caski, pages 267–281. Budapest: Akademiai Kiado, 1973.

[2] O. Alter, P. O. Brown, and D. Botstein. Singular value decomposition for genome-wide expression data processing and modeling. *Proc. Natl. Acad. Sci.*, 97:10101–10106, 2000.

[3] T.W. Anderson. *An Introduction to Multivariate Statistical Analysis.* John Wiley, 1984.

[4] L. Andersson, C.S. Haley, H. Ellegren, S.A. Knott, M. Johansson, K. Andersson, L. Andersson-Eklund, I. Edfors-Lilja, M. Fredholm, I. Hansson, J. Hakanson, and K. Lundstrom. Genetic mapping of quantitative trait loci for growth and fatness in pigs. *Science*, 263:1771–1774, 1994.

[5] J.S. Bailey, L. Grabowski-Boase, B.M. Steffy, T. Wiltshire, G.A. Churchill, and L.M. Tarantino. Identification of quantitative trait loci for locomotor activation and anxiety using closely related inbred strains. *Genes, Brain and Behavior*, 7(7):761–769, 2008.

[6] S. Bakin. Adaptive regression and model selection in data mining problems. PhD thesis, Australian National University.

[7] S. Banerjee, B. S. Yandell, and N.J. Yi. Bayesian quantitative trait loci mapping for multiple traits. *Genetics*, 179:2275–2289, 2008.

[8] J.T. Bensen, L.A. Lange, C. D. Langefeld, B. L. Chang, E. R. Bleecker, D.A. Meyers, and J Xu. Exploring pleiotropy using principal components. *BMC Genet*, 4:S53, 2003.

[9] J. O. Berger. *Statistical Decision Theory and Bayesian Analysis, second edition.* Springer-Verlag, New York, 1985.

[10] J.M. Bernardo and A.F.M. Smith. *Bayesian Theory.* John Wiley, 1993.

[11] J.P. Bidanel, D. Milan, N. Iannuccelli, Y. Amigues, M.Y. Boscher, F. Bourgeois, J.C. Caritez, J. Gruand, P. Le Roy, H. Lagant, R. Quintanilla, C. Renard, J. Gellin, L. Ollivier, and C. Chevalet. Detection of

quantitative trait loci for growth and fatness in pigs. *Genet. Sel. Evol.*, 33:289–309, 2001.

[12] M. Bidanel, J.P.and Rotschild. Current status of quantitative trait locus mapping in pigs. *Pig News Inf.*, 23:39N–53N, 2002.

[13] S. Biswas, J. D. Storey, and J. M. Akey. Mapping gene expression quantitative trait loci by singular value decomposition and independent component analysis. *BMC Bioinformatics*, 9:244, 2008.

[14] M. Bogdan, R. Doerge, and J.K. Ghosh. Modifying the Schwarz Bayesian information criterion to locate multiple interacting quantitative trait loci. *Genetics*, 167:989–999, 2004.

[15] M. Bogdan, F. Frommlet, P. Biecek, R. Cheng, J. K. Ghosh, and R.W. Doerge. Extending the modified bayesian information criterion (mBIC) to dense markers and multiple interval mapping. *Biometrics*, 64:1162–1169, 2008.

[16] V. L. Boyartchuk, K. W. Broman, R. E. Mosher, S. F. F. Dorazio, and M. N. et al Starnbach. Multigenetic control of *Listeria monocytogenes* susceptibility in mice. *Nat. Genet.*, 27:259–260, 2001.

[17] L. Boyd, S.and Vandenberghe. *Convex Optimization*. Cambridge University Press, Cambridge, 2004.

[18] P. Breheny and J. Huang. Coordinate descent algorithms for nonconvex penalized regression, with applications to biological feature selection. *The Annals of Applied Statistics*, 5(1):232, 2011.

[19] K. W. Broman. Quantitative trait locus mapping in the case of a spike in the phenotype distribution. *Genetics*, 163:1169–1175, 2003.

[20] K. W. Broman. The genomes of recombination inbred lines. *Genetics*, 169:1133–1146, 2005.

[21] K. W. Broman and S. Sen. *A Guide to QTL Mapping with R/qtl*. Springer-Verlag, 2009.

[22] K. W. Broman and T. Speed. A model selection approach for the identification of quantitative trait loci in experimental crosses. *J. Roy. Statist. Soc. B*, 64:641–656, 2002.

[23] R. J. Brooker. *Genetics, Analysis and Principles*. Addison-Wesley, New York, 1999.

[24] T. Calinski, Z. Kaczmarek, P. Krajewski, C. Frova, and M. Sari-Gorla. A multivariate approach to the problem of QTL localization. *Heredity*, 84:303–310, 2000.

[25] T. C. Carter and D. S. Falconer. Stocks for detecting linkage in the mouse and the theory of their design. *J. Genetics*, 50:307–323, 1951.

[26] G. Casella and E. I. George. Explaining the Gibbs sampler. *The American Statistician*, 46:167–174, 1992.

[27] J. Chen and Z. Chen. Extended Bayesian Information Criteria for Model Selection with Large Model Space. *Biometrika*, 95:759–771, 2008.

[28] J. Chen and Z. Chen. Extended BIC for small-n-large-P sparse GLM. *Statistica Sinica*, 22:555–574, 2012.

[29] Z. Chen. The full EM algorithm for the MLEs of QTL effects and positions and their estimated variances in multiple interval mapping. *Biometrics*, 61:474–480, 2005.

[30] Z. Chen and H. Chen. On some statistical aspects of the interval mapping for QTL detection. *Statistica Sinica*, 15:909–925, 2005.

[31] Z. Chen and J. Chen. Tournament screening cum EBIC for feature selection with high dimensional feature spaces. *Science in China, Series A*, 56(6):1327–1341, 2009.

[32] Z. Chen and J. Liu. Multiple interval mapping for ordinal traits using mixture GLIM. Manuscript.

[33] Z. Chen and J. Liu. Mixture generalized linear models for multiple interval mapping of quantitative trait loci in experimental crosses. *Biometrics*, 65:470–477, 2009.

[34] Z. Chen, G. Zheng, K. Ghosh, and Z. Li. Linkage disequilibrium mapping of quantitative-trait loci by selective genotyping. *Am. J. Hum. Genet.*, 77:661–669, 2005.

[35] V. G. Cheung and R. S. Spielman. Genetics of human gene expression: mapping DNA variants that influence gene expression. *Nature Reviews Genetics*, 10:595–604, 2009.

[36] A. P. Chiang, J.S. Beck, H.-J. Yen, M.K. Tayeh, T.E. Scheetz, R. E. Swiderski, D. Y. Nishimura, T. A. Braun, K-Y. Kim, J. Huang, K. Elbedour, R. Carmi, D. C. Slusarski, T.L. Casavant, E. M. Stone, and V. C. Sheffield. Homozygosity mapping with SNP arrays identifies TRIM32, an E3 ubiquitin ligase, as a BardetCBiedl syndrome gene (BBS11). *Proc. Nat. Acad. Sci.*, 103:6287–6292, 2006.

[37] S. Chib and E. Greenberg. Understanding the metropolis-hastings algorithm. *The American Statistician*, 49(4):327–335, 1995.

[38] H. Chun and S. Keles. Expression quantitative loci mapping with multivariate sparse partial least squares. *Genetics*, 182:267–278, 2009.

[39] H. Chun and S. Keles. Sparse partial least squares regression for simultaneous dimension reduction and variable selection. *J. R. Statist. Soc. B*, 72:3–25, 2010.

[40] G. A. Churchill and R. W. Doerge. Empirical threshold values for quantitative trait mapping. *Genetics*, 138:963–971, 1994.

[41] W. Cookson, L. Liang, G. Abecasis, M. Moffatt, and M. Lathrop. Mapping complex disease traits with global gene expression. *Nature Reviews Genetics*, 10:184–194, 2009.

[42] B.D. Cotter, S.F.and Rao and K. Engan, K.and Kreutz-Delgado. Sparse solutions to linear inverse problems with multiple measurement vectors. *IEEE Trans. Signal Process*, 53(7):2477–2488, 2005.

[43] A. P. Dempster, N. M. Laird, and D. B. Rubin. Maximum likelihood from incomplete data via the EM algorithm. *Journal of the Royal Statistical Society, Series B*, 39 (1):1–38, 1977.

[44] J. Dong and S. Horvath. Understanding network concepts in modules. *BMC Syst Biol*, 1:24, 2007.

[45] A. Druka1, E. Potokina, Z. Luo, N. Jiang, X. Chen, M. Kearsey, and R. Waugh. Expression quantitative trait loci analysis in plants. *Plant Biotechnology Journal*, 8:10–27, 2010.

[46] J. Dupuis and D. Siegmund. Statistical methods for mapping quantitative trait loci from a dense set of markers. *Genetics*, 151:373–386, 1999.

[47] C. W. Edwards, M. D.and Stuber and J. F. Wendel. Molecular-marker-facilitated investigation of quantitative-trait loci in maize. I. Numbers, genomic distribution and types of gene action. *Genetics*, 116:113–125, 1987.

[48] B. Efron. Large-scale simultaneous hypothesis testing: the choice of a null hypothesis. *J. Am. Stat. Assoc*, 99:96–104, 2004.

[49] B. Efron, T. Hastie, I. Johnstone, and R. Tibshirani. Least angle regression. *Annals of Statistics*, 32:407–499, 2004.

[50] B. Efron and R. J. Tibshirabi. *An Introduction to the Bootstrap*. Chapman & Hall, 1993.

[51] V. et al. Emilsson. Genetics of gene expression and its effect on disease. *Nature*, 452:423–428, 2008.

[52] L. Fahrmeir and G. Tutz. *Multivariate Statistical Modelling Based on Generalized Linear Models*. Springer-Verlag, New York, 1994.

[53] D. S. Falconer and T. F. C. Mackay. *Introduction to Quantitative Genetics*. Longman, London, 1996.

[54] J. Fan, Y. Feng, and R. Song. Nonparametric independence screening in sparse ultra-high-dimensional additive models. *Journal of the American Statistical Association*, 106(494):544–557, 2011.

[55] J. Fan and R. Li. Variable selection via nonconcave penalized likelihood and its oracle properties. *Journal of the American Statistical Association*, 96:1348–1360, 2001.

[56] J. Fan and J. Lv. Sure independence screening for ultrahigh dimensional feature space. *J. R. Statist. Soc. B*, 70:849–911, 2008.

[57] E. Feingold, P. O. Brown, and D. Siegmund. Guassian models for genetic linkage analysis using complete high-resolution maps of identity by descent. *Am. J. Hum. Genet.*, 53:234–251, 1993.

[58] J. Felsenstein. A Methematically tractable family of genetic mapping functions with different amounts of interference. *Genetics*, 91:769–775, 1979.

[59] M.E. Ferreira, J. Satagopan, B.S. Yandell, P.H. Williams, and T.C. Osborn. Mapping loci controlling vernalization requirement and flowering time in brassica napus. *Theoretical and Applied Genetics*, 90(5):727–732, 1995.

[60] D. Firth. Bias reduction of maximum likelihood estimates. *Biometrika*, 80:27–38, 1993.

[61] R. A. Fisher. *The Design of Experiments*, third edition, Oliver & Boyd Ltd., London, 1935.

[62] R. Foygel and M. Drton. Extended Bayesian information criteria for Gaussian graphical models. *Adv. Neural. Inf. Process.Syst.*, 23:2020–2028, 2010.

[63] I.E. Frank and J. H. Friedman. A statistical view of some chemometrics regression tools. *Technometrics*, 35:109–148, 1993.

[64] A. Frigyesi, S. Veerla, D. Lindgren, and M. Hoglund. Independent component analysis reveals new and biologically significant structures in micro array data. *BMC Bioinformatics*, 7:290–302, 2006.

[65] T. F. Fuller, A. Ghazalpour, J. E. Aten, T. A. Drake, A. J. Lusis, and et al. Weighted gene coexpression network analysis strategies applied to mouse weight. *Mamm Genome*, 18:463–472, 2007.

[66] A.E. Gelfand, S.E. Hills, A. Racine-Poon, and A.F.M. Smith. Illustration of Bayesian inference in normal data models using Gibbs sampling. *Journal of the American Statistical Association*, 85(412):972–985, 1990.

[67] A.E. Gelfand and A.F.M. Smith. Sampling-based approaches to calculating marginal densities. *Journal of the American statistical association*, 85(410):398–409, 1990.

[68] J. A. L. Gelfond, J. G. Ibrahim, and F. Zou. Proximity model for expression quantitative trait loci (eQTL) detection. *Biometrics*, 63:1108–1116, 2007.

[69] S. Geman and D. Geman. Stochastic Relaxation, Gibbs Distributions, and the Bayesian Restoration of Images. *IEEE Transactions on Pattern Analysis and Machine Intelligence*, 6:721–741, 1984.

[70] A. Ghazalpour, S. Doss, B. Zhang, S. Wang, C. Plaisier, and et al. Integrating genetic and network analysis to characterize genes related to mouse weight. *PLoS Genet*, 2:e130, 2006.

[71] Y. Gilad, S. A. Rifkin, and J. K. Pritchard. Revealing the architecture of gene regulation: the promise of eQTL studies. *Trends Genet.*, 24(8):408–415, 2008.

[72] H. Gilbert and P. Le Roy. Comparison of three multitrait methods for QTL detection. *Genet Sel Evol*, 35:281–304, 2003.

[73] M. Girvan and M. E. Newman. Community structure in social and biological networks. *Proc Natl Acad Sci U S A*, 99:7821–7826, 2002.

[74] L. Gonnick and M. Wheelis. *The Cartoon Guide to Genetics*. Harper Perennial, New York, 1991.

[75] H. H. Goring et al. Discovery of expression QTLs using large-scale transcriptional profiling in human lymphocytes. *Nature Genet.*, 39:1208–1216, 2007.

[76] J. Gray and N. Mcnaughton. *The Neuropsychology of Anxiety*. Oxford University Press, Oxford., 2000.

[77] P.J. Green. Reversible jump Markov chain Monte Carlo computation and Bayesian model determination. *Biometrika*, 82(4):711–732, 1995.

[78] C. A. Hackett and J. I. Weller. Genetic mapping quantitative trait loci for traits with ordinal distributions. *Biometrics*, 51:1252–1263, 1995.

[79] J. B. S. Haldane. The combination of linkage values and the calculation of distance between the loci of linked factors. *J. Genetics*, 8:299–309, 1919.

[80] J. B. S. Haldane and C. H. Waddington. Inbreeding and linkage. *Genetics*, 16:357–374, 1931.

[81] C.S. Haley and S. A. Knott. A simple regression method for mapping quantitative trait loci in line crosses using flanking markers. *Heredity*, 69:315–324, 1992.

[82] T. Hastie, R. Tibshirani, and J. Friedman. *The Elements of Statistical Learning*, second edition. New York: Springer, 2009.

[83] W.K. Hastings. Monte Carlo sampling methods using Markov chains and their applications. *Biometrika*, 57(1):97–109, 1970.

[84] Y. He and Z. Chen. Extended bic for linear interactive models with high or ultra-high feature space. *Preprint*, 2012.

[85] I. S. Helland. Partial least squares regression and statistical models. *Scand. J. Statist.*, 17:97–114, 1990.

[86] R. M. Henig. *The Monk in the Garden*. Houghton Mifflin Company, New York, 2000.

[87] I. Hoeschele and PM VanRaden. Bayesian analysis of linkage between genetic markers and quantitative trait loci. 1. Prior knowledge. *TAG Theoretical and Applied Genetics*, 85(8):953–960, 1993.

[88] I. Hoeschele and PM VanRaden. Bayesian analysis of linkage between genetic markers and quantitative trait loci. 2. Combining prior knowledge with experimental evidence. *TAG Theoretical and Applied Genetics*, 85(8):946–952, 1993.

[89] J. Huang, S. Ma, H. Xie, and C-H. Zhang. A group bridge approach for variable selection. *Biometrika*, 96 (2):339–355, 2009.

[90] J. Huang, S. Ma, and C.H. Zhang. Adaptive lasso for sparse high-dimensional regression models. *Statistica Sinica*, 18(4):1603, 2008.

[91] K.W. Hunter, K. W. Broman, T. Le Voyer, L. Lukes, and D. et al. Gozma. Predisposition to efficient mammary tumor metastatic progression is linked to the breast cancer metastasis suppressor gene Brms1. *Cancer Res.*, 61:8866–8872, 2001.

[92] W.Y. Hwang, H.H. Zhang, and S. Ghosal. First: Combining forward iterative selection and shrinkage in high dimensional sparse linear regression. *Stat. Interface*, 2:341–348, 2009.

[93] A. Hyvarinen and E. Oja. Independent component analysis: algorithms and applications. *Neural Networks*, 13:411–430, 2000.

[94] Rafael A. Irizarry, Bridget Hobbs, Francois Collin, Yasmin D. Beazer-Barclay, Kristen J. Antonellis, Uwe Scherf, and Terence P. Speed. Exploration, normalization, and summaries of high density oligonucleotide array probe level data. *Biostatistics*, 4(2):249–264, 2003.

[95] R. C. Jansen. Interval mapping of multiple quantitative trait loci. *Genetics*, 135:205–211, 1993.

[96] Z. Jia and S. Xu. Mapping quantitative trait loci for expression abundance. *Genetics*, 176:611–623, 2007.

[97] C. Jiang and Z. B. Zeng. Multiple trait analysis of genetic mapping for quantitative trait loci. *Genetics*, 140:11 11–1127, 1995.

[98] C. H. Kao and Z. B. Zeng. General formulas for obtaining the MLEs and the asymptotic variance covariance matrix in mapping quantitative trait loci when using the EM algorithm. *Biometrics*, 53:653–665, 1997.

[99] C. H. Kao and Z. B. Zeng. Modeling epistasis of quantitative trait loci using Cockerhams model. *Genetics*, 160:1243–1261, 2002.

[100] C. H. Kao, Z. B. Zeng, and R. D. Teasdale. Multiple interval mapping for quantitative trait loci. *Genetics*, 152:1203–1216, 1999.

[101] C. M. Kendziorski, M. Chen, M. Yuan, H. Lan, and A. D. Attie. Statistical methods for expression quantitative trait loci (eQTL) mapping. *Biometrics*, 62:19–27, 2006.

[102] C. M. Kendziorski, M. A. Newton, H. Lan, and M. N. Gould. On parametric empirical Bayes methods for comparing multiple groups using replicated gene expression profiles. *Statistics in Medicine*, 22:3899–3914, 2003.

[103] B. W. Kernighan and S. Lin. An efficient heuristic procedure for partitioning graphs. *Bell System Tech. J.*, 49:291–307, 1970.

[104] Y. Kim, H. Choi, and H.S. Oh. Smoothly clipped absolute deviation on high dimensions. *Journal of the American Statistical Association*, 103(484):1665–1673, 2008.

[105] Y. Kim, J. Kim, and Y. Kim. Blockwise sparse regression. *Statist. Sinica*, 16 (2):375–390, 2006.

[106] S. A. Knott and C. S. Haley. Aspects of maximum likelihood methods for the mapping of quantitative trait loci in line crosses. *Genetical Research*, 60:139–151, 1992.

[107] D. D. Kosambi. The estimation of the map distance from recombination values. *Ann. Eugen.*, 12:172–175, 1944.

[108] L. Kruglyak and E. S. Lander. A nonparametric approach for mapping quantitative trait loci. *Genetics*, 139:1421–1428, 1995.

[109] H. Lan, C. M. Kendziorski, J. D. Haag, L. A. Shepel, M. A. Newton, and M. N. Gould. Genetic loci controlling breast cancer susceptibility in the Wistar-Kyoto rat. *Genetics*, 157:331–339, 2001.

[110] H. Lan, J.P. Stoehr, S.T. Nadler, K.L. Schueler, B.S. Yandell, and A.D. Attie. Dimension reduction for mapping mRNA abundance as quantitative traits. *Genetics*, 164:1607–1614, 2003.

[111] Hong Lan, Meng Chen, Jessica B. Flowers, Brian S. Yandell, Donnie S. Stapleton, Christine M. Mata, Eric Ton-Keen Mui, Matthew T. Flowers, Kathryn L. Schueler, Kenneth F. Manly, et al. Combined expression trait correlations and expression quantitative trait locus mapping. *PLoS Genetics*, 2(1):e6, 2006.

[112] E. S. Lander and D. Botstein. Mapping Mendelian factors underlying quantitative traits using RELP linkage maps. *Genetics*, 121:185–199, 1989.

[113] K. Lange. *Mathematical and Statistical Methods for Genetic Analysis*. Springer-Verlag, 2002.

[114] S. Lee and S. Batzoglou. Application of independent component analysis to microarrays. *Genome Biology*, 4:R76, 2003.

[115] E. C. Lehmann. *Testing Statistical Hypotheses*, second edition. John Wiley, New York, 1986.

[116] H. Li, G. Ye, and J. Wang. A modified algorithm for the improvement of composite inteval mapping. *Genetics*, 175:361–374, 2007.

[117] J. Li, S. Wang, and Z. Zeng. Multiple-interval mapping for ordinal traits. *Genetics*, 173:1649–1663, 2006.

[118] W. Li and Z. Chen. Multiple interval mapping for quantitative trait loci with a spike in the trait distribution. *Genetics*, 182:337–342, 2009.

[119] Y. Lin and H.H. Zhang. Component selection and smoothing in multivariate nonparametric regression. *Ann. Statist.*, 34 (5):2272–2297, 2006.

[120] S. E. Lincoln, M. J. Daly, and E. S. Lander. *Constructing Genetic Linkage Maps with Mapmaker/Exp Version 3.0*. The Whitehead Institue, Cambridge, MA, 1993.

[121] B. H. Liu. *Statistical Genomics*. CRC Press, New York, 1998.

[122] J. Liu, Y. Liu, X. Liu, and H.-W. Deng. Bayesian mapping of quantitative trait loci for multiple complex traits with the use of variance components. *Am. J. Hum. Genet.*, 81:304–320, 2007.

[123] S. Luo and Z. Chen. Selection consistency of EBIC for GLIM with non-canonical links and diverging number of parameters. *arXiv preprint arXiv:1112.2815*, 2011.

[124] S. Luo and Z. Chen. Sequential lasso for feature selection with ultra-high dimensional feature space. *arXiv preprint arXiv:1107.2734*, 2011.

[125] S. Luo and Z. Chen. Extended BIC for linear regression models with diverging number of parameters and high or ultra-high feature spaces. *Journal of Statistical Planning and Inference*, 143:494–504, 2013.

[126] S. Luo, Y. He, and Z. Chen. Feature selection for sparse high-dimensional generalized linear models by profile marginal score search. *Preprint*, 2013.

[127] T. F. Mackay and J. D. Fry. Polygenic mutation in *Drosophila melanogaster*: Genetic interactions between selection lines and candidate quantitative trait loci. *Genetics*, 144:671–688, 1996.

[128] T. F. C Mackay, E. A. Stone, and J. F. Ayroles. The genetics of quantitative traits: challenges and prospects. *Nature Reviews Genetics*, 10:565–577, 2009.

[129] D. Malioutov, M. Cetin, and A.S Willsky. A sparse signal reconstruction perspective for source localization with sensor arrays. *IEEE Trans. Signal Process*, 53(8):3010–3022, 2005.

[130] O.C. Martin and F. Hospital. Two- and three-locus tests for linkage analysis using recombinant inbred lines. *Genetics*, 173:451–459, 2006.

[131] O. Martinez and R. N. Curnow. Estimating the locations and the sizes of the effects of quantitative trait loci using flanking markers. *Theor. Appl. Genet*, 85:480–488, 1992.

[132] P. McCullagh and J. A. Nelder FRS. *Generalized Linear Models, 2nd ed.* Chapman and Hall, 1990.

[133] L. Meier, S. van de Geer, and P. Bhlmann. The group lasso for logistic regression. Technical Report, Eidgenssische Technische Hochschule.

[134] X.M. Meng and D. B. Rubin. Maximum likelihood estimation via the ECM algorithm: a general framework. *Biometrika*, 80:267–278, 1993.

[135] N. Metropolis, A.W. Rosenbluth, M.N. Rosenbluth, A.H. Teller, and E. Teller. Equation of state calculations by fast computing machines. *The Journal of Chemical Physics*, 21:1087, 1953.

[136] T. H. Meuwissen and M. E. Goddard. Mapping multiple QTL using linkage disequilibrium and linkage analysis information and multitrait data. *Genet. Sel. Evol.*, 36:261–279, 2004.

[137] J. J. Michaelson, S. Loguercio, and A. Beyer. Detection and interpretation of expression quantitative trait loci (eQTL). *Methods*, 48:265–276, 2009.

[138] P. Naik and C.-L. Tsai. Partial least squares estimator for single-index models. *J. R. Statist. Soc. B*, 62:763–771, 2000.

[139] M. E. Newman. Modularity and community structure in networks. *Proc. Natl. Acad. Sci. USA*, 103:8577–8582, 2006.

[140] M. E. Newman and M. Girvan. Finding and evaluating community structure in networks. *Phys. Rev. E Stat. Nonlin. Soft Matter Phys.*, 69:026–113, 2004.

[141] M. A. Newton, A. Noueiry, D. Sarkar, and P. Ahlquist. Detecting differential gene expression with a semiparametric hierarchical mixture method. *Biostatistics*, 5:155–176, 2004.

[142] R package version 0.2-2. http://cran.r-project.org/web/packages/parcor/index.html.

[143] M.Y. Park and T. Hastie. L_1-regularization path algorithm for generalized linear models. *J. R. Statist. Soc. B*, 69:659–677, 2007.

[144] T. Park, M.Y.and Hastie. Regularization path algorithms for detecting gene interactions. Technical Report, Stanford University.

[145] Erzsébet Ravasz, Anna Lisa Somera, Dale A Mongru, Zoltán N Oltvai, and A-L Barabási. Hierarchical organization of modularity in metabolic networks. *science*, 297(5586):1551–1555, 2002.

[146] S. Raychaudhuri, J. M. Stuart, and R. B. Altman. Principal components analysis to summarize microarray experiments: application to sporulation time series. *Pacific Symposium on Biocomputing*, pages 455–466, 2000.

[147] A. Rebai, B. Goffinet, and B. Mangin. Approximate thresholds of interval mapping tests for QTL detection. *Genetics*, 138:235–240, 1994.

[148] A. Rebai, B. Goffinet, and B. Mangin. Comparing power of different methods for QTL detection. *Biometrics*, 51:87–99, 1995.

[149] Alvin C. Rencher. *Methods of Multivariate Analysis*. John Wiley, New York, 1995.

[150] M. V. Rockman and L. Kruglyak. Genetics of global gene expression. *Nature Reviews Genetics*, 7:862–872, 2006.

[151] J. Ruan and W. Zhang. Identification and evaluation of weak community structures in networks. In *Proceedings of the Twenty-First National Conference on Artificial Intelligence*, page 470–475, 2006.

[152] J.M. Satagopan and B.S. Yandell. Estimating the number of quantitative trait loci via Bayesian model determination. In *Special Contributed Paper Session on Genetic Analysis of Quantitative Traits and Complex Disease, Biometric Section, Joint Statistical Meetings, Chicago*. Citeseer, 1996.

[153] J.M. Satagopan, B.S. Yandell, M.A. Newton, and T.C. Osborn. A Bayesian approach to detect quantitative trait loci using Markov chain Monte Carlo. *Genetics*, 144(2):805–816, 1996.

[154] K. Sax. The association of size differences with seed-coat pattern and pigmentation in *Phaseolus vulgaris*. *Genetics*, 8:552–560, 1923.

[155] E. E. et al. Schadt. Mapping the genetic architecture of gene expression in human liver. PLoS Biol. 6, e107 (2008).

[156] S. Schadt, E.and Monks, T. A. Drake, and et al. Genetics of gene expression surveyed in maize, mouse and man. *Nature*, 422:297–302, 2003.

[157] Juliane Schäfer and Korbinian Strimmer. Learning large-scale graphical Gaussian models from genomic data. In *AIP Conference Proceedings*, volume 776, page 263, 2005.

[158] T.E. Scheetz, K.-Y.A. Kim, A.R. Swiderski, R.E.and Philip1, T.A. Braun, K.L. Knudtson, A.M. Dorrance, et al. Regularization of gene expression in the mammalian eye and its relevance to eye disease. *Proc. Nat. Acad. Sci.*, 103:14429–14434, 2006.

[159] G. Schwarz. Estimating the dimension of a model. *Ann. Statist.*, 6:461–464, 1978.

[160] L. A. Shepel, H. Lan, J. D. Haag, G. M. Brasic, J. S. Gheen, M. E.and Simon, P. Hoff, M. A. Newton, and M. N. Gould. Genetic identification of multiple loci that control breast cancer susceptibility in the rat. *Genetics*, 149:289–299, 1998.

[161] D. Siegmund. Model selection in irregular problems: Application to mapping quantitative trait loci. *Biometrika*, 91:785–800, 2004.

[162] D. Siegmund and B. Yakir. *The Statistics of Gene Mapping*. Springer-Verlag, 2007.

[163] M.J. Sillanpää and E. Arjas. Bayesian mapping of multiple quantitative trait loci from incomplete inbred line cross data. *Genetics*, 148(3):1373–1388, 1998.

[164] M.J. Sillanpää and E. Arjas. Bayesian mapping of multiple quantitative trait loci from incomplete outbred offspring data. *Genetics*, 151(4):1605–1619, 1999.

[165] B. W. Silverman. *Density Estimation for Statistics and Data Analysis*. Chapman and Hall, 1986.

[166] T. Simila and J. Tikka. Multiresponse sparse regression with application to multidimensional scaling. In *International Conference on Artificial Neural Networks (ICANN)*, pages 97–102. Warsaw, Poland, 2005.

[167] T. Simila and J. Tikka. Common subset selection of inputs in multiresponse regression. In *IEEE International Joint Conference on Neural Networks*, pages 1908–1915. Vancouver, Canada, 2006.

[168] T. Simila and J. Tikka. Input selection and shrinkage in multiresponse linear regression. *Computational Statistics & Data Analysis*, 52:406–422, 2007.

[169] M. Slatkin. Disequilibrium mapping of a quantitative-trait locus in an expanding population. *Am. J. Hum. Genet.*, 64:1764–1772, 1999.

[170] M. Soller and T. Brody. On the power of experimental designs for the detection of linkage between marker loci and quantitative loci in crosses between inbred lines. *Theor. Appl. Genet.*, 47:35–39, 1976.

[171] T. P. Speed. What is a genetic mapping function? In *Genetic Mapping and DNA Sequencing*. Springer-Verlag, New York, 1995.

[172] P. Stoica and T. Soderstorom. Partial least squares: a first-order analysis. *Scand. J. Statist.*, 25:17–24, 1998.

[173] E. A. Stone and J. F. Ayroles. Modulated modularity clustering as an exploratory tool for functional genomic inference. *PLoS Genetics*, 5:1–13, 2009.

[174] M. Stone. Cross-validatory choice and assessment of statistical predictions (with Discussion). *J. Roy. Statist. Soc. B*, 39:111–147, 1974.

[175] T. Sun and C.H. Zhang. Scaled sparse linear regression. *arXiv preprint arXiv:1104.4595*, 2011.

[176] S. D. Tanksley, H. Medina-Filho, and C. M. Rick. Use of naturally-occurring enzyme variation to detect and map genes controlling quantitative traits in an interspecific backcross of tomato. *Heredity*, 49:11–25, 1982.

[177] M.A. Tanner and W.H. Wong. The calculation of posterior distributions by data augmentation. *Journal of the American Statistical Association*, 82(398):528–540, 1987.

[178] F. Teuscher, V. Guiard, P. E. Rudolph, and G. A. Brockmann. The map expanson obtained with recombinant inbred strains and intermated recombinant inbred populations for finite generation designs. *Genetics*, 170:875–879, 2005.

[179] G. Thaller and I. Hoeschele. A Monte Carlo method for Bayesian analysis of linkage between single markers and quantitative trait loci. I. Methodology. *TAG Theoretical and Applied Genetics*, 93(7):1161–1166, 1996.

[180] G. Thaller and I. Hoeschele. A Monte Carlo method for Bayesian analysis of linkage between single markers and quantitative trait loci. II. A simulation study. *TAG Theoretical and Applied Genetics*, 93(7):1167–1174, 1996.

[181] R. Tibshirani. Regression shrinkage and selection via the lasso. *J. Roy. Statist. Soc. Ser. B.*, 58:267–288, 1996.

[182] R. Tibshirani, G. Walther, and T. Hastie. Estimating the number of clusters in a dataset via the gap statistic. *Journal of the Royal Statistical Society, Series B*, 32(2):411–423, 2001.

[183] N.A. Tinker, D.E. Mather, B.G. Rossnagel, K.J. Kasha, A. Kleinhofs, P.M. Hayes, D.E. Falk, T. Ferguson, L.P. Shugar, W.G. Legge, et al. Regions of the genome that affect agronomic performance in two-row barley. *Crop Science*, 36(4):1053–1062, 1996.

[184] J.A. Tropp. Greed is good: algorithmic results for sparse approximation. *IEEE Transactions on Information Theory*, 50:1–21, 2004.

[185] J.A. Tropp. Algorithms for simulataneous sparse approximation. Part II: convex relaxation. *Signal Processing*, 86:589–602, 2006.

[186] J.A. Tropp, A. C. Gilbert, and M. J. Strauss. Algorithms for simultaneous sparse approximation. Part I: greedy pursuit. *Signal Processing*, 86:572–588, 572-588.

[187] J.A. Tropp and A.C. Gilbert. Signal recovery from random measurements via orthogonal matching pursuit. *IEEE Transactions on Information Theory*, 53:4655–4666, 2007.

[188] B.A. Turlach. On homotopy algorithms in statistics. in: Symposium on Optimisation and Data Analysis. Canberra, Australia, http://www.maths.uwa.edu.au/~berwin/.

[189] B.A. Turlach and S.J. Venables, W.N.and Wright. Simultaneous variable selection. *Technometrics*, 47:349–363, 2005.

[190] P. Uimari, G. Thaller, and I. Hoeschele. The use of multiple markers in a Bayesian method for mapping quantitative trait loci. *Genetics*, 143(4):1831–1842, 1996.

[191] P. M. Visscher, C. S. Haley, and S. A. Knott. Mapping QTLs for binary traits in backcross and F-2 populations. *Genetical Research*, 68:55–63, 1996.

[192] C. Wallace, J.M. Chapman, and D. G. Clayton. Improved power offered by a score test for linkage disequilibrium mapping of quantitative-trait loci by selective genotyping. *Am. J. Hum. Genet.*, 78:498–504, 2006.

[193] H. Wang. Forward regression for ultra-high dimensional variable screening. *J. Amer. Statist. Assoc.*, 104:1512–1524, 2009.

[194] H. Wang, Y. C. M. Zhang, X. Li, G. L. and Mohan S. Masinde, et al. Bayesian shrinkage estimation of quantitative trait loci parameters. *Genetics*, 170:465–480, 2005.

[195] Y. Wang, Y. Fang, and S. Wang. Clustering and principal-components approach based on heritability for mapping multiple gene expressions. *BMC Proceedings*, page S121, 2007.

[196] S. Weisberg. *Applied Linear Regression*. John Wiley, New York, 1980.

[197] J.I. Weller, G.R. Wiggans, P.M. VanRaden, and M. Ron. Application of a canonical transformation to detection of quantitative trait loci with the aid of genetic markers in a multi-trait experiment. *Theor. Appl. Genet.*, 92:998–1002, 1996.

[198] C. R. Winkler, N. M. Jensen, M. Cooper, D. W. Podlich, and O.S. Smith. On the determination of recombination rates in intermated recombinant inbred populations. *Genetics*, 164:741–745, 2003.

[199] H. Wittenburg, F. Lammert, D. Q. Wang, G. A. Churchill, and R. et al. Li. Interacting QTLs for cholesterol gallstones and gallbladder mucin in AKR and SWR strains of mice. *Physiol. Genomics*, 8:67–77, 2002.

[200] H. Wold. *Estimation of Principal Components and Related Models by Iterative Least Squares*. Academic Press, New York, 1966.

[201] S. Wright. An analysis of variability in number of digits in an inbred strain of guinea pigs. *Genetics*, 19:506–536, 1934.

[202] R. Wu, C-X. Ma, and G. Casella. *Statistical Genetics of Quantitative Traits: Linkage, Maps and QTL*. Springer-Verlag, 2007.

[203] M. Xiong, R. Fan, and L. Jin. Linkage disequilibrium mapping of quantitative trait loci under truncation selection. *Hum. Hered.*, 53:158–172, 2002.

[204] S. Xu. Iteratively reweighted least squares mapping of quantitative trait loci. *Behavior Genetics*, 28:341–355, 1998.

[205] S. Xu and W. R. Atchley. Mapping quantitative trait loci for complex binary diseases using line crosses. *Genetics*, 143:1417–1424, 1996.

[206] S. Xu, N. Yonash, R. L. Vallejo, and H. H. Cheng. Mapping quantitative trait loci for complex binary traits using a heterogeneous residual variance model: An application to Mareks disease susceptibility in chickens. *Genetica*, 104:171–178, 1998.

[207] B. S. Yandell, T. Mehta, S. Banerjee, R. Shriner, D. et al. R/qtlbim: QTL with Bayesian interval mapping in experimental crosses. *Bioinformatics*, 23(5):641–643, 2007.

[208] R. Yang and S. Xu. Bayesian shrinkage analysis of quantitative trait loci for dynamic traits. *Genetics*, 176:1169–1185, 2007.

[209] M. Yeager, N. Orr, R. B. Hayes, K. B. Jacobs, P. Kraft, S. Wacholder, M. Minichiello, et al. Genome-wide association study of prostate cancer identifies a second risk locus at 8q24. *Nature Genetics*, 39:645–649, 2002.

[210] K.Y. Yeung and W. L. Ruzzo. Principal component analysis for clustering gene expression data. *Bioinformatics*, 17:763–774, 2001.

[211] N. Yi and S. Xu. Bayesian mapping of quantitative trait loci for complex binary traits. *Genetics*, 155(3):1391–1403, 2000.

[212] N. Yi, S. Xu, and D.B. Allison. Bayesian model choice and search strategies for mapping interacting quantitative trait loci. *Genetics*, 165(2):867–883, 2003.

[213] N. Yi, S. Xu, et al. Mapping quantitative trait loci with epistatic effects. *Genetical Research*, 79(2):185–198, 2002.

[214] N. Yi, B. S. Yandell, G.A. Churchill, D.B. Allison, E. J. Eisen, et al. Bayesian model selection for genomewide epistatic quantitative trait loci analysis. *Genetics*, 170:1333–1344, 2005.

[215] A. M. Yip and S. Horvath. Gene network interconnectedness and the generalized topological overlap measure. *BMC Bioinformatics*, 8:22, 2007.

[216] M. Yuan and Y. Lin. Model selection and estimation in regression with grouped variables. *J. Roy. Statist. Soc. Ser. B*, 68 (1):49–67, 2006.

[217] A. Zellner. An efficient method of estimating seemingly unrelated regressions and tests for aggregation bias. *J. Am. Stat. Assoc.*, 57:348–368, 1962.

[218] Z-B. Zeng. Precision mapping of quantitative trait loci. *Genetics*, 136:1457–1468, 1994.

[219] B. Zhang and S. Horvath. A general framework for weighted gene coexpression network analysis. *Stat. Appl. Genet. Mol. Biol.* 4, 17, 2005.

[220] C-H. Zhang. Nearly unbiased variable selection under minimax concave penalty. *The Annals of Statistics*, 38:894–942, 2010.

[221] H. Zhao and T. P. Speed. On genetic map functions. *Genetics*, 142:1369–1377, 1996.

[222] J. Zhao and Z. Chen. A two-stage penalized logistic regression approach to case-control genome-wide association studies, Article ID 642403, doi:10.1155/2012/642403. *Journal of Probability and Statistics*, 2012:1–15, 2012.

Index